OXFORD CLASSIC TEXTS IN THE PHYSICAL SCIENCES

ALGEBRAIC PROJECTIVE GEOMETRY

BY

J. G. SEMPLE

AND

G. T. KNEEBONE

CLARENDON PRESS · OXFORD

This book has been printed digitally and produced in a standard specification in order to ensure its continuing availability

OXFORD
UNIVERSITY PRESS

Great Clarendon Street, Oxford OX2 6DP

Oxford University Press is a department of the University of Oxford.
It furthers the University's objective of excellence in research, scholarship,
and education by publishing worldwide in

Oxford New York

Auckland Cape Town Dar es Salaam Hong Kong Karachi
Kuala Lumpur Madrid Melbourne Mexico City Nairobi
New Delhi Shanghai Taipei Toronto

With offices in

Argentina Austria Brazil Chile Czech Republic France Greece
Guatemala Hungary Italy Japan South Korea Poland Portugal
Singapore Switzerland Thailand Turkey Ukraine Vietnam

Published in the United States
by Oxford University Press Inc., New York

Oxford is a registered trade mark of Oxford University Press
in the UK and in certain other countries

Published in the United States
by Oxford University Press Inc., New York

Reprinted 2005

ISBN 0-19-850363-6

PREFACE

THIS book is intended primarily for the use of students reading for an honours degree in mathematics, and our aim in writing it has been to give a rigorous and systematic account of projective geometry, which will enable the reader without undue difficulty to grasp the fundamental ideas of the subject and to learn to apply them with facility.

Projective geometry is a subject that lends itself naturally to algebraic treatment, and we have had no hesitation in developing it in this way—both because to do so affords a simple means of giving mathematical precision to intuitive geometrical concepts and arguments, and also because the extent to which algebra is now used in almost all branches of mathematics makes it reasonable to assume that the reader already possesses a working knowledge of its methods. We have accordingly taken for granted acquaintance with the elements of linear algebra and the calculus of matrices, and, except in one instance, we have not gone into the proofs of purely algebraic theorems. The exception is a theorem which is fundamental in our system but is possibly not met with in quite the same form outside geometry, and this theorem we have proved in the Appendix.

In spite, however, of treating geometry algebraically, we have tried never to lose sight of the synthetic approach perfected by such geometers as von Staudt, Steiner, and Reye. If one consideration has been more prominent in our minds than any other it is that of giving precedence to the geometrical content of the system and the geometrical way of thinking about it. Nothing, in our opinion, could be more undesirable than that this traditionally elegant subject should be allowed to take on the appearance of being merely a dressing-room in which algebra is decked out in geometrical phraseology. We have, therefore, tried to show that although the basis of the formal structure is algebraic, the structure itself is thoroughgoing geometry, inasmuch as its concepts, its methods, and its results are all essentially dependent on geometrical ideas.

The book is divided into two parts. Part I consists of two chapters, of an historical and introductory character, which are intended both to serve as a link with elementary coordinate geometry and to prepare the reader for looking at projective geometry from a more advanced point of view. Part II is devoted to the formal theory, which is developed afresh from a new beginning so as to be logically independent of previous geometrical knowledge. Projective spaces of one, two, and three dimensions are considered in succession, and in the final chapter a short introduction is given to the geometry of higher space. In addition to discussing the usual topics of homographies, conics, quadrics, twisted cubics, and line geometry, we have given considerable space to collineations and linear transformations generally, since the fundamental importance of these, and particularly of the geometrical approach to them, is now universally recognized. Throughout the book, moreover, we have laid considerable emphasis on euclidean and affine specializations of projective results. This hardly needs justification. It is the experience of both the authors, and of most of those whom they have taught, that the exhibition of concrete instances of the rather formal projective concepts and results always endows these with a new prestige and stimulates interest in them as nothing else could at this stage.

Numerous exercises have been provided throughout the book, both interspersed in the text and collected at the ends of the chapters. Very many of the problems are taken from recent papers set in London University at the examination for B.A. Honours and B.Sc. Special Mathematics, and a few from the Cambridge Mathematical Tripos and the Mathematical Moderations at Trinity College, Dublin. The authors' grateful acknowledgement is due for permission to draw upon these sources.

Acknowledgement is also due to Mr. A. E. Ingleton for having read the manuscript and made many useful suggestions for its improvement, and to Miss A. S. Dennis for criticism of the early chapters. Finally, we would like to record our thanks to the officers of the Clarendon Press for their unfailing courtesy and helpfulness and for the excellence of the printing.

J. G. S.

April 1952 G. T. K.

CONTENTS

PART I

THE ORIGINS AND DEVELOPMENT OF GEOMETRICAL KNOWLEDGE

PART II

ABSTRACT PROJECTIVE GEOMETRY

PART I

THE ORIGINS AND DEVELOPMENT OF GEOMETRICAL KNOWLEDGE

> 'That all our knowledge begins with experience, there is indeed no doubt . . . but although all our knowledge originates *with* experience, it does not all arise *out of* experience.'
>
> KANT: *Critique of Pure Reason*

CHAPTER I

THE CONCEPT OF GEOMETRY

OUR main purpose in this book is to construct and develop a systematic theory of projective geometry, and in order to make the system both rigorous and easily comprehensible we have chosen to build it on a purely algebraic foundation. In adopting such a course, however, we may run the risk of appearing to reduce our subject to an ingenious manipulation of symbols in accordance with certain arbitrarily prescribed rules. Although the axiomatic form is the proper one in which to present a mathematical theory, we must not lose sight of the fact that an abstract system can only be fully appreciated when seen in relation to a more concrete background; and this is the reason why we have prefaced the formal development of projective geometry with two introductory chapters of a more informal character. The present chapter is devoted to a rather general consideration of the nature of mathematics and, more specifically, of geometry, while Chapter II contains an outline of the intuitive treatment of projective geometry from which the axiomatic theory has gradually been disentangled by progressive abstraction.

The growth of geometrical knowledge in the past has been marked by a gradual shifting away from empirical observation towards rational deduction; and we shall begin by looking for a moment at this process.

Geometry is commonly regarded as having had its origins in ancient Egypt and Babylonia, where much empirical knowledge was acquired through the experience of surveyors, architects, and builders; but it was in the Greek world that this knowledge took on the characteristic form with which we are now familiar. The Greek geometers were not only interested in the facts as such, but were intensely interested in exploring the logical connexions between them. In other words, they wished to raise the status of mathematics from that of a mere catalogue to that of a deductive science—and the *Elements* of Euclid is an embodiment of this ideal. In the *Elements* we have the systematic derivation of a large body of geometrical theorems by strict deduction from a small number of axioms. The system, as is now known, is not altogether perfect, and modern mathematicians have shown how it needs to be amended; but the modifications required are comparatively slight, and there is perhaps no easier way for a student to learn to appreciate mature mathematical reasoning than by studying the first book of Euclid and observing the way in which it is constructed.

Now for the Greeks, we must remember, geometry meant study of the space of ordinary experience, and the truth of the axioms of geometry was guaranteed by appeal to self-evidence. This view persisted for a very long time, and was still accepted without question at the end of the eighteenth century—when Kant, for example, made it an integral part of his philosophy. But about that time mathematicians were already beginning to see their subject in a new light, as a branch of study not directly dependent on experience, and this change of outlook was encouraged by the discovery, early in the nineteenth century, of the non-euclidean geometries, systems consistent within themselves but incompatible with Euclid's system. Since then it has become a commonplace that the mathematician is free to study the consequences of any axioms that interest him, whether or not they have any application in experience, provided only that they are not mutually contradictory.

We see, then, that in the period which elapsed between the first beginnings of mathematics and the conscious adoption of the modern axiomatic method, two major revolutions took place in mathematical thinking. First, the mere collecting of useful or interesting facts gave place to the rational deduction of theorems; and then, much later, mathematicians began to detach themselves from experience and to concentrate on the study of *formal* axiomatic systems. Neither of the revolutions came about suddenly, and the second is in a sense still in progress.

Mathematics, as conceived today, is fundamentally the study of structure. Thus, although arithmetic is ostensibly about numbers and geometry about points and lines, the real objects of study in these branches of mathematics are the *relations* which exist between numbers and between geometrical entities. As mathematics develops, so it becomes more abstract, until at last it is seen to be concerned with networks of formal relations only, and not with any particular sets of entities between which the relations hold. The process of abstraction whereby the formal structure is by degrees detached from the concrete systems in which it is exhibited is of so great importance to the understanding of the nature of mathematics as to justify closer examination of the manner in which it takes place.

One of the simplest illustrations of the process is provided by the evolution of the concept of number. Our first rudimentary idea of number is arrived at by simple abstraction from the processes of counting and measuring ordinary objects, and this idea is adequate at the level of school arithmetic. At a more advanced stage, numbers are seen to require redefinition in purely logical terms, and several alternative definitions have, in fact, been given. In whatever way numbers are defined, however, they obey the same formal 'laws of algebra'—the associative law of addition $(a+b)+c = a+(b+c)$, the distributive law $a(b+c) = ab+ac$, etc. —and many of the standard theorems of arithmetic and algebra can be deduced directly from these laws, without any need to specify further the nature of the numbers that are represented by the symbols a, b, etc. But this is not all. When studying elementary algebra one soon becomes aware of the close analogy that exists between the algebra of polynomials and the arithmetic of whole numbers; and it is now easy to account for this analogy by pointing out that polynomials, as well as numbers, satisfy the 'laws of algebra'. This is tantamount to saying that the system of numbers and the system of polynomials have a common structure; and when once this fact is recognized it is a natural step to undertake the study of an abstract system whose nature is unspecified beyond the fact that it has this particular structure. Such a system is known in algebra as a *ring*. If, on the other hand, we apply a similar process of abstraction to the system of rational numbers or the system of rational functions, we arrive at the abstract system known as a *field*.

There is no need for recognition of structural similarity to come to an end, even at this stage. Thus we might observe, for instance, that addition of rational numbers and multiplication of non-zero rational numbers obey similar laws; and we could then verify that the additive structure of a field and its multiplicative structure (when the element zero is excluded) are formally alike. Carrying the process of abstraction one stage farther, we could now introduce the abstract system known as a *group*.

Mathematics, then, is concerned with abstract systems of various kinds, each defined by a suitable set of axioms, which serves to characterize its structure. But although, from the point of view of pure mathematics, each structure is regarded as self-contained, the mathematical scheme usually has one or more concrete *realizations*; that is to say, the structure is usually to be found (possibly only to a certain degree of approximation) in a more concrete system. Abstract euclidean geometry of three dimensions, for instance, has as one of its realizations the structure of ordinary space. Indeed this is what led to its discovery, as well as what makes it so much more interesting than other systems which are logically of equal status with it. We do not, of course, always have to go all the way back to everyday experience for a realization of a mathematical formalism, since one is usually provided, as in the arithmetical example already considered, by a more concrete part of mathematics itself. One of the most important instances is the widespread occurrence of the group structure, which is found not only in additive and multiplicative groups of numbers, but also in groups of transformations and groups of matrices. Since this type of structure pervades much of mathematics, we may say that it is especially *significant*.

In this book we shall study the structure of projective geometry which, as is well known, is closely associated with certain simple algebraic structures, and with linear algebra particularly. Since the relevant algebra is part of every mathematician's essential equipment, we shall take it for granted that the reader is already familiar with it.

What we have said so far about the nature of mathematics holds quite generally, but when we limit the discussion to geometry we are able to be rather more specific. The structures studied in this branch of mathematics occur in experience as spatial structures, and from this alone we can infer something of their general character.

If, in fact, we turn back once again to Greek geometry, we may recall that the geometrical knowledge with which the Greeks began was derived ultimately from measurements made upon rigid bodies, and was therefore essentially a knowledge of shapes. Now the shape of a body can be conceived as determined by those relations between its parts which remain unaltered when the body is moved about in space. Whenever one body can be made in this way to take the place of another, the two bodies have the same shape; and they are then equivalent as regards their geometrical properties, or, in the language of elementary geometry, 'equal in all respects'. It will be remembered that in order to prove that certain sets of conditions are sufficient to ensure the congruence of two triangles Euclid showed that, if the conditions are satisfied, one triangle may be placed so as to bring it into coincidence with the other.

The idea of studying those properties of bodies which remain unaltered when the bodies are displaced in any way is most suggestive to a modern mathematician. In the language now in use, we would say that the geometrical (or, more accurately, the euclidean) properties of a body are invariant with respect to the operation of displacement in space; and invariance with respect to a certain *kind* of operation at once suggests the existence of an underlying *group* of operations. In the present instance the appropriate group is not far to seek. The totality of all displacements in space is a group of transformations; two bodies are congruent if and only if one can be made to take the place of the other by an operation of the group; and the shape of a body is determined by those of its spatial characteristics which are invariant with respect to the whole group. This, then, is the nature of euclidean geometry—it is the invariant-theory of the group of displacements.

Euclidean geometry, however, is not the whole of geometry. Early in the nineteenth century it was realized that other systematic collections of geometrical properties are possible besides that of Euclid, and in 1822 Poncelet published his *Traité des propriétés projectives des figures*, the first systematic treatise on projective geometry. In constructing this system Poncelet was fully conscious that his classification of geometrical theorems was based upon a new kind of fundamental operation, namely conical projection. A projective property of a figure is, in fact, simply a property that is invariant with respect to projection, and this enables us easily to identify the associated group of transformations. Confining ourselves, for simplicity, to two-dimensional geometry, we may consider the totality of all those transformations of the plane into itself which can be resolved into finite chains of projections from one plane on to another; and it is clear that this totality of transformations is a group and that it has plane projective geometry as its invariant-theory. Since the euclidean group, consisting of all displacements of the plane, may be shown to be a proper subgroup of the projective group, it follows at once that every projectively invariant property is also a euclidean invariant, whereas not every euclidean property is projective.

If we were now to take any arbitrarily chosen group of transformations of the plane into itself (containing the group of displacements as a subgroup) we could use this group in order to *define* an associated system of geometry; and all such systems are, mathematically speaking, of equal status. This was the general principle laid down by Klein in his famous *Erlangen Programme* of 1872.† Some of the geometries that can be obtained in this way, such as euclidean geometry, affine geometry, and projective geometry, are very well known; others, such as inversive geometry (which arises from the group of all transformations that can be resolved into finite sequences of inversions with respect to circles) are known but not usually studied in much detail; and yet others are presumably ignored altogether.

We shall confine our attention to the three geometries first mentioned—the geometries of the *projective hierarchy*—and since this restriction is somewhat arbitrary from a purely mathematical point of view, we should perhaps give some indication of why we choose to impose it. In the first place, euclidean geometry is of particular interest on account of its close connexion with the space of common experience, and this alone is sufficient to single it out for special attention. It so happens, however, that euclidean geometry is complicated; and we can appreciate it better when we relate it to projective geometry, where the structure is very much simpler. Projective geometry is more symmetrical than euclidean, by virtue both of the existence of a principle of duality and also of the fact that it may be handled by means of homogeneous coordinates. When homogeneous coordinates are used

† Klein: *Vergleichende Betrachtungen über neuere geometrische Forschungen* (Erlangen, 1872). Reprinted in *Mathematische Annalen*, **43** (1893).

for this purpose, the algebra has the merit of being either already linear or else readily made so. Thus the system of projective geometry is easy to work out and equally easy to comprehend when it has been worked out. Furthermore, projective transformations have the property of transforming conics into conics; and this means that the conic takes its place as naturally in projective geometry as does the circle in euclidean geometry. Finally, the essentials of euclidean geometry may be treated projectively by the simple artifice of introducing the line at infinity and the circular points. We thus have two geometries, projective geometry and euclidean geometry, which fit naturally together and which between them include most of the classical geometrical theorems. It is convenient to take in conjunction with them affine geometry, an intermediate geometry that is more general than euclidean but less so than projective; and the projective hierarchy is then complete.

What has been said so far concerns the subject-matter of our book, and it still remains for us to say something of the kind of approach that we shall use. It is customary to distinguish between two modes of reasoning in geometry, commonly referred to as synthetic and analytical. In a synthetic treatment we argue directly about geometrical entities (points, lines, etc.) and geometrical relations between them, whereas in an analytical treatment we first represent the geometrical entities by coordinates or equations, in order to be able to use the technique of algebraic manipulation. Since the discussion of projective geometry which follows in Part II is to be analytical, we shall conclude this chapter by touching upon the use of coordinates; but it should be realized, nevertheless, that we are under no logical compulsion to introduce a coordinate system at all. In the *Elements*, as in all Greek treatises, euclidean geometry is treated synthetically, and synthetic treatments of projective geometry are to be found in a number of modern books on the subject.†

Coordinates were first introduced into geometry by Descartes, in the seventeenth century, and the fruitfulness of the innovation soon became apparent. The older method of labelling figures was by letters of the alphabet, as in 'the triangle ABC', but such labels

† The first work of this kind was von Staudt's *Geometrie der Lage* (Nuremberg, 1847). A standard text-book, written in a similar spirit, is Veblen and Young's *Projective Geometry* (Boston, 1910).

were in fact no more than arbitrarily assigned names. Descartes's new technique of coordinates, on the other hand, made use of a system of labels which itself possesses a mathematical structure capable of reflecting the structure of the system labelled. This method of labelling has since become indispensable in mathematics, and the domain in which it can be applied now extends far beyond that originally envisaged by Descartes. In geometry itself, not only points but also lines and other entities can be represented by sets of coordinates; and in dynamics—to take an instance of another kind—the configuration of a system is ordinarily specified by n coordinates $q_1, q_2, \ldots, q_n$.

We have now seen how mathematics may be looked upon as a study of formal structure, and how geometry may be fitted into the general scheme. What has been said so far has been of a rather general character, and we must now turn more specifically to the details of the geometries of the projective hierarchy. This will be the topic of the second chapter of Part I, in which our purpose will be to recall enough of the elementary treatment of projective geometry to enable the reader to appreciate the process of abstraction which leads to the formal system of Part II.

CHAPTER II

THE ANALYTICAL TREATMENT OF GEOMETRY

THIS chapter is devoted, for the most part, to a discussion of the basic ideas involved in projective geometry and the apparatus of coordinates which allows them to be handled algebraically, and the point of view adopted is essentially elementary. The whole account is to be regarded as introductory, and in Part II a completely fresh beginning will be made. The formal system to be presented there is wholly abstract and independent of all previous geometrical knowledge; but even so, an elementary treatment such as that given in the present chapter is necessary as a psychological though not a logical presupposition of the more advanced theory. It alone can give body to the abstract formalism.

This chapter is not meant to be more than a summary, and the reader who desires a fuller account of the subjects touched upon in it is referred to Graustein: *Introduction to Higher Geometry* (New York, 1930).

§ 1. The Projective Hierarchy

We have already referred in Chapter I to the three geometries of the projective hierarchy and the possibility of defining them in terms of certain groups of transformations. It will be convenient, before proceeding further, to make these ideas more precise by giving a few details of each of the geometries; and once again we shall confine ourselves to the geometry of the (real) plane.

Euclidean geometry

The underlying group (p. 5) is the group of all displacements in the plane. The simplest invariant of this group is *length*, or the distance between two points. *Angle* is another invariant, and it follows from a theorem on congruent triangles (Euclid, I. 4) that angles may be characterized by suitably chosen lengths.

Among the figures appropriately studied in euclidean geometry is the *circle*, or locus of a variable point whose distance from a fixed point is constant. The theorems which properly belong to euclidean geometry include most of those in the *Elements*.

Analytically, euclidean geometry is best handled by means of rectangular cartesian coordinates, since, by virtue of the theorem of Pythagoras, the expression for the distance between two points then has a particularly simple form. Euclidean geometry may also be handled by vectors, the length of a vector being expressed in terms of the scalar product.

Projective geometry

The underlying group consists of all finite chains of projections that begin and end on the given plane. Relations of *incidence*, *collinearity*, and *tangency* are all projectively invariant, and *cross ratio* (cf. p. 17) is an invariant quantity.

A figure that is appropriately studied in projective geometry is the conic, since every conic is obtainable by projection from a circle.

Analytically, projective geometry is best handled by means of projective coordinates, which will be defined in § 5. These coordinates are expressible in terms of cross ratios. Vectors, as ordinarily defined in elementary books, have no application in projective geometry proper.

Affine geometry

Affine geometry occupies an intermediate position between euclidean geometry and projective geometry. The underlying group is generated by all *parallel* projections in space. The simplest invariant quantity for this group is the *position ratio* AP/PB of a point P with respect to two points A, B with which it is collinear. All projective properties are *a fortiori* affine properties; and when we pass from the projective group to the more restricted affine group, *parallelism* is introduced as a new invariant property.

Among the figures entering appropriately into affine geometry are the *parallelogram* and the separate kinds of conic, the *ellipse*, *hyperbola*, and *parabola*. The theorems which belong to affine geometry include the theorem on the concurrence of the medians of a triangle, Ceva's theorem, and the theorems on diameters of conics.

The coordinates which are suitable for handling affine geometry are oblique cartesian coordinates (perpendicularity of the axes in this case producing no essential simplification) or areal coordinates (see p. 25 below). Vectors may also be used; and since the scalar product is not involved, only linear vector algebra is required.

§ 2. The Modern Approach to Geometry

Although it is over two thousand years since Euclid compiled his treatise, our conception of geometry still continues to be moulded by the tradition which goes back to the Greek geometers; and the geometry that is taught in schools is essentially that of the *Elements*, though modified to some extent for educational reasons. Nevertheless, in the interval that has elapsed since the system was constructed mathematical thought has had a long time to mature, and we now do many things as a matter of course that might have seemed startling or even incomprehensible to a Greek mathematician. These new habits of thought have had a profound effect upon our conception of geometry, and it is worth while to consider them briefly in the present context.

For Euclid, a segment was simply a portion of a line intercepted between two of its points and, as such, it had no essential characteristic apart from its length. Nowadays we know the value of taking into account the sense of the segment as well as the length, i.e. we give the directed segment $\overrightarrow{PQ}$ precedence over the undirected segment PQ. The advantage of doing this is that we are enabled to formulate more comprehensive statements than would otherwise be possible. Thus, for instance, if P, Q, R are any three collinear points then

$$\overrightarrow{PR} = \overrightarrow{PQ}+\overrightarrow{QR},$$

irrespective of the order of the points in their line; whereas with undirected segments we should have to distinguish a number of different cases.

Again in contrast to Euclid, we introduce coordinates whenever convenient. Having chosen a pair of perpendicular lines OX, OY as axes, we say that a point P has as its coordinates (X, Y) the projections of the vector $\overrightarrow{OP}$ along the positively-directed axes. It is partly because we cannot avoid taking sign into account when we use coordinates that we instinctively do the same even when coordinates are not introduced.

In Euclid's system we often find ourselves distinguishing different cases in what appears to be essentially the same configuration or the same theorem, and having to apply somewhat different arguments to each case. This is rather disturbing, and we often try to avoid it by means of some kind of unifying device like that used above in connexion with the equation $PR = PQ+QR$. Sometimes, when we work in terms of coordinates, an appropriate artifice is suggested by the algebra.

To take a simple example, a straight line may either cut a circle in two distinct points, or it may touch it, or it may lie wholly outside it. Trying to find the coordinates of the points of intersection leads us to a quadratic equation, and in the three cases the roots are respectively real and distinct, real and coincident, and conjugate complex. We are accordingly led to say that a secant cuts a circle in two distinct real points, a tangent cuts it in two coincident real points, and a line that lies entirely outside it cuts it in two conjugate complex points. There are two new notions involved here—coincident points and complex points—and both of them are of considerable service in making it possible to state theorems in a more comprehensive form. We can say quite simply, for instance, that every line cuts every circle in two points; and the distinction between the possible cases is converted into a more manageable distinction between different kinds of point-pair.

In complex points we have our first example of *ideal elements* in geometry; for such points, having no 'real' existence, can only be thought of as artificial entities which we find convenient to adjoin to the actual plane. We are justified in treating them as if they were actual points by the fact that they can be represented by complex coordinates—and complex numbers behave algebraically in the same way as real numbers.

A rather different way in which we are led to introduce formal artifices unknown to the Greeks is by thinking of geometrical figures as capable of continuous variation. The most familiar examples of this are our ways of thinking of tangency and parallelism, which were two awkward concepts in Euclid's system.

A tangent to a curve may be regarded as a limiting position of a chord, in which one of the end-points tends towards coincidence with the other; and in the case of the circle this fits in with what we have just said about a tangent meeting the curve in two coincident points. The crucial importance of this new way of looking at tangency is that it enables us to find the equations of tangents by a simple application of the differential calculus. In a similar way, under suitable restrictions, we can define the circle of curvature of a curve at a point as the limiting position of the circle which passes through three points of the curve, two of which tend to coincidence with the third; and the differential calculus may then be applied to this also. The way is thus made open for the development of another subject—differential geometry. This branch of geometry, however, important though it is, does not fall within the scope of our discussion.

A second limiting process in geometry is connected with the relation of parallelism between straight lines. Euclid, as we know, distinguished between two different kinds of line-pair in the plane, namely intersecting lines and parallel lines. His distinction was absolute, but we can make it less sharp in the following way. Suppose that, being given two lines l and m which meet in P, we choose a point A of l and then rotate l about this point. As we do so, P moves farther and farther away along m until, when l is parallel to m, it disappears altogether. If we go on turning beyond this position, the point of intersection reappears and moves along m from the other end. Using a natural metaphor, therefore, we may say that the two lines have a point of intersection even when they are parallel, but that this point is then at infinity. Every line possesses a single point at infinity, which may be approached by travelling along the line in either direction, and all lines parallel to the given line meet it in this same point.

So far this use of the term 'point at infinity' is an expressive metaphor but nothing more, and if it remained a metaphor we should only have obscured an important distinction by introducing it; but fortunately we are able to give a precise mathematical meaning to the notion and so to make genuine use of it. One way of doing this is to identify 'point at infinity' with unsensed or absolute direction. Consider the following three properties of straight lines:

(i) there is a unique line which passes through two given points;
(ii) there is a unique line which passes through one given point and has a given absolute direction;
(iii) two lines, which do not coincide, either have a unique common point or have the same absolute direction.

If we agree to replace the phrase 'absolute direction' by 'point at infinity', and if we class together all the points at infinity as a single *line at infinity*, the three properties can be replaced by one, namely: *There is a unique line through any two distinct points, and any two distinct lines have a unique point in common.* In this way it is possible to construct a rigorous and serviceable theory of points at infinity.

When we introduced complex points (on p. 12) we explained that they are to be regarded as ideal points adjoined to the euclidean plane; and the same is now true of points at infinity. These too are ideal points, but since they have not been introduced by way of the coordinate representation it is not immediately obvious how they can be handled algebraically. This can, in fact, be managed quite easily, as we shall now show, by making the coordinates homogeneous.

When a pair of axes has been chosen, all actual (i.e. non-ideal) points of the plane admit of unique and unexceptional representation by pairs of real numbers (X, Y). Two lines, whose equations are

$$aX+bY+c = 0 \quad \text{and} \quad a'X+b'Y+c' = 0,$$

meet, if they are not parallel, in the actual point whose coordinates are given by

$$X : Y : 1 = bc'-b'c : ca'-c'a : ab'-a'b.$$

Now whether or not the lines are parallel, supposing only that they are distinct, the ratios

$$bc'-b'c : ca'-c'a : ab'-a'b$$

are always determinate, in the sense that at least one of the quantities concerned is different from zero. Our object, then, must be to represent points by such sets of ratios rather than by pairs of numbers which are liable to become infinite.

Instead, therefore, of representing points by pairs of coordinates (X, Y), we propose to represent them by triads of homogeneous coordinates (x, y, z), connected with the ordinary coordinates by the relations $X = x/z$, $Y = y/z$ whenever the point concerned is actual. We agree, furthermore, that (i) proportional triads shall always represent the same point, and (ii) the special triad $(0, 0, 0)$ shall be excluded. Plainly now, whenever $z \neq 0$ the triad (x, y, z) represents an actual point, while every triad for which $z = 0$ represents unequivocally a definite point at infinity.

A homogeneous linear equation $ax+by+cz = 0$ represents an actual line unless $a = b = 0$, namely the line $aX+bY+c = 0$; and the sole exceptional equation of this type, the equation $z = 0$, represents a unique ideal line—the *line at infinity*—which contains every point at infinity. We have here a justification for our earlier convention of regarding the set of all points at infinity as a line.

Although it was the consideration of a limiting process that originally led us to the notion of points at infinity, we now see that such points can be treated purely algebraically. When once we have introduced the representation of the plane by *homogeneous* cartesian coordinates x, y, z, we can dispense with limits in this context, and parallelism is made to depend on a distinction between ordinary and ideal elements in our extended geometrical system.

Besides affording a simple algebraic representation of points at infinity, homogeneous coordinates also have an incidental merit that is by no means negligible. In terms of such coordinates, the equation of every algebraic locus is homogeneous—e.g. for the straight line we have $ax+by+cz = 0$ in place of $aX+bY+c = 0$ —and the greater symmetry of equations of this kind makes the algebra very much easier to work with.

§ 3. Conical Projection and Projective Equivalence

Adoption of the modern approach to mathematics instead of the classical Greek approach enables us, as we have just seen in § 2, to simplify euclidean geometry and to give it greater homogeneity. But to transform the subject in this manner is to alter its character, and we may wonder whether we should not rather have reorganized geometry even more radically by constructing an entirely new system, more in accordance with our new way of thinking. Such a system is in fact possible—namely the projective geometry inaugurated by Poncelet—and it is to this that we shall now turn.

The ideal points at infinity that were introduced on p. 13 fit into the projective scheme even more naturally than into the euclidean. Any two figures that can be transformed into each other by a series of projections are to be regarded as *projectively equivalent*, so that from the projective point of view they are two instances of the same figure (as congruent figures are 'equal in all respects' in Euclid). This means, in particular, that there is no *projective* distinction between a pair of intersecting lines and a pair of parallel lines, the actual point of intersection of the first and the ideal point of intersection of the second having exactly the same status. Indeed, when we project from a plane π on to another plane π'

that is not parallel to π, the line at infinity in π may be said to project into an actual line of π'.

We shall now discuss various geometrical concepts in order to see which of them are projective.

In the first place, straight lines project into straight lines, and this means that collinearity of points is a projective property. Similarly, concurrence of lines is also projective.

The length of a segment is not invariant over projection; any finite segment can, in fact, be projected into a portion of a line that extends to infinity (by projecting a point I of the segment into a point at infinity).

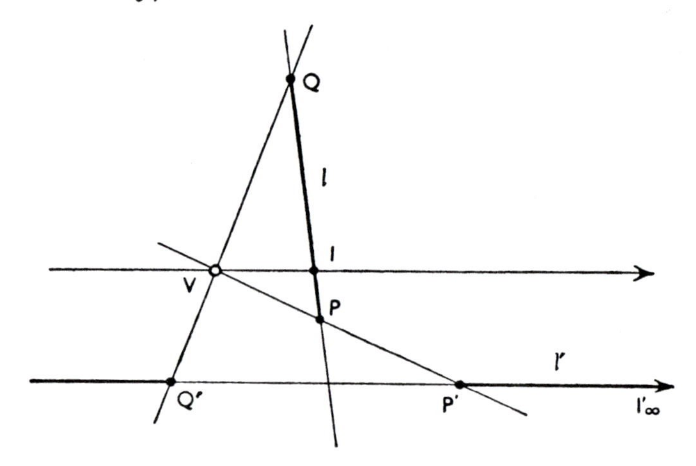

Since any two (distinct) points can be transformed projectively into any other two (distinct) points, a point-pair can have no projective individuality. In the same way, a triad of distinct points has no invariant property beyond its collinearity or non-collinearity.

EXERCISE. Show that any three collinear points can be transformed into any other three collinear points by a chain of at most three projections, and any three non-collinear points can be transformed into any other three non-collinear points by a chain of at most four projections.

When we come to a figure consisting of four collinear points we have for the first time a metrical characteristic—the cross ratio of the points—which is a projective invariant and which enables us to discriminate projectively between different collinear tetrads. This invariant plays as fundamental a part in projective geometry as does length in euclidean geometry. The precise definition of cross ratio is as follows.

DEFINITION: Let A, B, C, D be four collinear points. Then the cross ratio $\{A, B; C, D\}$ of the ordered pair (C, D) with respect to the ordered pair (A, B) is $\dfrac{AC}{CB}\Big/\dfrac{AD}{DB}$.

Thus the cross ratio is the quotient of the position ratios of C and D with respect to A and B. All segments are, of course, to be treated vectorially in the expression given in the definition.

EXERCISE. Prove that cross ratio is symmetrical in the two ordered pairs, i.e. $\{A, B; C, D\} = \{C, D; A, B\}$.

The projective invariance of cross ratio is established in the following theorem.

THEOREM 1. *If four collinear points A, B, C, D are projected from a vertex V into four collinear points A', B', C', D', then*

$$\{A, B; C, D\} = \{A', B'; C', D'\}.$$

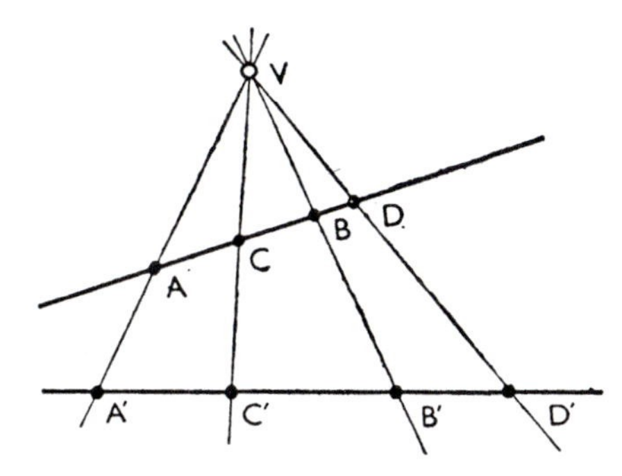

Proof. Considering magnitudes only,

$$\begin{aligned}\frac{AC}{CB} &= \left(\frac{AC}{VA}\Big/\frac{CB}{VB}\right)\frac{VA}{VB}\\ &= \left(\frac{\sin AVC}{\sin VCA}\Big/\frac{\sin CVB}{\sin VCB}\right)\frac{VA}{VB}\\ &= \frac{VA}{VB}\frac{\sin AVC}{\sin CVB}.\end{aligned}$$

Hence
$$\frac{AC}{CB}\Big/\frac{AD}{DB} = \frac{\sin AVC}{\sin CVB}\Big/\frac{\sin AVD}{\sin DVB},$$

and it follows that the two cross ratios have the same expression in terms of the angles at V. Further, $\{A, B; C, D\}$ is positive or negative according as A and B are not separated or are separated by C and D; and, since the relation of separation of pairs is evidently projectively invariant, the two cross ratios are equal in sign as well as in magnitude.

COROLLARY. *Any ordered set of four concurrent lines is cut by a variable transversal in a set of points with a constant cross ratio.*

This corollary makes possible the definition of cross ratio of two ordered pairs of *lines* through a common point. If the lines VA, VB, VC, VD just considered are denoted by a, b, c, d, we define their cross ratio as

$$V\{A,B;\,C,D\} \equiv \{a,b;\,c,d\} = \{A,B;\,C,D\}.$$

An important special case occurs when the value of the cross ratio $\{A,B;\,C,D\}$ is -1. In this case C and D divide the segment AB internally and externally in the same ratio, and we have the special mode of division known as *harmonic section*. The two pairs of points (A,B) and (C,D) are then said to be pairs of *harmonic conjugates* with respect to each other. If, in particular, C and D divide AB internally and externally in the ratio $1:1$, one of them is the mid-point of the segment AB and the other is the point at infinity on the line AB. Thus two points are always harmonic with respect to the mid-point of the segment which they determine and the point at infinity on their line.

Up to the present we have considered only tetrads of collinear points, but four points of the plane which are such that no three of them are collinear also give rise to an interesting configuration, known as the *complete quadrangle*.

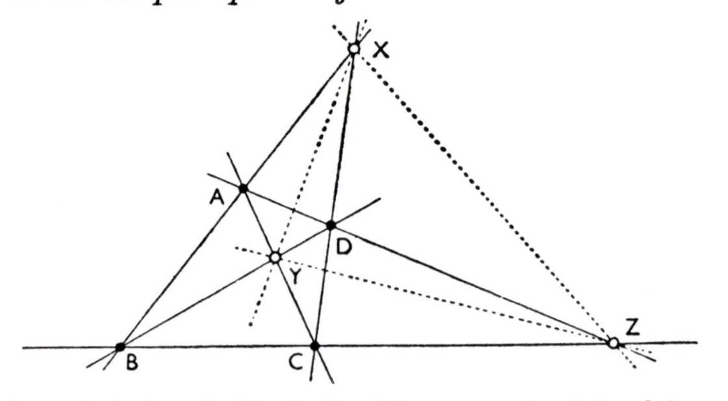

Let the points be A, B, C, D. They may be joined in pairs in three ways, giving the three pairs of opposite sides of the quadrangle: (AB,CD), (AC,BD), (AD,BC). Each pair of opposite sides has a point of intersection, and we thus have three new points X, Y, Z, the vertices of the diagonal triangle of the quadrangle.

There is also a corresponding figure, the *complete quadrilateral*, which is built up from lines and points as the quadrangle is built up from points and lines. The two figures possess important harmonic properties, as we shall now show.

THEOREM 2. *The two sides of the diagonal triangle of a quadrangle which meet in any diagonal point are harmonic with respect to the two sides of the quadrangle which meet in that point.*

The two vertices of the diagonal trilateral of a quadrilateral which lie on any diagonal line are harmonic with respect to the two vertices of the quadrilateral which lie on that line.

Proof. To prove the first part we project the figure so that ZX becomes the line at infinity. Then A, B, C, D become the vertices of a parallelogram, and the points where CD is met by YC, YD, YZ, YX become respectively two vertices of the parallelogram, the mid-point of the side on which they lie, and the point at infinity on this side. But these points form a harmonic range, and hence the points of the original figure of which they are the projections also form a harmonic range. The lines YC, YD, YZ, YX, which join Y to the points, accordingly form a harmonic pencil; and this is sufficient to prove the first part of the theorem. The second part may be proved by a similar argument.

We give this well-known proof here because it illustrates a powerful method of argument which has many applications in mathematics. Suppose, in fact, that we have (i) a class of mathematical objects of some specified kind, (ii) a set of transformations such that any one of the given objects may be changed into any other by a suitable transformation of the set, and (iii) a property of objects which is left invariant by every transformation of the set. Then we are able to show that the property holds for all the objects simply by showing that it holds for a particular one—and for this one we may be able to take an especially simple object, as we did in choosing the parallelogram in the proof above.

§ 4. Geometry of the Conic

Since, as we have already remarked, projective geometry provides the natural setting for the geometry of the conic, it is appropriate at this juncture to say something about this special type of curve. The classical definition of a *conic section* or *conic* is as a plane section of a circular cone. In other words, every conic is a projection of a circle and every projection of a circle is a conic. The whole theory of conics may be derived from this definition, and there are various standard treatments in which this is done.†

The classical definition leads, in particular, to the focus-directrix property, and this in turn yields the standard equations

$$X^2/a^2 \pm Y^2/b^2 = 1 \quad \text{and} \quad Y^2 = 4aX.$$

The theorems on conics that can be proved by elementary co-ordinate geometry are therefore all valid for the conic as defined above.

Besides the properties that are peculiar to the different types of conic, there are many properties that are common to all conics—the polar properties, for example. These are, in fact, the projective properties, and they may be established by the general method referred to at the end of § 3; for since every conic is projectively equivalent to a circle, we have only to write down the projective properties of the circle and restate them in terms of the general conic.

Some projective properties of the circle are readily deducible from the propositions in the third book of Euclid. An important example of this is provided by the theory of pole and polar, which can be treated by euclidean methods although it was not discussed by Euclid himself. The theory may be worked out by using the simple fact that the polar of a point P with respect to a circle whose centre is O is the line which passes through the inverse point P' and is perpendicular to OP'. Here we make use of a notion (that of inverse points) which does not occur in projective geometry proper; but we use this notion to prove theorems on the circle, in the formulation of which it does not appear, and then we generalize the *theorems* by the argument that they are projective and therefore valid equally for every conic. A key theorem that may be proved in this way is Chasles's Theorem:

THEOREM 3. *If A, B, C, D are four fixed points of a conic s and P is a variable point of s, and if a, b, c, d, p are the tangents at A, B, C, D, P respectively, then* (i) *the cross ratio $P\{A, B;\ C, D\}$ is constant;* (ii) *the cross ratio $p\{a, b;\ c, d\}$ is constant; and* (iii) *the two constant values are the same.*

† For the classical presentation, see Apollonius of Perga, *Treatise on Conic Sections* (Cambridge, 1896). For a modern account, see Askwith, *A Course of Pure Geometry* (Cambridge, 1917).

The first part is an immediate consequence of Euclid's theorem on the angle in a segment, and the remainder follows from the fact that the polars of four collinear points are four concurrent lines with the same cross ratio.

§ 5. Duality and Projective Coordinates

In the development of projective geometry we need to make systematic use of points at infinity, and for this reason we naturally work with homogeneous cartesian coordinates (x, y, z) rather than with the non-homogeneous coordinates (X, Y) from which they are derived. In terms of the new coordinates, every locus is represented, as we have seen, by a homogeneous equation $f(x, y, z) = 0$; and more particularly, every line has an equation of the form $ux+vy+wz = 0$. Thus the lines of the plane are represented algebraically in the simplest possible way, by homogeneous linear equations, and this makes it easy for us to bring out the duality between points and lines that is characteristic of plane projective geometry.

We have already noticed various ways in which the projective properties of points appear to be duplicated in analogous properties of lines. Thus, for example:

(i) when points at infinity are taken into account, two distinct points always determine a unique line and two distinct lines always determine a unique point;

(ii) we can define cross ratio not only for collinear points but also for concurrent lines;

(iii) the complete quadrangle and the complete quadrilateral have strictly analogous properties.

The correspondence between properties of points and properties of lines that is exemplified here runs through the whole of plane projective geometry, and to every theorem involving points and lines there corresponds a *dual* theorem involving lines and points. This duality of the projective plane may be exhibited very clearly, as we shall now show, by means of homogeneous cartesian coordinates.

Instead of treating the points of the plane as the primary geometrical entities, and the lines as *loci* or sets of points, we may fix our attention from the outset upon the lines, regarding these as simple entities. If we do this, we can define points in terms of lines, characterizing a point by the complete set of lines which pass through it; and every point now appears as the *envelope* of a variable line. When we adopt this second point of view, we naturally wish to represent lines by suitable sets of coordinates, and a simple way of doing this readily suggests itself. Each line is represented, in the original coordinate system, by an equation of the form

$$ux+vy+wz = 0,$$

and this equation is uniquely determined by the ratios between its coefficients. We may therefore take the three numbers u, v, w as *homogeneous cartesian coordinates of the line*; and a line is then determined by its coordinates (u, v, w) just as a point is determined by its coordinates (x, y, z). If we wish for a geometrical interpretation of the line-coordinates, which will enable us to plot a line when its coordinates are known, it is sufficient to note that the ratios u/w and v/w are the negative reciprocals of the intercepts made by the line on the axes OX and OY.

If u, v, w are fixed, the equation

$$ux+vy+wz = 0$$

means that the variable point (x, y, z) lies on a fixed line, but if x, y, z are fixed, the same equation means that the variable line (u, v, w) passes through the fixed point (x, y, z). Thus when a point (x, y, z) is regarded as the common point of the system of all lines through it, the equation $ux+vy+wz = 0$, with u, v, w variable, may be called its equation in line-coordinates or its *line-equation*. The coefficients in this equation are the homogeneous coordinates of the point—and there is thus complete duality between the representation of points and lines in terms of point-coordinates, on the one hand, and the representation of lines and points in terms of line-coordinates, on the other.

But although homogeneous cartesian coordinates reflect the duality of projective geometry, and although we can often use them to good effect in proving projective theorems, they are still not the most appropriate coordinates to use. In every cartesian coordinate system the line at infinity has the same equation $z = 0$; and since this line enjoys no special status in projective geometry, but can be transformed projectively into any other line, the restriction imposed in this way is quite arbitrary. We can remove it by defining a more general type of coordinate representation—

the projective coordinate system—which includes the cartesian representation as a particular case.

Projective coordinates may be introduced in different ways: either directly, in terms of cross ratios of certain pencils of lines, or indirectly, as numbers arrived at by algebraic transformation of cartesian coordinates. We shall here discuss the algebraic method, which will be found in the end to lead to an interpretation of the new coordinates in terms of cross ratios.

We begin by choosing once and for all a pair of rectangular axes OX, OY, and using these to define a system of homogeneous cartesian coordinates x, y, z. We write

$$\begin{aligned} x' &= a_{11}x+a_{12}y+a_{13}z, \\ y' &= a_{21}x+a_{22}y+a_{23}z, \\ z' &= a_{31}x+a_{32}y+a_{33}z, \end{aligned} \tag{A}$$

where the expressions on the right are three fixed linearly independent linear forms in x, y, z, and the determinant $|a_{rs}|$ is accordingly different from zero. The equations (A) determine x', y', z' uniquely when x, y, z are given, and vice versa; if x, y, z are not all zero, neither can x', y', z' all be zero; and if x, y, z are all multiplied by a factor λ, then x', y', z' are multiplied by the same factor. It is reasonable, therefore, to regard x', y', z' as homogeneous coordinates in the plane; and we say that *any transformation* (A) *of the above form defines a system of projective coordinates* (x', y', z') *for the points of the plane.*

Plainly every homogeneous rectangular cartesian coordinate system is projective, since it is related to the original system (x, y, z) by equations

$$\begin{aligned} x' &= x\cos\alpha - y\sin\alpha + az, \\ y' &= \pm(x\sin\alpha + y\cos\alpha) + bz, \\ z' &= z, \end{aligned}$$

which are of the form (A); and it can be shown that the oblique systems are also included. Since, further, the product of two transformations of type (A) is another such transformation, we deduce that (i) all projective coordinate systems can be derived as above from *any* one cartesian system, and (ii) the equations of transformation from one projective coordinate system to another are also of the form (A).

The equation of any line in the new coordinate system is of the form $u'x'+v'y'+w'z' = 0$; and (u', v', w') will be taken as the projective coordinates of the line in this system.

The transformation of coordinates is determined by the ratios of the nine coefficients a_{rs} in equations (A) and has thus eight degrees of freedom. If the old and new coordinates of a point are given, this yields two homogeneous linear relations between the a_{rs}; and the transformation is therefore uniquely determined when the projective coordinates of four points are assigned (assuming that the eight conditions so imposed are compatible and independent). In particular, the new system may be fixed† by means of the four points whose projective coordinates are to be $(1, 0, 0)$, $(0, 1, 0)$, $(0, 0, 1)$, and $(1, 1, 1)$. The first three points are called the vertices of the triangle of reference XYZ, and the fourth is called the unit point E.

Now let P be a general point of the plane, with projective coordinates (x', y', z'). By transforming back to cartesian coordinates and evaluating the cross ratio directly from its definition, the reader may verify that

$$X\{E, P; Y, Z\} = y'/z'.$$

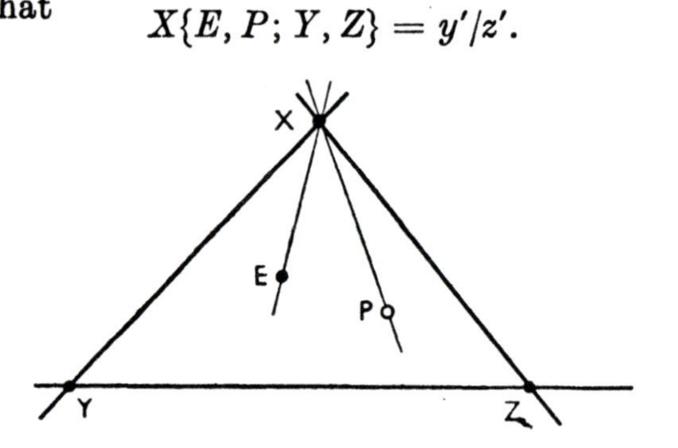

This means that the ratios between the coordinates of P may be identified with the cross ratios of certain pencils of lines determined by P and the four fundamental points X, Y, Z, E which characterize the coordinate system. Projective coordinates are thus related in a projectively invariant manner to the frame of reference.

By taking special tetrads of points as X, Y, Z, E we are able to define various projective coordinate systems with special properties:

(i) If XY is the line at infinity and E lies on the bisector of the angle XZY, the coordinate system is a homogeneous

† We omit the formal details, since the same result appears in Part II as Theorem 1 of Chapter IV.

cartesian system (in general oblique) with Z as origin, ZX, ZY as axes, and a scale of measurement determined by the unit point E.

(ii) If XYZ is an actual triangle and E is its centroid, the coordinates x', y', z' are *areal coordinates*. They are proportional to the areas of the triangles PYZ, PZX, PXY, with suitable signs.

(iii) If XYZ is an actual triangle and E is its incentre, x', y', z' are *trilinear coordinates*. They are proportional to the distances of P from YZ, ZX, XY, again with suitable signs.

In projective geometry proper we do not distinguish these special systems.

If two points Q_1, Q_2 of the plane have projective coordinates (x'_1, y'_1, z'_1), (x'_2, y'_2, z'_2) respectively, the coordinates of a general point of the line $Q_1 Q_2$ may be written as

$$(\lambda x'_1+\mu x'_2,\ \lambda y'_1+\mu y'_2,\ \lambda z'_1+\mu z'_2),$$

and the ratio $\lambda:\mu$ is uniquely determined. This is an algebraic consequence of the fact that the line has a linear equation. We often write $\theta = \mu/\lambda$ (allowing θ to take the improper value ∞ at Q_2) and we then call θ a *projective parameter* for the line. The parameter is uniquely determined when the *coordinates* of Q_1 and Q_2 are specified or, what comes to the same thing, when the *points* Q_1, Q_2 are fixed and a third point Q is also given, whose parameter is 1.

The projective parameter is of fundamental importance in projective geometry, as the invariant $\{P_1, P_2; P_3, P_4\}$ of four points of the line $Q_1 Q_2$ admits of a simple expression in terms of the parameters of the points.

THEOREM 4. *If P_1, P_2, P_3, P_4 are four points of the line $Q_1 Q_2$, with parameters θ_1, θ_2, θ_3, θ_4, then*

$$\{P_1, P_2; P_3, P_4\} = \frac{\theta_1-\theta_3}{\theta_2-\theta_3}\Big/\frac{\theta_1-\theta_4}{\theta_2-\theta_4}.$$

Proof. Take a cartesian representation (X, Y) of the plane, with Q_1 as origin, $Q_1 Q_2$ as axis of X, and Q_2 as the point $(1, 0)$. Then, if P_i is the point $(X_i, 0)$ $(i = 1, 2, 3, 4)$, it follows immediately from the definition of cross ratio that

$$\{P_1, P_2; P_3, P_4\} = \frac{X_1-X_3}{X_2-X_3}\Big/\frac{X_1-X_4}{X_2-X_4}.$$

In terms of homogeneous cartesian coordinates, the points Q_1, Q_2, P_i may be represented respectively by the coordinates $(0,0,1)$, $(1,0,1)$, $(X_i, 0, 1)$; and for the cartesian system—a special projective system—the projective parameter of P_i with respect to Q_1, Q_2 is $X_i/(1-X_i)$. If we now pass, by a transformation of type (A) above, from the cartesian coordinates (x,y,z) to the projective coordinates (x',y',z'), the new coordinates of any point

$$(x_1+\theta x_2,\ y_1+\theta y_2,\ z_1+\theta z_2)$$

will be $(x_1'+\theta x_2',\ y_1'+\theta y_2',\ z_1'+\theta z_2')$; and the value of the parameter θ will remain unaltered. In our particular case, therefore, we have $\theta_i = X_i/(1-X_i)$ $(i = 1,2,3,4)$, and a simple calculation shows that

$$\frac{\theta_1-\theta_3}{\theta_2-\theta_3}\bigg/\frac{\theta_1-\theta_4}{\theta_2-\theta_4} = \frac{X_1-X_3}{X_2-X_3}\bigg/\frac{X_1-X_4}{X_2-X_4}.$$

This completes the proof of the theorem.

It follows from Theorem 4 that the expression

$$\frac{\theta_1-\theta_3}{\theta_2-\theta_3}\bigg/\frac{\theta_1-\theta_4}{\theta_2-\theta_4}$$

depends only on the points P_1, P_2, P_3, P_4, and is independent of the choice of parametric representation of the line (in particular, of the choice of Q_1 and Q_2). It is called the *cross ratio of the four numbers* θ_i, taken in the proper order, and is denoted by $\{\theta_1,\theta_2;\ \theta_3,\theta_4\}$. The properties of this rational function of four variables will be discussed in Chapter III. Using the new terminology, we may now restate Theorem 4 in the form: *When the points of a line are represented by means of a projective parameter, the cross ratio of two given pairs of points is equal to the cross ratio of the corresponding pairs of parameters.*

It may further be shown that if (u_1', v_1', w_1'), (u_2', v_2', w_2') are the projective coordinates of two lines then the coordinates of any line through their point of intersection are expressible uniquely in the form

$$(u_1'+\theta u_2',\ v_1'+\theta v_2',\ w_1'+\theta w_2').$$

The parameter θ is called a projective parameter for the pencil of lines; and the cross ratio of two pairs of lines of the pencil is equal to the cross ratio of the corresponding pairs of parameters. The range of points and the pencil of lines are, in fact, completely dual to each other, and this duality is fully reflected in their algebraic representations.

We have now at our disposal a coordinate system, valid for both points and lines, which makes possible an algebraic treatment of the whole of plane projective geometry. One of the most important applications of this formal apparatus is to the study of the general conic, which has a simple representation in terms of projective coordinates. The cartesian equation of any proper conic, referred to suitably chosen rectangular axes, may easily be shown to be quadratic and irreducible; and since the equations of transformation (A) are linear, it follows that the equation of a conic in projective coordinates is always an irreducible quadratic equation. It may also be shown that, conversely, any (real) curve represented by such an equation is a proper conic. This means that the projective properties of the general conic may all be derived from algebraic properties of the general quadratic equation

$$ax^2+by^2+cz^2+2fyz+2gzx+2hxy = 0.$$

Since an equivalent discussion will form part of the formal theory of Chapter V, we shall not pursue this subject further at the present stage.

Besides, however, being algebraically deducible from the general quadratic equation, the projective properties of the conic can also be arrived at synthetically; and this alternative mode of derivation is often more suggestive and satisfying.

A suitable starting-point for a synthetic treatment of the general conic is provided by Chasles's Theorem, which was stated on p. 20. It follows from this theorem that if A and B are two fixed points of a proper conic s and P is a variable point of s, then the cross ratio of any four rays AP, taken in any order, is equal to the cross ratio of the four associated rays BP, taken in the corresponding order. Now this property has an important converse:

THEOREM 5 (*Steiner's Theorem*). *If p and p' describe pencils of lines, with vertices A and B respectively, and if the rays of the two pencils are associated in pairs in such a way that the cross ratio of any four rays p is equal to the cross ratio of the corresponding rays p', then the locus of the point of intersection of corresponding rays is a conic through A and B.*

Taken together, the theorems of Chasles and Steiner permit us to identify conics with loci of the points of intersection of corresponding rays of two pencils which are associated in the manner described. Since all the concepts involved in the definition of such a locus are projective, we have here a projective generation of the conic; and it is instructive to compare it with Euclid's generation of the circle. The projective generation of the conic, as we shall see, opens the way to a synthetic treatment of the projective properties of this curve.

§ 6. Homographic Correspondences

The idea of a correspondence that preserves cross ratio, which we meet for the first time in connexion with the projective generation of the conic, is very fruitful in projective geometry, and such correspondences—usually referred to as *homographic correspondences* or *homographies*—have been studied in considerable detail. The concept may be defined precisely as follows:

A one–one correspondence between two ranges of points, two pencils of lines, or a range of points and a pencil of lines, is said to be homographic when the cross ratio of any four elements of either system is equal to the cross ratio of the four corresponding elements, taken in the corresponding order, of the other system.

Let us consider the case of a homographic correspondence between two ranges. If we take projective parameters θ, θ' for the ranges, we can represent the homography algebraically by means of an equation connecting the parameters of corresponding points. To do this it is only necessary to take three fixed corresponding pairs (θ_1, θ_1'), (θ_2, θ_2'), (θ_3, θ_3'), chosen once for all, and a variable corresponding pair (θ, θ'). Then we have, by Theorem 4,

$$\{\theta, \theta_1; \theta_2, \theta_3\} = \{\theta', \theta_1'; \theta_2', \theta_3'\};$$

and this relation can be written in the form

$$a\theta\theta' + b\theta + c\theta' + d = 0.$$

Now conversely, as may easily be verified directly, every equation of this form, with $ad - bc \neq 0$, defines a homographic correspondence (see Theorem 2 of Chapter III). We can say, therefore, that *a one–one correspondence between two ranges is homographic if and only if it is associated with a bilinear relation between projective parameters.*

The representation of homographies by bilinear equations is the starting-point for the algebraic investigation of the properties of such correspondences; but this is another subject that we prefer to leave until we meet it again in the more formal setting of the next chapter.

§ 7. PROJECTIVE TRANSFORMATIONS

Plane projective geometry has been interpreted so far as the invariant theory of the group of projective transformations—a projective transformation of a plane into itself being defined (as on p. 6) as a transformation which can be resolved into a finite sequence of projections from one plane to another, beginning and ending with the given plane. This definition is stated in purely geometrical language, and we need to translate it into the language of algebra. This is made possible by the following theorem.

THEOREM 6. *When the plane is referred to a fixed system of projective coordinates, every projective transformation is represented algebraically by a non-singular linear transformation; and conversely, every such transformation defines a projective transformation of the plane into itself.*

This means that every projective transformation of the plane into itself is equivalent to an algebraic transformation

$$\begin{aligned} \rho x' &= a_{11}x+a_{12}y+a_{13}z, \\ \rho y' &= a_{21}x+a_{22}y+a_{23}z, \\ \rho z' &= a_{31}x+a_{32}y+a_{33}z, \end{aligned}$$

where $|a_{rs}| \neq 0$, and ρ is an arbitrary factor of proportionality which arises from the fact that the coordinates are homogeneous. In practice we usually omit the factor ρ, making the tacit reservation that two transformations are to be regarded as identical if their coefficients are proportional.

In order to prove the above fundamental theorem we need a number of lemmas.

LEMMA 1. *There exists a projective transformation which transforms four given points, no three of which are collinear, into four other given points, some or all of which may coincide with the first four, and no three of which are collinear.*

Proof. Let π_1 be the plane which contains the given tetrads of points. We begin by applying a preliminary transformation, projecting the second tetrad from an arbitrary vertex, not in π_1, on to a general plane π_2. We then have two tetrads, A_1, B_1, C_1, D_1 in π_1 and A_2, B_2, C_2, D_2 in π_2, and it is sufficient to show how to transform the second of these into the first by a chain of projections.

Since π_2 has been chosen generally, $A_1 A_2$ does not lie in π_2;

and we may choose a vertex of projection U on $A_1 A_2$ and then project π_2 on to a plane π_3, distinct from π_1, which passes through A_1. Let the projections of A_2, B_2, C_2, D_2 be A_1, B_3, C_3, D_3.

If we call $A_1 B_1 . C_1 D_1$ the point E_1 and $A_1 B_3 . C_3 D_3$ the point E_3, the lines $A_1 B_1 E_1$ and $A_1 B_3 E_3$ are coplanar, and $B_1 B_3$, $E_1 E_3$ therefore intersect, in V say. We now take V as vertex of projection and project π_3 on to a plane π_4, distinct from π_1, which passes through $A_1 B_1$. Let the projections of A_1, B_3, C_3, D_3, E_3 be A_1, B_1, C_4, D_4, E_1.

Since C_3, D_3, E_3 are collinear, so also are C_4, D_4, E_1; and the line $C_4 D_4 E_1$ is clearly coplanar with the line $C_1 D_1 E_1$. It follows that $C_1 C_4$ and $D_1 D_4$ intersect, in W say. Taking W as vertex of projection, we now project π_4 on to π_1, and A_1, B_1, C_4, D_4 project into A_1, B_1, C_1, D_1.

A chain of projections has now been defined which transforms the second of the given tetrads of points into the first. If, in a particular case, some of the points of the second set coincide with points of the first, it may be possible to simplify the construction; but this is of no importance in the present context.

LEMMA 2. *In terms of a fixed projective coordinate system, there is a unique non-singular linear transformation of the plane into itself which transforms four given points, no three of which are collinear, into four other given points, no three of which are collinear.*

This is a purely algebraic result, and it follows at once from Theorem 1 of the Appendix.

LEMMA 3. *In terms of a fixed projective coordinate system, every projective transformation of the plane into itself is a non-singular linear transformation.*

Proof. Suppose the given transformation changes the points X, Y, Z, E into points X', Y', Z', E'. Then no three of these transformed points are collinear, and they define a new system of projective coordinates x', y', z'. If P, P' are any two corresponding points, and the coordinates of P in the original system are (x, y, z), while those of P' in the new system are (x', y', z'), we have

$$y/z = \{E, P; Y, Z\} \quad \text{and} \quad y'/z' = \{E', P'; Y', Z'\};$$

and consequently, by the projective invariance of cross ratio, $y/z = y'/z'$. By symmetry, therefore, $x:y:z = x':y':z'$. But any two projective coordinate systems are connected, as we have seen, by a non-singular linear transformation, and this means that the coordinates of P and P' in the original system are connected by an algebraic transformation of this kind. The lemma is therefore proved.

Combining the results expressed in the above three lemmas, we now have Theorem 6, as enunciated on p. 29.

The significance of the theorem is that it leads to an algebraic characterization of projective properties of figures. A property is projective if it is invariant for every projective transformation of the plane into itself, and this is now seen to mean that *a property is projective if and only if its expression in terms of projective coordinates remains invariant when the coordinates are subjected to any non-singular linear transformation.*

We have now encountered two distinct geometrical interpretations of the algebraic transformation

$$\begin{aligned} x' &= a_{11}x+a_{12}y+a_{13}z, \\ y' &= a_{21}x+a_{22}y+a_{23}z, \\ z' &= a_{31}x+a_{32}y+a_{33}z, \end{aligned}$$

where $|a_{rs}| \neq 0$:

(i) as a change of *coordinates* from one projective system to another—(x,y,z) and (x',y',z') being coordinates of the same point in the two systems; and

(ii) as a projective transformation of *points*, referred to one and the same coordinate system—(x,y,z) and (x',y',z') being coordinates of the original point and the transformed point in this single system.

Since the algebra is the same in both cases, it makes no difference whether we define a projective property, as we did at first, as one which is invariant over projective transformation, or, alternatively, as one whose algebraic expression in every projective coordinate system is the same. We shall in fact find it convenient now to shift our ground, and throughout the formal development of projective geometry in Part II we shall adopt the second point of view.

In the same way we can redefine the euclidean properties of the plane—originally defined as invariants with respect to displacement—as those properties whose expression is the same in all right-handed rectangular cartesian systems with the same unit of measurement; for the equations which represent a transformation from one such coordinate system to another are of the same form as those which represent a displacement in terms of a fixed rectangular system.

More generally, we may define *similarity euclidean geometry* as consisting of those properties of configurations which are invariant, not only over displacement, but also over the operations of radial expansion about a point and reflection in a line, and this is equivalent to our defining it as consisting of those properties of configurations which are expressible in the same way in *all* rectangular cartesian coordinate systems, changes now being allowed both in the unit of measurement and in the senses of the axes. The equations of the typical transformation—of the plane or of coordinates, as the case may be—are now

$$\begin{aligned} x' &= \rho(x\cos\alpha - y\sin\alpha) + az, \\ y' &= \pm\rho(x\sin\alpha + y\cos\alpha) + bz, \\ z' &= z, \end{aligned} \qquad \text{(B)}$$

where ρ, α, a, b are arbitrary parameters, with $\rho > 0$, and the alternative sign is $-$ or $+$ according as the transformation of the plane does or does not involve reflection. In terms of change of coordinates, the corresponding distinction is between the case in which one of the two sets of axes is right-handed and the other left-handed and the case in which they are both of the same kind.

We now have before us two quite distinct systems of geometry, each defined in terms of a particular kind of algebraic transformation. Our next task will be to show that there is a simple relationship between the two systems, and that the whole of similarity euclidean geometry can readily be derived from projective geometry by means of a convenient device, that of the so-called 'circular points'.

§ 8. The Circular Points at Infinity

The circular points are two conjugate complex points, lying on the line at infinity, which have many remarkable properties. One of these is that the coordinates of the points are the same in every rectangular coordinate system.

Consider a general circle, whose equation in some fixed system of (homogeneous) rectangular cartesian coordinates is

$$x^2+y^2+2fyz+2gzx+cz^2 = 0.$$

The circle is met by the line at infinity $z = 0$ in the two points whose coordinates are given by the equations

$$x^2+y^2 = 0 = z,$$

i.e. the conjugate complex points $(1, \pm i, 0)$. These points are independent of the particular circle which served to define them—i.e. they are common to all circles of the plane—and they are known as the *circular points* I and J.

It is easily verified that the coordinates of I and J are left invariant by every transformation of type (B) above. More precisely, if the alternative sign in the equations is $+$, the triads of coordinates $(1, i, 0)$ and $(1, -i, 0)$ are individually invariant, while if the sign is $-$ they are interchanged by the transformation. Thus the circular points, taken as a pair, have the same representation in every rectangular coordinate system (as is also evident from the way in which they were defined). We can now establish the converse of this result, by showing that every projective coordinate system in which I and J have coordinates $(1, \pm i, 0)$ is rectangular cartesian. In the first place, if I and J have these coordinates, the equation of the line at infinity IJ is $z = 0$, so that $x/z = \lambda X$, $y/z = \mu Y$, where X, Y are cartesian coordinates and λ, μ are positive constants. If the axes of X and Y are inclined at angle ω, the equation $X^2+Y^2+2XY\cos\omega = 1$ represents a circle; and hence, in the projective coordinates, the circular points are given by

$$(x/\lambda)^2+(y/\mu)^2+2(x/\lambda)(y/\mu)\cos\omega = 0 = z.$$

Their coordinates are therefore $(1, \pm i, 0)$ only if $\cos\omega = 0$ and $\lambda = \mu$, i.e. only if the coordinates are rectangular cartesian.

A non-singular linear transformation of coordinates, then, changes any given rectangular system into another rectangular system if and only if it leaves the representation of the point-pair (I, J) invariant; and, in view of the twofold interpretation of the equations of transformation (B), we can infer from this that a projective transformation of the plane into itself preserves all similarity euclidean properties if and only if it leaves the point-pair (I, J) invariant. In other words, *similarity euclidean geometry may be interpreted as projective geometry relative to the point-pair* (I, J). The subclass of projective coordinate systems which characterize similarity euclidean geometry consists of those projective systems in which the point-pair (I, J) is represented by the equations

$$x^2+y^2 = 0 = z.$$

From this new point of view, the *properties* of configurations that enter specifically into similarity euclidean geometry must be interpreted as projective *relations* to the *absolute point-pair* (I, J). To mention only two examples: two lines AB, AC are perpendicular when they are separated harmonically by AI, AJ; and a point F is a focus of a conic s when the lines FI, FJ are tangents to s. We shall not go into further details here as the subject will be treated more fully in the formal theory of Part II (see the discussion in § 8 of Chapter IV).

Not only euclidean geometry but also affine geometry may be treated projectively in this manner. Affine geometry is concerned with those properties of figures which are invariant over affine transformation, and affine transformations are simply projective transformations which leave the line at infinity invariant. The affine properties of figures are accordingly those of which the expression is the same in every cartesian coordinate system, rectangular or oblique.

§ 9. Geometry in Spaces of Other Dimensionality

Throughout this chapter we have been thinking of geometry as primarily the study of plane figures; that is to say, our geometry has been two-dimensional. This restriction, however, was imposed merely for the sake of convenience, and we could equally well consider the geometry of configurations in spaces of other dimensionality. Before leaving this introductory survey, therefore, we shall refer briefly to spaces whose dimensionality is different from that of the plane.

In the first place, little need be said of one-dimensional geometry. When we confine ourselves to the points of a single line, the resulting geometry is of necessity very simple; and one-dimensional projective geometry reduces, in fact, to the theory of homographic correspondences on the line. This will be made clear in Chapter III. We might perhaps remark, in anticipation of later discussions, that in one dimension the analogue of the conic is the point-pair, given by a single quadratic equation $ax^2+2hxy+by^2 = 0$.

When once we are familiar with the geometry of the plane, the extension to three-dimensional space is not particularly troublesome. The geometrical structure is of course more elaborate, and the fact that figures cannot be drawn so easily makes three-dimensional geometry rather harder to visualize; but there is no difficulty of principle, and we shall accordingly pass over the details of the extension. A point that is perhaps worth mentioning, however, is that a projective property can now no longer be defined in terms of invariance over projection; for ordinary space has only three dimensions, and we cannot speak, in this connexion, of projection from one space to another. But we can introduce the idea of a projective coordinate representation of space (obtained from a fixed homogeneous cartesian representation by applying an arbitrary non-singular linear transformation) and then define a projective transformation of space into itself as one that is represented in any projective coordinate system by a non-singular set of linear equations. If we call a property projective when it is invariant with respect to every such transformation, we obtain a three-dimensional projective geometry that is strictly analogous to the two-dimensional projective geometry discussed in the earlier sections of this chapter. As soon, in fact, as projective geometry has been translated into the language of algebra, we are able without difficulty to increase the number of dimensions of the space under consideration from two to three, and beyond. So we arrive at n-dimensional projective geometry, in which a point is defined by a set of $n+1$ homogeneous coordinates $(x_0, x_1, \ldots, x_n)$, and a projective property is one that is invariant for every non-singular linear transformation of the coordinates. When n is greater than three the system has no concrete realization as a space of geometrical intuition, but it is adequately defined algebraically as an object of abstract mathematical thought.

§ 10. Conclusion

We have now reached the end of the preliminary survey of geometry which forms the first part of this book. In the compass of two chapters we have attempted to trace the evolution of the geometrical ideas that will be treated more systematically in Part II, and the reader should by this time be in a position to appreciate the structure of the formal system.

We have seen in the course of our discussion how axiomatic geometry had its origin in the work of Euclid and other Greek geometers, how the two subjects of geometry and algebra were fused together by the genius of Descartes, and how the modern approach to geometry ultimately found formal expression in Klein's *Erlangen Programme* of 1872. We have also examined in some detail

Poncelet's projective geometry, which we recognize to be one of the major creations of nineteenth-century mathematics. As the ancestral member of the projective hierarchy, projective geometry unifies and makes more intelligible a wide variety of classical geometrical theorems.

Besides discussing the concepts on which projective geometry is based, we have also shown how the cartesian method comes fully into its own in this sphere. The representation of the plane or three-dimensional space by projective coordinates is devised in the first instance as a technique ancillary to geometry proper, but it soon shows itself capable of fulfilling a higher function. When projective geometry is expressed in algebraic language the application of Klein's principles becomes remarkably simple and illuminating. We have seen, in fact, how projective geometry is rooted in the full linear group, i.e. the group of all non-singular linear transformations; and this fundamental connexion provides us with an alternative foundation for projective geometry, firmer and more fully under our control than geometrical intuition. We are thus enabled to take the final step in the process of successive abstraction outlined in Chapter I by rebuilding our subject from the beginning as a purely abstract formalism.

In § 7 we showed that a geometrical property is projective if and only if its expression in terms of projective coordinates remains invariant when the coordinates are subjected to any non-singular linear transformation. We now *define* the projective plane as a set of abstract entities which can be represented by homogeneous coordinates (x, y, z) and projective geometry as the study of those properties of the 'plane' whose expression in terms of the coordinates is left invariant by every non-singular linear transformation. In this way we set up an abstract axiomatic projective geometry, based on an algebraic foundation, which has intuitive projective geometry as a concrete realization.

EXERCISES ON CHAPTER II

1. Being given a quadrilateral $ABCD$ in a plane π, show how to choose a vertex of projection A and a plane of projection α so that $ABCD$ projects into (i) a parallelogram, (ii) a rectangle, (iii) a rhombus, or (iv) a square.

2. Two parallel planes π, π' being given, and also two points U, V not in either plane, a projective self-transformation of π is defined by the condition that points P, P' of this plane correspond when UP and VP' meet on π'. If UV meets π in O, prove that the correspondence between P and P'

is a radial expansion from O (i.e. that O, P, P' are in line and the ratio OP'/OP is constant) and find out under what conditions the correspondence reduces to reflection in the point O.

Show also that if UV is parallel to π the correspondence between P and P' is a translation.

3. Three planes α, β, γ meet in O, and a self-correspondence in α is set up by projecting α from a vertex L on to β, then projecting β from a vertex M on to γ, and finally projecting γ from a vertex N on to α. Show that O is one self-corresponding point in α and that, in general, there are two others on the line of intersection of α with the plane LMN.

Discuss the special case in which L, M, N are collinear.

4. Prove that a transformation of the plane into itself which is compounded of a displacement and a radial expansion about a point, with or without reflection in a fixed line, has equations (referred to any given rectangular axes) of the form

$$X' = \rho(X\cos\alpha - Y\sin\alpha) + a,$$
$$Y' = \pm\rho(X\sin\alpha + Y\cos\alpha) + b,$$

the $-$ or $+$ sign being taken according as a reflection has or has not been included.

Show that the result of applying successively any number of transformations of this general type is another transformation of the same type.

5. Show that every transformation of the type considered in the preceding exercise expands or contracts all distances in the same fixed ratio $\rho : 1$.

6. Show that, in rectangular cartesian coordinates, the equations of any projective transformation of the plane into itself are of the form

$$X' = \frac{a_1 X + b_1 Y + c_1}{a_3 X + b_3 Y + c_3}, \qquad Y' = \frac{a_2 X + b_2 Y + c_2}{a_3 X + b_3 Y + c_3},$$

where the determinant $(abc)_{123}$ is not zero.

Find the equations of the projective transformation which leaves the origin and the point $(1, 1)$ invariant and transforms the points $(1, 0)$ and $(0, 1)$ respectively into the points at infinity on the axes OX and OY.

Find all the circles which are transformed into circles by this transformation.

7. If X, Y are cartesian coordinates and x, y, z are the corresponding homogeneous coordinates, such that $X = x/z$, $Y = y/z$, prove that the point $(x_1 + \lambda x_2, y_1 + \lambda y_2, z_1 + \lambda z_2)$ divides the line joining (x_1, y_1, z_1) and (x_2, y_2, z_2) in the ratio $\lambda z_2 : z_1$.

Show that this line is divided harmonically by the conic $ax^2 + by^2 + cz^2 = 0$ if and only if $ax_1 x_2 + by_1 y_2 + cz_1 z_2 = 0$.

8. A, B, C, D are the four points $(1, 1)$, $(-1, 1)$, $(-1, -1)$, $(1, -1)$ respectively. Find the locus of a variable point P such that the cross ratio

$$P\{A, C;\ B, D\}$$

has a constant value k.

9. X', Y', Z', E' are the reference points and unit point of a system of projective coordinates x', y', z', and in a given system of homogeneous cartesian coordinates x, y, z the coordinates of the four points are $(-1, -1, 1)$,

$(1, -1, 1)$, $(0, 2, 1)$, and $(0, 0, 1)$ respectively. Obtain equations expressing the ratios of x', y', z' in terms of x, y, z.

10. If u, v, w and u', v', w' are homogeneous line-coordinates in the cartesian and projective systems of the preceding exercise, express u', v', w' in terms of u, v, w, and vice versa.

11. By plotting a sufficient number of lines, find out what envelopes are represented by the following equations in rectangular cartesian line-coordinates:

(i) $3u^2+4v^2-w^2 = 0$;
(ii) $u^2-2v^2+w^2 = 0$;
(iii) $v^2-wu = 0$.

Check your conclusions algebraically.

12. If $ABCD$ is a given quadrangle, show that a projective coordinate system can be defined in which the points A, B, C, D all have coordinates of the form $(1, \pm 1, \pm 1)$.

Find the coordinates of the sides, diagonal points, and sides of the diagonal triangle of the quadrangle; and hence give an algebraic proof of Theorem 2.

13. Show that parallelism may be handled by means of ideal points at infinity in space as well as in the plane. Obtain an analytical representation of such points by introducing homogeneous cartesian coordinates x, y, z, t; and show that

(i) the points at infinity are those points for which $t = 0$;
(ii) the totality of points at infinity may be regarded as forming a plane at infinity ι;
(iii) two planes are parallel if and only if they meet ι in the same line, and two lines are parallel if and only if they meet ι in the same point;
(iv) the line at infinity in any given plane π is the line in which π is met by ι;
(v) all spheres meet ι in the same (virtual) conic.

14. If (X, Y) and (X^*, Y^*) are two conjugate complex points of the plane, prove that there is one and only one real line whose equation is satisfied by the coordinates (X, Y) and that the equation of this same line is also satisfied by the coordinates (X^*, Y^*).

Show that the line is the radical axis of a system of real coaxal circles, with real limiting points and with the given complex points as common points, and find the equation of the line of centres of this coaxal system.

15. Find the two points at infinity on the hyperbola whose equation in homogeneous cartesian coordinates is

$$x^2+5xy+4y^2+6xz-2yz+3z^2 = 0.$$

Find the asymptotes of the hyperbola.

16. A variable tangent to an ellipse k meets the tangents at the ends of the major axis in U, V and those at the ends of the minor axis in U', V'. Prove that the circles on UV and $U'V'$ as diameters describe orthogonal coaxal systems, of which the first has the foci of k as common points. Find the (conjugate complex) common points of the second coaxal system.

17. OX and OY are respectively the transverse and conjugate axes of a central conic k, c is any circle touching k at its intersections with a line p

parallel to OY, and d is any circle touching k at its intersections with a line q parallel to OX. If P is any point of k, prove that

$$PT^2 = e^2PM^2 \quad \text{and} \quad PT'^2 = e'^2PN^2,$$

where PT, PT' are tangents from P to c, d, PM, PN are perpendiculars from P to p, q, e is the eccentricity of k, and e' is a number which satisfies the equation $1/e^2+1/e'^2 = 1$.

Show that when p is a directrix of k (meeting k in a pair of conjugate complex points) the circle c reduces to a point-circle at the corresponding focus.

18. A conic k is the section of a right circular cone κ by a plane π. Show that the foci of k are the points at which π is touched by two spheres inscribed in κ, the corresponding directrices being the lines in which π is met by the planes of contact of κ with the spheres (Dandelin's Theorem).

19. If two curves touch at a point P, prove that the ratio of their curvatures at P is a projective invariant. [*Hint.* Use Newton's expression for curvature.]

20. If P and P' are inverse points for the circle c, whose equation is $X^2+Y^2 = k^2$, show that their coordinates are connected by the equations

$$X' = \frac{k^2X}{X^2+Y^2}, \qquad Y' = \frac{k^2Y}{X^2+Y^2}.$$

Establish the following properties of the self-transformation of the plane in which every point P goes over into its inverse point P' (i.e. *inversion* with respect to c):

(i) a circle which does not pass through the centre O of c transforms into a circle;

(ii) a circle through O transforms into a straight line parallel to the tangent to the original circle at O;

(iii) inversion preserves tangency;

(iv) inversion preserves the (unsensed) angle of intersection of two curves;

(v) a circle k and a pair of points A, B, inverse for k, transform into a circle k' and a pair of points A', B', inverse for k'.

21. Prove that the transformations of inversion defined by two circles c_1 and c_2 are commutative with each other if and only if c_1 and c_2 cut orthogonally.

22. Prove that inversion with respect to a circle c transforms a general conic into a bicircular quartic, i.e. a curve of the fourth order which passes twice through each of the circular points.

Investigate the effect of inversion in a circle whose centre is at the origin on the lemniscate whose polar equation is $r^2 \cos 2\theta = a^2$.

PART II
ABSTRACT PROJECTIVE GEOMETRY

CHAPTER III
PROJECTIVE GEOMETRY OF ONE DIMENSION

§ 1. INTRODUCTION. FORMAL DEFINITION OF PROJECTIVE GEOMETRY IN GENERAL

THE somewhat informal discussion of projective geometry in Part I was intended to introduce the reader to the intuitive geometrical notions which it is our present purpose to formalize and to treat by rigorous mathematical argument. We shall now begin afresh and work out the details of a system of abstract projective geometry based on a foundation of pure algebra. The algebra that will be presupposed is essentially elementary, consisting mainly of linear algebra and the simplest portions of the theory of groups. Two key theorems, which will be applied again and again in the course of the development, are given in the appendix at the end of the book.

Although the formal system of projective geometry is logically self-contained and independent of geometrical intuition, the full significance of the formal steps is only to be grasped by bearing in mind all the time the concrete realization of the abstract model that has already been sketched in Part I. In order to help the reader to do this we shall occasionally insert remarks that form no part of the formal system.

The n-dimensional projective domain

Our plan is to discuss projective spaces of one, two, and three dimensions, in this order, and the present chapter will be devoted to one-dimensional projective geometry. Before we begin this systematic treatment, however, it is desirable to give once for all a general definition of projective space of n dimensions, which covers all the particular cases at the same time. In order to do this we first make precise the notion of a coordinate representation.

Let C be a given class of entities (whose nature may remain unspecified). We say that C admits of an n-dimensional representation by homogeneous coordinates belonging to a given field K if a correspondence can be set up between the elements of C and the ordered $(n+1)$-tuples $(x_0, x_1, \ldots, x_n)$ of elements of K—i.e. the vectors $\mathbf{x}$ in the vector space $V_{n+1}(K)$—in such a way that (i) to every vector $\mathbf{x}$, other than the null vector $\mathbf{0}$, corresponds a unique element of C, and (ii) two vectors $\mathbf{x}$, $\mathbf{y}$ are associated with the same element of C if and only if $\mathbf{y} = \lambda\mathbf{x}$, where λ is a non-zero element of K.

The field K from which the coordinates x_i are drawn is usually referred to as the ground field. The choice of ground field is quite arbitrary as far as the abstract concept of a coordinate representation is concerned, but when once a particular field has been selected the choice must be adhered to.

A coordinate representation of C of the kind just defined gives rise to a derived representation, by non-homogeneous coordinates, of all those elements of C for which $x_0 \neq 0$. If, in fact, we put $X_i = x_i/x_0$ $(i = 1, 2, \ldots, n)$, each element of C for which $x_0 \neq 0$ is represented uniquely by the n non-homogeneous coordinates $(X_1, X_2, \ldots, X_n)$. Conversely, if we begin with a non-homogeneous representation of a class C', we can derive an associated homogeneous representation if we adjoin to C' new elements which correspond to coordinate vectors with $x_0 = 0$—just as we have already adjoined points at infinity to the affine plane.

DEFINITIONS. An *n-dimensional projective domain over the ground field K*, or *projective space* $S_n(K)$, is a set of entities (usually called the *points* of the space) that admits of a certain class $(\mathscr{R})$ of *allowable representations* by homogeneous coordinates $(x_0, x_1, \ldots, x_n)$ in K, this class being such that, if $\mathscr{R}_0$ is any one allowable representation, the whole class $(\mathscr{R})$ consists of all those representations that can be obtained from $\mathscr{R}_0$ by non-singular linear transformation

$$x_i' = \sum_{k=0}^{n} a_{ik} x_k \quad (i = 0, 1, \ldots, n).$$

The *projective properties* of $S_n(K)$ are those properties of which the expression in every allowable coordinate system $\mathscr{R}$ is the same. The totality of all these properties is the *projective geometry* of $S_n(K)$, or *n-dimensional projective geometry over K*.

It will be seen from the above definitions that the allowable representations $\mathscr{R}$ of $S_n(K)$ are connected by a group of non-singular linear transformations. The group is called the *projective group* $PGL(n; K)$, and from the algebraic point of view n-dimensional projective geometry over K is simply the invariant theory of this group.

Many of the projective properties of $S_n(K)$ are the same whatever field is taken as K, but others are dependent on the particular choice of ground field. *When nothing is said to the contrary, we shall assume that K is the field of all complex numbers.* This choice of ground field is particularly convenient in view of the fact that the complex field is algebraically closed, every equation with complex coefficients having just as many roots as its degree indicates. When we wish to confine our attention to real projective geometry we must take as ground field the field of all real numbers; and since the real field is a subfield of the complex field we can do this by picking out the real points in complex space and considering them alone.

We can, if we choose, take as ground field one of the very special Galois fields, which have only a finite number of elements, and then we obtain a projective space with a finite number of points. For details of the finite geometries see, for example, Veblen and Young, *Projective Geometry*, I, 201 and the references there given.

§ 2. Projective Geometry of One Dimension

We now take up the systematic treatment of the simplest projective space, namely the projective line $S_1(K)$, which we get by taking n to be 1. Since we are now supposing that the ground field is the field of complex numbers, we may suppress explicit reference to K, and we shall accordingly denote the complex projective line by S_1.

In any allowable representation $\mathscr{R}$ of S_1, each point is given by a pair of homogeneous coordinates (x_0, x_1), and for every non-zero value of λ the coordinates $(\lambda x_0, \lambda x_1)$ represent the same point.

THEOREM 1. *If the coordinates of three points of the line are specified, the representation $\mathscr{R}$ is uniquely determined; and the coordinates of the three points may be chosen arbitrarily, as long as no two pairs of coordinates are proportional.*

Proof. Take a fixed allowable representation $\mathscr{R}_0$ of S_1; and let A, B, C be three (distinct) points of S_1, represented in $\mathscr{R}_0$ by coordinate vectors $\mathbf{a}_0$, $\mathbf{b}_0$, $\mathbf{c}_0$. Then every two of these three vectors are linearly independent. If, now, $\mathbf{a}$, $\mathbf{b}$, $\mathbf{c}$ are three given coordinate vectors, no two of which differ only by a scalar factor, the vectors $\mathbf{a}$, $\mathbf{b}$, $\mathbf{c}$ are linearly independent in pairs, and it follows from Theorem 1 of the Appendix that there is a unique non-singular linear transformation which transforms $\mathbf{a}_0$, $\mathbf{b}_0$, $\mathbf{c}_0$ into $\mathbf{a}$, $\mathbf{b}$, $\mathbf{c}$.

This transformation then transforms $\mathscr{R}_0$ into an allowable representation $\mathscr{R}$ in which A, B, C are represented respectively by the vectors **a**, **b**, **c**, and $\mathscr{R}$ is clearly the only representation with this property.

Remarks

(i) The points whose coordinates are $(1, 0)$ and $(0, 1)$ are called the *reference points* X_0, X_1, and the point whose coordinates are $(1, 1)$ is called the *unit point* E for the representation $\mathscr{R}$. It follows that $\mathscr{R}$ may be defined by specifying these three points.

EXERCISE. Find the equations of transformation from a given coordinate representation $\mathscr{R}_0$ to a new representation $\mathscr{R}$ if the reference points and unit point of $\mathscr{R}$ are represented in $\mathscr{R}_0$ by the coordinates $(3, 2)$, $(2, 1)$, $(-1, 3)$ respectively.

(ii) By choosing the representation $\mathscr{R}$ suitably, we can arrange for any three distinct points to have any assigned linearly independent coordinates. This means that no triad of points of S_1 is projectively different from any other.

§ 3. The Projective Parameter

The algebra that is involved in one-dimensional projective geometry is so simple that the elaborate apparatus of suffix notation is frequently an encumbrance. It is useful to formulate general results in this language in order to exhibit their relation to the systems of projective geometry of higher dimensionality, but when we are handling specific problems we usually do better to work with a single non-homogeneous coordinate $\theta = x_0/x_1$. We call θ the *projective parameter* associated with the coordinate representation $\mathscr{R}$.

The point whose projective parameter is θ is simply the point $(\theta, 1)$, and by letting θ run through all complex values we obtain all the points of S_1 with the single exception of the reference point X_0, whose coordinates are $(1, 0)$. In order to include this point as well, we have to allow θ to take the improper value ∞, which is not a number of the complex field. Infinite values of the parameter can be avoided altogether by going back to the pair of homogeneous coordinates; but with a little experience one discovers how to perform equivalent formal manipulations with the symbol ∞ itself, and this often simplifies the algebraic working.

The results that we have already obtained may now be restated in the language of projective parameters.

(i) When the representation is changed from $\mathscr{R}$ to $\mathscr{R}'$, the projective parameter undergoes a transformation of the form

$$\theta' = \frac{a\theta+b}{c\theta+d},$$

where $ad-bc \neq 0$; and conversely, every such transformation may be interpreted as the transformation from $\mathscr{R}$ to a second allowable representation.

(ii) An allowable representation of the points of S_1 by a projective parameter is uniquely determined when the parameters ∞, 0, 1 are assigned to three points. These points are the reference points and unit point of the representation.

(iii) More generally, any three distinct values θ_1, θ_2, θ_3 may be assigned to three chosen points, and the representation is then fixed.

EXAMPLE. An ordinary euclidean line, completed by its point at infinity I, provides a concrete realization of the projective space $S_1(R)$, R being the field of real numbers. If we choose a point O of the line as origin and a segment OE as unit segment, every point P has a uniquely defined cartesian coordinate X, given by $\overrightarrow{OP} = X.\overrightarrow{OE}$. X is, of course, a projective parameter, I and O being the reference points and E the unit point.

If X_0, X_1 are any two finite points of the line we can easily define a projective parameter for which X_0 and X_1 are reference points—namely the position ratio $\overrightarrow{X_1 P}/\overrightarrow{PX_0}$. In this representation of the line, the unit point is the mid-point of the segment $X_1 X_0$.

If we wish to obtain the most general allowable representation of $S_1(R)$ we must find a way of choosing the unit point arbitrarily as well as the reference points, and this can be done by taking as parameter θ a fixed multiple of the position ratio:

$$\theta = k\frac{\overrightarrow{X_1 P}}{\overrightarrow{PX_0}}.$$

If a given point E is to be the unit point, then we must have

$$1 = k\frac{\overrightarrow{X_1 E}}{\overrightarrow{EX_0}}.$$

The parameter θ is therefore given by

$$\begin{aligned}\theta &= \frac{\overrightarrow{X_1 P}}{\overrightarrow{PX_0}}\Big/\frac{\overrightarrow{X_1 E}}{\overrightarrow{EX_0}} \\ &= \{P, E; X_1, X_0\} \\ &= \{E, P; X_0, X_1\}.\end{aligned}$$

Thus the most general projective parameter for the euclidean line is not position ratio but a cross ratio.

§ 4. Cross Ratio

The cardinal role already assumed by cross ratio in our informal treatment of projective geometry is an indication that this concept will also be important in the formal theory. Before proceeding farther, therefore, we shall define cross ratio algebraically and establish its fundamental properties.

DEFINITION. If (x_1, x_2) and (x_3, x_4) are two ordered pairs of elements of a field K, the rational function

$$\frac{x_1-x_3}{x_2-x_3}\Big/\frac{x_1-x_4}{x_2-x_4}$$

is called the *cross ratio* of the second pair with respect to the first.

Since the rational function is unaltered if the two ordered pairs are interchanged, it is also the cross ratio of the first pair with respect to the second, and we may refer to it simply as the cross ratio of the two ordered pairs. We shall denote it by $\{x_1, x_2; x_3, x_4\}$.

It may easily be verified that, although the value of the cross ratio depends upon the order of each pair, if the orders of both pairs are changed simultaneously the value of the cross ratio is unaltered, i.e.

$$\{x_1, x_2; x_3, x_4\} = \{x_2, x_1; x_4, x_3\}.$$

THEOREM 2. *Cross ratio is invariant over non-singular bilinear transformation; i.e. if*

$$x_i = \frac{ax'_i+b}{cx'_i+d} \quad (i = 1, 2, 3, 4),$$

where $ad-bc \neq 0$, *then*

$$\{x_1, x_2; x_3, x_4\} = \{x'_1, x'_2; x'_3, x'_4\}.$$

Proof. By direct substitution, and cancellation of the non-zero factor $ad-bc$.

COROLLARY. *The unique bilinear transformation which is defined by the three corresponding pairs* (x_i, x'_i) $(i = 1, 2, 3)$ *may be written*

$$\{x, x_1; x_2, x_3\} = \{x', x'_1; x'_2, x'_3\}.$$

EXERCISE. Verify that this equation may be reduced to the form

$$x = \frac{ax'+b}{cx'+d}.$$

THEOREM 3. *If P_1, P_2, P_3, P_4 are four given points of S_1, and θ_1, θ_2, θ_3, θ_4 are their parameters in a representation $\mathcal{R}$, the value of the cross ratio $\{\theta_1, \theta_2; \theta_3, \theta_4\}$ is independent of the particular choice of $\mathcal{R}$.*

Proof. Since a change of the representation $\mathcal{R}$ is equivalent to a bilinear transformation of the parameter, this theorem is an immediate consequence of Theorem 2.

Theorem 3 states, in effect, that $\{\theta_1, \theta_2; \theta_3, \theta_4\}$ is a projective characteristic of the two ordered pairs of points (P_1, P_2) and (P_3, P_4), and this makes legitimate the following definition.

DEFINITION. If (P_1, P_2), (P_3, P_4) are two ordered pairs of points in S_1, their *cross ratio* is defined as

$$\{P_1, P_2; P_3, P_4\} = \{\theta_1, \theta_2; \theta_3, \theta_4\},$$

where θ_1, θ_2, θ_3, θ_4 are the parameters of P_1, P_2, P_3, P_4 in some allowable representation $\mathcal{R}$.

THEOREM 4. *The cross ratio $\{P_1, P_2; P_3, P_4\}$ is equal to the ratio x_0/x_1, where (x_0, x_1) are coordinates of P_2 in the representation $\mathcal{R}$ of S_1 for which P_3, P_4 are the reference points and P_1 is the unit point.*

Proof. If, in some allowable representation, the coordinates of P_i are $(x_0^{(i)}, x_1^{(i)})$ $(i = 1, 2, 3, 4)$, the value of the cross ratio is

$$\frac{\dfrac{x_0^{(1)}}{x_1^{(1)}} - \dfrac{x_0^{(3)}}{x_1^{(3)}}}{\dfrac{x_0^{(2)}}{x_1^{(2)}} - \dfrac{x_0^{(3)}}{x_1^{(3)}}} \div \frac{\dfrac{x_0^{(1)}}{x_1^{(1)}} - \dfrac{x_0^{(4)}}{x_1^{(4)}}}{\dfrac{x_0^{(2)}}{x_1^{(2)}} - \dfrac{x_0^{(4)}}{x_1^{(4)}}} = \frac{x_0^{(1)}x_1^{(3)} - x_1^{(1)}x_0^{(3)}}{x_0^{(2)}x_1^{(3)} - x_1^{(2)}x_0^{(3)}} \Big/ \frac{x_0^{(1)}x_1^{(4)} - x_1^{(1)}x_0^{(4)}}{x_0^{(2)}x_1^{(4)} - x_1^{(2)}x_0^{(4)}}.$$

Substituting the coordinates $(1, 1)$, (x_0, x_1), $(1, 0)$, $(0, 1)$ we obtain the value

$$\frac{1.0 - 1.1}{x_0.0 - x_1.1} \Big/ \frac{1.1 - 1.0}{x_0.1 - x_1.0},$$

i.e. x_0/x_1.

Remark. We have given the calculation in full in order to show how difficulties with the improper number ∞ can be avoided by working with homogeneous coordinates. In practice we should compute the cross ratio less rigorously as follows:

$$\{P_1, P_2; P_3, P_4\} = \{1, \theta; \infty, 0\} = \frac{1-\infty}{\theta-\infty} \Big/ \frac{1-0}{\theta-0} = 1 \Big/ \frac{1}{\theta} = \theta.$$

§ 5. THE SIX CROSS RATIOS OF FOUR POINTS

The cross ratio $\{P_1, P_2; P_3, P_4\}$ has been defined as a function of the two ordered pairs (P_1, P_2) and (P_3, P_4). If, now, we take an unordered tetrad of points of S_1, we can arrange the points as two ordered pairs in various ways, and we then have a number of different cross ratios associated with the same tetrad. There are twenty-four possible arrangements of the four points P_i, and therefore twenty-four possible cross ratios $\{P_\alpha, P_\beta; P_\gamma, P_\delta\}$. But since the cross ratio is unaltered if the two pairs are interchanged or if the points of both pairs are transposed simultaneously, each of the cross ratios is equal to a cross ratio of the form $\{P_\alpha, P_\beta; P_\gamma, P_4\}$, with P_4 in the last place. Thus we have at most six distinct cross ratios: $\{P_1, P_2; P_3, P_4\}$, $\{P_1, P_3; P_2, P_4\}$, $\{P_2, P_1; P_3, P_4\}$, $\{P_2, P_3; P_1, P_4\}$, $\{P_3, P_1; P_2, P_4\}$, $\{P_3, P_2; P_1, P_4\}$. When the tetrad of points is general the six values are in fact distinct, for if the first is denoted by λ the others are respectively $1-\lambda, \frac{1}{\lambda}, 1-\frac{1}{\lambda}, \frac{1}{1-\lambda}, \frac{\lambda}{\lambda-1}$.

The set of six rational functions of λ that we have just written down is of considerable interest, as all six of the functions may be obtained from any one of them by carrying out alternately the two operations of subtracting from unity and forming the reciprocal. It follows from this fact that any symmetric function of the six cross ratios is a projective invariant of the unordered tetrad of points $\{P_1, P_2, P_3, P_4\}$. Many such symmetric functions are constants—for example, the product of the six cross ratios—but there are non-trivial ones. The sum of the squares of the six cross ratios is such a function. (Cf. p. 69, Ex. 18.)

The six expressions $\lambda, 1-\lambda, \frac{1}{\lambda}, 1-\frac{1}{\lambda}, \frac{1}{1-\lambda}$, and $\frac{\lambda}{\lambda-1}$ are distinct functions of λ, but when a numerical value is given to λ the values of the functions need not all be different. This means that if the four points P_1, P_2, P_3, P_4 happen to be related in a suitable way some of the six cross ratios formed from them may be equal. A complete catalogue of the special cases may easily be drawn up by equating λ successively to the various expressions given above, solving the equations, and interpreting the results geometrically. It is found that there are three and only three such special cases.

(i) The six values, in some order, are $1, 1, 0, 0, \infty, \infty$. This case arises when two of the four points coincide.

(ii) The six values are -1, -1, $\frac{1}{2}$, $\frac{1}{2}$, 2, 2. This is the important case of harmonic separation, to be discussed in detail immediately below.

(iii) The six values are $-\omega$, $-\omega$, $-\omega$, $-\omega^2$, $-\omega^2$, $-\omega^2$, where $\omega = e^{2\pi i/3}$. In this case the four points, which cannot all have real parameters, are said to form an *equianharmonic tetrad*.

It will be noticed that in every case all the values of the cross ratio occur the same number of times in the full set of twenty-four—four times in the general case, eight in the first two special cases, and twelve in the last case.

§ 6. The Harmonic Relation

We say that the pairs of points (P_1, P_2) and (P_3, P_4) are *harmonic* if $\{P_1, P_2; P_3, P_4\} = -1$. This means that there exists a representation $\mathscr{R}$ in which the four points have parameters 0, ∞, 1, -1 respectively.

If $\{P_1, P_2; P_3, P_4\} = -1$, then

$$\{P_1, P_2; P_4, P_3\} = \{P_2, P_1; P_3, P_4\} = \{P_3, P_4; P_1, P_2\} = -1.$$

Thus if the pairs (P_1, P_2) and (P_3, P_4) are harmonic, so also are the pairs (P_1, P_2) and (P_4, P_3), (P_2, P_1) and (P_3, P_4), and (P_3, P_4) and (P_1, P_2). In other words, *the harmonic relation is independent of the order of the points within each pair and of the order of the pairs.* We may say that the unordered pairs (P_1, P_2) and (P_3, P_4) separate each other harmonically; and this is a projective relation between them.

Exercise. Interpret geometrically the harmonic relation between two pairs of points on the euclidean line, using the parameter defined on p. 44. Show that if two pairs are harmonic then they separate each other in the ordinary sense.

The harmonic relation is a fundamental relation in projective geometry, and in view of its importance we shall now give three different formulations of the criterion for two pairs to be harmonic.

Condition 1. The pairs (P_1, P_2) and (P_3, P_4) are harmonic if and only if $\{P_1, P_2; P_3, P_4\} = \{P_1, P_2; P_4, P_3\}$.

Condition 2. If, in some representation $\mathscr{R}$, the parameters of P_1, P_2, P_3, P_4 are θ_1, θ_2, θ_3, θ_4, the pairs (P_1, P_2) and (P_3, P_4) are harmonic if and only if

$$(\theta_1+\theta_2)(\theta_3+\theta_4) = 2(\theta_1\theta_2+\theta_3\theta_4).$$

Condition 3. If the parameters of (P_1, P_2) and (P_3, P_4) are given respectively by the quadratic equations $a\theta^2+2h\theta+b = 0$ and $a'\theta^2+2h'\theta+b' = 0$, then the pairs are harmonic if and only if

$$2hh'-ab'-a'b = 0.$$

(This condition is the polarized form of the condition for the first quadratic equation to have equal roots, namely $h^2-ab = 0$.)

The reader should verify that, when the four points P_1, P_2, P_3, P_4 are distinct, the above three conditions are equivalent to each other and to the original definition of the harmonic relation.

When coincidences occur among the points, we encounter difficulties in trying to use the original definition, for the cross ratio may then become an indeterminate expression. If we use a continuity argument we can show that if P_3 coincides with P_2 the harmonic relation holds if and only if P_4 or P_1 also coincides with P_2 and P_3, which means that two pairs can only be harmonic if either the four points are distinct or at least three of them coincide. Since, however, we are basing our system of projective geometry on an algebraic foundation, the appeal to continuity is not legitimate, and we must try to dispense with it. We can do so by turning to Condition 2 or Condition 3. Both these conditions are satisfied in the two cases just mentioned, and in these cases only. If, therefore, we take either of these criteria as our definition of the harmonic relation, the new definition is equivalent to the old one taken in conjunction with the argument in terms of continuity. *We shall accordingly regard the harmonic relation from now on as defined by the equation*

$$(\theta_1+\theta_2)(\theta_3+\theta_4) = 2(\theta_1\theta_2+\theta_3\theta_4).$$

The above reference to continuity is of some interest. We can eliminate analytical ideas from projective geometry and make the subject purely algebraic if we take care to avoid quotients and rational functions and to deal always with polynomials. We have already seen in Chapter I how points at infinity can be introduced either analytically, by means of a limiting process, or algebraically, by making the coordinates homogeneous. If we are to carry our programme through consistently, we must dispense with cross ratio, which is a rational function, in order to be able to deal with coincident points; and this is the reason why we have had to modify the definition of the harmonic relation.

Exercise. If $\{\theta_1,\theta_2;\theta_3,\theta_4\} = -1$, show that

(a) if $\theta_4 = 0$, then $\dfrac{1}{\theta_1}+\dfrac{1}{\theta_2} = \dfrac{2}{\theta_3}$;

(b) if $\theta_4 = \infty$, then $\theta_1+\theta_2 = 2\theta_3$.

Interpret these results geometrically when θ is an ordinary cartesian coordinate on a euclidean line.

§ 7. Homographies

The only projective property of S_1 that we have met so far is the fact that any two ordered pairs of points have a cross ratio—so that, in particular, the harmonic relation may possibly hold between them. In order to develop further the geometry of S_1 we need to introduce an altogether new idea, namely that of homographic transformation or correspondence.

DEFINITION. A non-singular linear transformation of the line into itself is called a *homographic transformation* or *homography*.

Let ϖ be a given homographic transformation. If a representation $\mathscr{R}$ is chosen, the coordinates of any two corresponding points will be connected by fixed equations

$$\begin{aligned} \rho x_0' &= a_{00}x_0+a_{01}x_1 \\ \rho x_1' &= a_{10}x_0+a_{11}x_1 \end{aligned} \quad \left(\begin{vmatrix} a_{00} & a_{01} \\ a_{10} & a_{11} \end{vmatrix} \neq 0\right),$$

or, as we usually write,

$$\rho x_i' = \sum_{k=0}^{1} a_{ik}x_k \quad (i = 0, 1),$$

where $|a_{rs}| \neq 0$ and ρ is an arbitrary factor of proportionality. Since the coordinates are homogeneous, we still have the same transformation of points when the factor ρ is suppressed, and we shall always suppose this done. The transformation ϖ may thus be written algebraically as

$$x_i' = \sum_{k=0}^{1} a_{ik}x_k \quad (i = 0, 1)$$

or, in matrix notation,

$$\mathbf{x}' = \mathbf{A}\mathbf{x} \quad (|\mathbf{A}| \neq 0),$$

where the coordinate vector $\mathbf{x}$ is the 2×1 matrix $\begin{pmatrix} x_0 \\ x_1 \end{pmatrix}$.

If, on the other hand, we choose to work with the projective parameter $\theta = x_0/x_1$ we can write the same transformation in the equivalent form

$$\theta' = \frac{a_{00}\theta+a_{01}}{a_{10}\theta+a_{11}} \quad (|a_{rs}| \neq 0).$$

It is usually more convenient to drop the suffixes, writing simply

$$\theta' = \frac{\alpha\theta+\beta}{\gamma\theta+\delta} \quad (\alpha\delta-\beta\gamma \neq 0).$$

The equation may be solved for θ in terms of θ', and we have

$$\theta = \frac{\delta\theta'-\beta}{-\gamma\theta'+\alpha}.$$

This is again a non-singular transformation, the inverse transformation ϖ^{-1}.

The two transformations ϖ and ϖ^{-1} together give a (1, 1) correspondence of the line with itself. The equation of this correspondence may be written in the symmetrical form

$$\gamma\theta\theta'-\alpha\theta+\delta\theta'-\beta = 0$$

or

$$a\theta\theta'+b\theta+c\theta'+d = 0 \quad (ad-bc \neq 0).$$

In terms of the coordinates x_0, x_1 this becomes

$$ax_0 x_0'+bx_0 x_1'+cx_0' x_1+dx_1 x_1' = 0.$$

If the condition $\alpha\delta-\beta\gamma \neq 0$ is not satisfied, the equation

$$\gamma\theta\theta'-\alpha\theta+\delta\theta'-\beta = 0$$

may be written as

$$\beta\gamma\theta\theta'-\alpha\beta\theta+\beta\delta\theta'-\beta^2 = 0,$$

i.e.

$$\alpha\delta\theta\theta'-\alpha\beta\theta+\beta\delta\theta'-\beta^2 = 0,$$

i.e.

$$(\alpha\theta+\beta)(\delta\theta'-\beta) = 0.$$

When this is the case, every value of θ gives rise to the same value β/δ of θ', and the transformation is degenerate.

There are thus several different ways of writing the equation of a homography, all of which are useful on different occasions. We still have to show, however, that the definition of homography is itself legitimate, i.e. that homography is a projective concept. This is the content of the next theorem.

THEOREM 5. *If two allowable representations $\mathscr{R}$, $\overline{\mathscr{R}}$ are connected by the equation*

$$\mathbf{x} = \mathbf{P}\bar{\mathbf{x}} \quad (|\mathbf{P}| \neq 0),$$

and if a transformation ϖ is represented in $\mathscr{R}$ by an equation of the form

$$\mathbf{x}' = \mathbf{A}\mathbf{x} \quad (|\mathbf{A}| \neq 0),$$

then ϖ is represented in $\overline{\mathscr{R}}$ by an equation of the same form

$$\bar{\mathbf{x}}' = \overline{\mathbf{A}}\bar{\mathbf{x}} \quad (|\overline{\mathbf{A}}| \neq 0).$$

Proof.

If

$$\mathbf{x}' = \mathbf{A}\mathbf{x},$$

then

$$\mathbf{P}\bar{\mathbf{x}}' = \mathbf{A}\mathbf{P}\bar{\mathbf{x}},$$

and hence
$$\bar{\mathbf{x}}' = \mathbf{P}^{-1}\mathbf{A}\mathbf{P}\bar{\mathbf{x}}$$
$$= \bar{\mathbf{A}}\bar{\mathbf{x}}, \text{ say.}$$

Then
$$|\bar{\mathbf{A}}| = |\mathbf{P}^{-1}\mathbf{A}\mathbf{P}|$$
$$= |\mathbf{P}^{-1}|.|\mathbf{A}|.|\mathbf{P}|$$
$$= |\mathbf{P}|^{-1}.|\mathbf{A}|.|\mathbf{P}|$$
$$= |\mathbf{A}|$$
$$\neq 0.$$

Since a homographic transformation ϖ, referred to some fixed representation $\mathscr{R}$, is of the same algebraic form as the transformation from one allowable representation to another, our earlier theorems on choice of the representation can be reformulated as theorems on homographies. In view of the fundamental importance of these simple properties of homographies we shall state the results explicitly.

THEOREM 6. *Every homographic transformation leaves cross ratio invariant.*

THEOREM 7. *There exists a unique homography ϖ which carries over three given distinct points P_i into three given distinct points P'_i, some or all of which may coincide with the P_i.*

COROLLARY 1. *The homography determined by the three pairs (θ_i, θ'_i) $(i = 1, 2, 3)$ has the equation*
$$\{\theta', \theta'_1; \theta'_2, \theta'_3\} = \{\theta, \theta_1; \theta_2, \theta_3\}.$$

COROLLARY 2. *Four given pairs of points (P_i, P'_i) $(i = 1, 2, 3, 4)$ correspond in some homography if and only if*
$$\{P'_1, P'_2; P'_3, P'_4\} = \{P_1, P_2; P_3, P_4\}.$$

COROLLARY 3. *If a $(1, 1)$ transformation of the line into itself leaves cross ratio invariant, then it is a homography.*

THEOREM 8. *The set of all homographic transformations of the line into itself is a group.*

Proof. Let ϖ_i $(i = 1, 2)$ be a homography given by
$$\varpi_i(\mathbf{x}) = \mathbf{A}_i.\mathbf{x} \quad (|\mathbf{A}_i| \neq 0).$$

Then
$$\varpi_1\{\varpi_2(\mathbf{x})\} = \varpi_1(\mathbf{A}_2\,\mathbf{x})$$
$$= \mathbf{A}_1.\mathbf{A}_2\,\mathbf{x}$$
$$= \mathbf{A}_1\mathbf{A}_2.\mathbf{x} \quad (|\mathbf{A}_1\mathbf{A}_2| \neq 0).$$

Thus when ϖ_2 and ϖ_1 are applied in succession, the product transformation $\varpi_1 \varpi_2$ is itself a homography with matrix $\mathbf{A}_1 \mathbf{A}_2$.

Corresponding to the unit matrix $\mathbf{I} = (\delta_{rs}) = \begin{pmatrix} 1 & 0 \\ 0 & 1 \end{pmatrix}$ we have a homography ϵ; and this clearly satisfies the equation

$$\epsilon\varpi = \varpi\epsilon = \varpi$$

for every ϖ.

Finally, every homography ϖ has an inverse ϖ^{-1}, already defined above, such that

$$\varpi^{-1}\varpi = \varpi\varpi^{-1} = \epsilon.$$

Thus the set of all homographies is a group in which ϵ, the *identical homography*, is the identity element.

EXERCISES

(i) If ϖ has matrix $\mathbf{A}$, show that ϖ^{-1} has matrix $\mathbf{A}^{-1}$.

(ii) Show, by means of an example, that the group of homographies is non-abelian, i.e. that there exist homographies ϖ_1, ϖ_2 such that

$$\varpi_1 \varpi_2 \neq \varpi_2 \varpi_1.$$

§ 8. Repetition of a Homographic Transformation

Consider a fixed homography ϖ, given, in terms of a chosen representation $\mathscr{R}$, by

$$\theta' = \varpi(\theta) = \frac{\alpha\theta+\beta}{\gamma\theta+\delta} \quad (\alpha\delta-\beta\gamma \neq 0).$$

The transformation may be represented diagrammatically as follows:

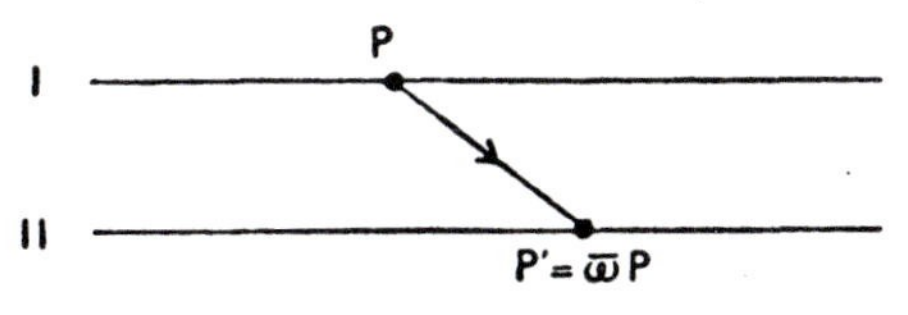

the line being drawn twice because it functions in two different capacities.

Now suppose we choose a point P and transform it repeatedly by ϖ. In this way we obtain a sequence of points: P, $P' = \varpi P$, $P'' = \varpi^2 P$,...; and there are two possible cases to be distinguished. The points may all be different, so that P gives rise to an infinite sequence of distinct points, or we may return to P after a finite number of steps. In the second case the sequence of points

$$P, P', P'', \ldots$$

consists of a finite set of points recurring cyclically; and since this situation is of considerable geometrical interest we shall examine in detail how it can arise.

(i) Suppose $P' = P$. In this case P is said to be a *self-corresponding* or *united point* of ϖ. If $\varpi = \epsilon$, every point of the line is self-corresponding; but if $\varpi \neq \epsilon$ there are just two self-corresponding points, which may possibly coincide. Their parameters are the roots of the equation

$$\theta = \frac{\alpha\theta+\beta}{\gamma\theta+\delta},$$

i.e.

$$\gamma\theta^2+(\delta-\alpha)\theta-\beta = 0.$$

If $(\delta-\alpha)^2+4\beta\gamma = 0$, the two self-corresponding points coincide; and in this case ϖ is said to be an *elation*.

If $(\delta-\alpha)^2+4\beta\gamma \neq 0$, there are two distinct self-corresponding points, which we usually denote by M and N.

(ii) Suppose $P' \neq P$, $P'' = P$.

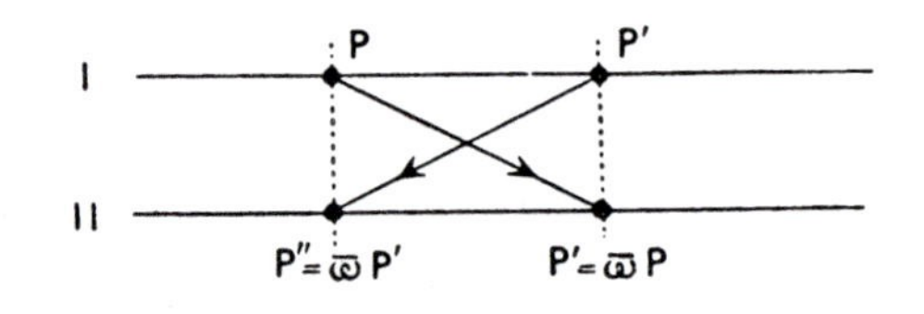

In this case $$\varpi^2 P = P$$
and hence

$$\varpi^2 P' = \varpi^2 . \varpi P = \varpi^3 P = \varpi . \varpi^2 P = \varpi P = P'.$$

Thus P and P' are united points of the homography ϖ^2. But ϖ has at least one united point M, necessarily distinct from P and P', and M is a united point of ϖ^2 also. The homography ϖ^2 thus has three distinct pairs (P, P), (P', P'), (M, M) in common with ϵ, and hence, by Theorem 7, $\varpi^2 = \epsilon$.

This relation may also be written $\varpi = \varpi^{-1}$, and it means that ϖ is self-inverse or, as we usually say, *involutory*. We have therefore the result:

THEOREM 9. *If there exists one point P, not a united point of ϖ, such that $\varpi^2 P = P$, then ϖ is involutory. The relation $\varpi^2 P = P$ then holds for every position of P.*

Taking the transformations ϖ and ϖ^{-1} together as a single (1, 1) correspondence, we may say that if a homographic correspondence has one symmetrical pair of distinct elements then all its pairs are symmetrical. Such a symmetrical homographic correspondence is called an *involution.*

THEOREM 10. *A homographic correspondence (other than the identical correspondence) represented by*

$$a\theta\theta'+b\theta+c\theta'+d = 0$$

is an involution if and only if $b = c$.

Proof. If $b = c$, the equation is symmetric, and the correspondence defined by it is clearly involutory.

If, conversely, the correspondence has an involutory pair (θ, θ') of distinct elements, then

$$a\theta\theta'+b\theta+c\theta'+d = 0$$

and $$a\theta\theta'+b\theta'+c\theta+d = 0.$$

By subtraction, $$(b-c)(\theta-\theta') = 0,$$

and hence $$b-c = 0.$$

(iii) We could now go on to consider the cases $P' \neq P$, $P'' \neq P$, $P''' = P$, and so on, but these are not of sufficient geometrical interest at this level to justify detailed study. The main results are given in Exercise 23 at the end of this chapter.

EXERCISE. Show that, if a homography ϖ has a cyclic triad of distinct points P, P', P'', such that $\varpi P = P'$, $\varpi P' = P''$, $\varpi P'' = P$, then $\varpi^3 = \epsilon$, and every point of the line belongs to a unique cyclic triad.

§ 9. Canonical Equations of Homographies

In any allowable representation $\mathscr{R}$, the equation of a given homography ϖ is necessarily of the bilinear form

$$a\theta\theta'+b\theta+c\theta'+d = 0.$$

If, instead of an arbitrary frame of reference, we take one that is specially related to the correspondence, we can simplify the equation by making some of the coefficients take special values, and in this way we obtain various canonical forms of the equation of a homography. The most obvious way to simplify the equation is by relating the reference points X_0, X_1 to the united points of ϖ.

Consider first the case in which the united points M, N are distinct, so that they may be taken as the points X_0, X_1 with parameters ∞, 0 respectively. If the equation of ϖ is

$$a\theta\theta'+b\theta+c\theta'+d = 0,$$

$0, \infty$ are the roots of the quadratic equation

$$a\theta^2+(b+c)\theta+d = 0.$$

Thus $a = d = 0$, and the equation of ϖ reduces to

$$b\theta+c\theta' = 0,$$

or

$$\theta' = k\theta.$$

This is the *canonical form of the equation of a homography with distinct united points.*

If $\{P, P')$ is a general corresponding pair, then

$$\{M, N; P, P'\} = \{\infty, 0; \theta, k\theta\} = k.$$

This gives us the important result:

THEOREM 11. *If ϖ has distinct united points M, N, and if (P, P') is a variable corresponding pair of distinct points, the value of the cross ratio $\{M, N; P, P'\}$ is constant.*

Remarks

(i) The constant k is called the *modulus* of ϖ. Whether the modulus is k or $1/k$ depends upon the order that is assigned to the united points.

(ii) If ϖ has united points M, N and modulus k, ϖ^{-1} has united points M, N and modulus $1/k$.

(iii) ϖ is completely determined by the ordered pair of united points and the modulus.

(iv) $\varpi = \epsilon$ if and only if $k = 1$, and $\varpi = \varpi^{-1}$ if and only if $k = \pm 1$.

(v) Since the modulus k is interpreted as a cross ratio of points, it is a projective characteristic of ϖ, independent of the choice of $\mathscr{R}$.

Now consider the other possible case, in which the united points of ϖ coincide, at M say. We may take this point as X_0, with $\theta = \infty$, and then the roots of the equation

$$a\theta^2+(b+c)\theta+d = 0$$

are both infinite. Thus $a = b+c = 0$, and the equation of ϖ reduces to

$$b\theta-b\theta'+d = 0,$$

or

$$\theta' = \theta+\alpha.$$

This is the *canonical form of the equation of a homography with coincident united points.*

Remarks

(i) Since this last equation is not homogeneous, the constant α, unlike the modulus k in the previous case, is not a projective invariant. For if we change the representation $\mathscr{R}$ by putting

$$x_0 = \lambda\bar{x}_0, \qquad x_1 = \bar{x}_1,$$

so that

$$\theta = \lambda\bar{\theta},$$

the equation of ϖ becomes

$$\lambda\bar{\theta}' = \lambda\bar{\theta}+\alpha,$$

i.e.

$$\bar{\theta}' = \bar{\theta}+\frac{\alpha}{\lambda}.$$

(ii) A homography with coincident united points is uniquely determined by the united point M and one other pair of corresponding points.

EXERCISE. Show that every non-singular bilinear equation

$$\theta' = \frac{\alpha\theta+\beta}{\gamma\theta+\delta}$$

may be written in the form

$$\frac{\theta'-\mu_1}{\theta'-\mu_2} = k\frac{\theta-\mu_1}{\theta-\mu_2},$$

or the form

$$\frac{1}{\theta'-\mu} = \frac{1}{\theta-\mu}+\alpha,$$

according as the associated homography has distinct united points with parameters μ_1, μ_2 or coincident united points with parameter μ. Reduce to one or other of these forms each of the equations

$$\theta' = \frac{4-5\theta}{2\theta-7}, \qquad \theta' = -\frac{11\theta+12}{3\theta+1}.$$

We now have at our disposal all the equipment that is needed in order to establish a large number of properties of homographies, namely the following notations and properties:

(i) the matrix representation $\mathbf{x}' = \mathbf{A}\mathbf{x}$;
(ii) the canonical forms $\theta' = k\theta$, $\theta' = \theta+\alpha$;
(iii) the fact that the set of all homographies constitutes a group;
(iv) the further fact that a homography is uniquely determined by three corresponding pairs.

The theorem which follows is a typical result in the theory of homographies. It concerns cyclic homographies, or homographies ϖ such that $\varpi^m = \epsilon$ for some positive integer m.

THEOREM 12. *Every cyclic homography has distinct united points. There exist cyclic homographies of every period, and the modulus of a cyclic homography of period m is a primitive mth root of unity.*

Proof. If ϖ is given by the equation

$$\theta' = \theta+\alpha \quad (\alpha \neq 0),$$

then ϖ^r is given by $\theta' = \theta+r\alpha,$

and therefore $\varpi^r = \epsilon$ if and only if $r = 0$. Thus if ϖ has coincident united points it cannot be cyclic.

If ϖ is given by $\theta' = k\theta,$

then ϖ^r is given by $\theta' = k^r\theta;$

and $\varpi^r = \epsilon$ if and only if $k^r = 1$. In this case there exists a smallest positive integer m for which $k^m = 1$, and k is then a primitive mth root of unity. Conversely, if k is a primitive mth root of unity, the homography given by

$$\theta' = k\theta$$

is cyclic of period m.

§ 10. Properties of Involutions

The involution, or cyclic homography of period 2, has a particularly important place in the general theory of homographies, and we shall accordingly discuss the special properties of involutions in greater detail.

It follows from Theorem 12 that the united points of an involution are necessarily distinct, and that the equation of an involution can be put in the canonical form

$$\theta' = -\theta.$$

The modulus of the involution is -1, and if M, N are the united points the pairs of the involution are simply the pairs of harmonic conjugates with respect to M and N. There is now no need to distinguish between the pairs (P, P') and (P', P), and we often refer to a pair of corresponding points as a *pair of mates* in the involution.

THEOREM 13. *An involution is uniquely determined by two corresponding pairs.*

Proof. The equation

$$a\theta\theta'+b(\theta+\theta')+d = 0$$

involves two effective coefficients; and the involution determined by (θ_1, θ'_1) and (θ_2, θ'_2) is given by

$$\begin{vmatrix} \theta\theta' & \theta+\theta' & 1 \\ \theta_1\theta'_1 & \theta_1+\theta'_1 & 1 \\ \theta_2\theta'_2 & \theta_2+\theta'_2 & 1 \end{vmatrix} = 0.$$

THEOREM 14. *Three given pairs* (P_i, P'_i) $(i = 1, 2, 3)$ *with* $P_1 \neq P'_1$, *correspond in a common involution if and only if*

$$\{P_1, P'_1; P_2, P_3\} = \{P'_1, P_1; P'_2, P'_3\}.$$

Proof. The three pairs determine a unique homography ϖ.

(i) Suppose ϖ is an involution. Then, since $\varpi P_1 = P'_1$, it follows that $\varpi P'_1 = P_1$; and hence ϖ carries over P_1, P'_1, P_2, P_3 into P'_1, P_1, P'_2, P'_3 respectively. Therefore

$$\{P_1, P'_1; P_2, P_3\} = \{P'_1, P_1; P'_2, P'_3\}.$$

(ii) Suppose the two cross ratios are equal. Then there is a homography which transforms P_1, P'_1, P_2, P_3 respectively into P'_1, P_1, P'_2, P'_3, and this homography is the homography ϖ defined by the three pairs (P_i, P'_i). But then ϖ has the involutory pair (P_1, P'_1) and, by Theorem 9, it is an involution. This completes the proof of the theorem.

Alternative forms of the equation of an involution

We obtain a second canonical form for the equation of an involution τ by taking as reference points not the united points of τ but an arbitrary pair of mates. Suppose this is done, and the equation of τ is

$$a\theta\theta' + b(\theta+\theta') + d = 0,$$

i.e.

$$ax_0 x'_0 + b(x_0 x'_1 + x'_0 x_1) + dx_1 x'_1 = 0.$$

Since the points (1, 0) and (0, 1) are corresponding points, $b = 0$; and the equation of τ accordingly reduces to

$$a\theta\theta' + d = 0.$$

The equation may be written as

$$\theta\theta' = m,$$

and this is the required *second canonical form.*

EXERCISES

(i) Show that the constant m is not a projective invariant.

(ii) Why is there no second canonical form for the equation of a non-involutory homography ?

Besides the two canonical equations, we have yet another way of representing an involution algebraically, by using ideas of a rather different kind. An involution is a simply-infinite system of unordered pairs of points, and as such it is represented quite naturally by a symmetric bilinear equation

$$a\theta\theta'+b(\theta+\theta')+d = 0.$$

A single pair of points, with parameters θ_0, θ_0', may, however, be represented by the quadratic equation

$$a\theta^2+2h\theta+b = 0$$

whose roots are θ_0 and θ_0', and we may ask what is the form of the quadratic equation which represents a variable or generic pair of a given involution τ. The answer is very simple, namely that the coefficients in the equation depend linearly on a parameter.

THEOREM 15. *In any fixed representation $\mathscr{R}$, the pairs of a given involution τ may be represented, for varying λ, by a quadratic equation of the form*

$$S+\lambda S' = 0,$$

where $S \equiv a\theta^2+2h\theta+b$ and $S' \equiv a'\theta^2+2h'\theta+b'$. Conversely, such a quadratic equation represents in general the pairs of an involution.

Proof. Take two pairs of the given involution τ, represented respectively by the quadratic equations

$$S \equiv a\theta^2+2h\theta+b = 0$$

and

$$S' \equiv a'\theta^2+2h'\theta+b' = 0,$$

and let the pair of united points (M, N) of τ be represented by the equation

$$a_1\theta^2+2h_1\theta+b_1 = 0.$$

Then, by the harmonic property of the involution,

$$2hh_1-ab_1-a_1b = 0$$

and

$$2h'h_1-a'b_1-a_1b' = 0,$$

and hence

$$2(h+\lambda h')h_1-(a+\lambda a')b_1-a_1(b+\lambda b') = 0.$$

Thus the pair of points represented by $S+\lambda S' = 0$ is a pair of τ for every λ. By choosing the value of λ suitably we can make this pair contain any assigned point, and the set of point-pairs therefore makes up the whole involution τ.

Since both an involution and a system of point-pairs $S+\lambda S' = 0$ are uniquely determined by two given pairs of points, it follows that every general system $S+\lambda S' = 0$ is an involution.

THEOREM 16. *Two homographies have, in general, two pairs of corresponding points in common. If the homographies are involutions, these two common pairs coincide, i.e. two involutions have, in general, a unique pair of mates in common.*

Proof. Suppose ϖ_1, ϖ_2 are two given homographies, and (P, P') is a common corresponding pair. Then

$$\varpi_1 P = \varpi_2 P$$

and hence

$$\varpi_2^{-1}\varpi_1 P = P.$$

Thus P is a united point of the homography $\varpi_2^{-1}\varpi_1$, and unless $\varpi_2^{-1}\varpi_1 = \epsilon$ there are two possible positions of P, distinct or coincident. Hence, if $\varpi_1 \neq \varpi_2$, the homographies have exactly two common pairs, which may possibly coincide.

If ϖ_1 and ϖ_2 are both involutions, and if (P, P') is a common pair, then (P', P) is also a common pair. The two common pairs are thus accounted for, and as order is immaterial we may say that two involutions have a unique pair of mates in common.

THEOREM 17. *If τ_1, τ_2 are involutions with (M, N) as a common pair of mates, then $\tau_1\tau_2$ is a homography with M and N as united points. Conversely, any homography with distinct united points M and N can be expressed in infinitely many ways as a product of two involutions with (M, N) as a common pair of mates.*

Proof. (i) If τ_1, τ_2 are involutions, each with (M, N) as a pair of mates, then $\tau_1\tau_2 M = \tau_1 N = M$ and $\tau_1\tau_2 N = \tau_1 M = N$. Thus the homography $\tau_1\tau_2$ has M and N as united points.

(ii) Let ϖ be a homography with united points M and N. If we choose an arbitrary involution τ_1 with (M, N) as a pair of mates, then $\tau_1\varpi M = \tau_1 M = N$, and similarly $\tau_1\varpi N = M$. Thus $\tau_1\varpi$ is a homography with (M, N) as an involutory pair, i.e. an involution τ_2. Then $\varpi = \tau_1^{-1}\tau_2 = \tau_1\tau_2$, and τ_1, τ_2 are involutions with (M, N) as a common pair of mates.

Remark. The above resolution of ϖ into two involutions is easily arrived at algebraically. Let ϖ be represented by its canonical equation

$$\varpi(\theta) = k\theta.$$

Then if τ_1 is an involution with the reference points M, N as a pair of mates, the equation of τ_1 assumes the canonical form

$$\theta\tau_1(\theta) = m,$$

and m can be assigned arbitrarily.

It now follows that

$$\tau_1 \varpi(\theta) = \tau_1(k\theta) = \frac{m}{k\theta},$$

so that $\tau_1 \varpi$ is an involution τ_2.

EXERCISE. Verify that Theorem 17 holds also when M and N coincide.

§ 11. PERMUTABILITY OF HOMOGRAPHIES

THEOREM 18. *If ϖ_1, ϖ_2 are two homographies, both distinct from ϵ, then $\varpi_1 \varpi_2 = \varpi_2 \varpi_1$ if and only if either (a) ϖ_1 and ϖ_2 have the same united points, distinct or coincident, or (b) ϖ_1 and ϖ_2 are involutions whose pairs of united points separate each other harmonically.*

Proof. (i) Suppose $\varpi_1 \varpi_2 = \varpi_2 \varpi_1$, and let M be a united point of ϖ_1. Then $\varpi_1 \varpi_2 M = \varpi_2 \varpi_1 M = \varpi_2 M$, and therefore $\varpi_2 M$ is a united point of ϖ_1. Thus any united point of either homography is transformed by the other homography into a united point (the same or another) of the original homography.

Now suppose the united points of ϖ_1 coincide at M. Then necessarily $\varpi_2 M = M$, and M is therefore a united point of ϖ_2 also. Suppose, if possible, that ϖ_2 has a second united point N, distinct from M. Then $\varpi_1 N$ is a united point of ϖ_2, and therefore $\varpi_1 N = M$ or $\varpi_1 N = N$.

If $\varpi_1 N = M$, then $N = \varpi_1^{-1} M = M$.

If $\varpi_1 N = N$, N is a united point of ϖ_1, and again $N = M$. Thus ϖ_2 has no united point distinct from M, and in this case, therefore, ϖ_1 and ϖ_2 have the same united points M, M.

We have now to consider the case in which one of the two homographies, and therefore the other as well, has distinct united points. Suppose, then, that the united points of ϖ_1 are M and N. By what was proved at the beginning, $\varpi_2 M$ and $\varpi_2 N$ are united points of ϖ_1, and hence either (a) $\varpi_2 M = M$, $\varpi_2 N = N$, or (b) $\varpi_2 M = N$, $\varpi_2 N = M$. In case (a) M and N are united points of ϖ_2, and the two homographies have the same united points. In case (b) ϖ_2 has an involutory pair (M, N). It is therefore an involution with (M, N) as a pair of mates, and its united points are separated harmonically by M and N. In this case, by symmetry, ϖ_1 is also an involution. This completes the proof of the necessity of the condition for ϖ_1 to commute with ϖ_2. We now consider its sufficiency.

(ii) Suppose ϖ_1 and ϖ_2 have the same united points, distinct or coincident. Then their equations may be reduced simultaneously either to

$$\varpi_1(\theta) = k_1\theta, \qquad \varpi_2(\theta) = k_2\theta$$

or to

$$\varpi_1(\theta) = \theta+\alpha_1, \qquad \varpi_2(\theta) = \theta+\alpha_2;$$

and in either case

$$\varpi_1\varpi_2(\theta) \equiv \varpi_2\varpi_1(\theta).$$

If, on the other hand, ϖ_1 and ϖ_2 are involutions whose pairs of united points separate each other harmonically, the united points of ϖ_1 are mates in ϖ_2; and if these points are taken as reference points the equations of ϖ_1 and ϖ_2 will be reduced simultaneously to

$$\varpi_1(\theta) = -\theta, \qquad \varpi_2(\theta) = \frac{m}{\theta}.$$

Once again, therefore,

$$\varpi_1\varpi_2(\theta) \equiv \varpi_2\varpi_1(\theta).$$

The theorem is now completely proved.

COROLLARY. *If τ_1 and τ_2 are involutions, both distinct from ϵ, then $\tau_2\tau_1$ is an involution if and only if the pairs of united points of τ_1 and τ_2 separate each other harmonically. If this condition is satisfied, then $\tau_1\tau_2 = \tau_2\tau_1 = \tau_3$, say; and the product, in either order, of any two of the involutions τ_1, τ_2, τ_3 is the third.*

Proof. A necessary and sufficient condition for the homography $\tau_2\tau_1$ to be an involution is

$$\tau_2\tau_1 = (\tau_2\tau_1)^{-1},$$

i.e.

$$\tau_2\tau_1 = \tau_1^{-1}\tau_2^{-1} = \tau_1\tau_2.$$

Since τ_1 and τ_2 cannot have the same united points unless they are the same involution, it follows from Theorem 18 that their pairs of united points must be harmonic.

If the condition is satisfied, and $\tau_1\tau_2 = \tau_2\tau_1 = \tau_3$, then

$$\tau_2\tau_3 = \tau_2^2\tau_1 = \tau_1$$

and

$$\tau_3\tau_2 = \tau_1\tau_2^2 = \tau_1,$$

and similarly

$$\tau_3\tau_1 = \tau_1\tau_3 = \tau_2.$$

This proves the second part of the corollary.

Remark. If τ_1, τ_2, τ_3 are three involutions, related in the manner just described, the homographies ϵ, τ_1, τ_2, τ_3 by themselves form a multiplicative group of four elements. This group has the multiplication table

ϵ	τ_1	τ_2	τ_3
τ_1	ϵ	τ_3	τ_2
τ_2	τ_3	ϵ	τ_1
τ_3	τ_2	τ_1	ϵ

It is isomorphic with the group formed by the four operations of permuting four given points in pairs (i.e. those permutations which leave the cross ratio invariant). Both these concrete groups are, in fact, realizations of the abstract group known as Klein's four-group.

THEOREM 19. *If ϖ, ϖ_1, ϖ_2 are three homographies such that ϖ commutes with ϖ_1 and ϖ_2, then*

(i) *ϖ commutes with $\varpi_1 \varpi_2$;*

(ii) *ϖ commutes with ϖ_1^{-1}.*

Proof. (i) Since $\varpi\varpi_1 = \varpi_1\varpi$ and $\varpi\varpi_2 = \varpi_2\varpi$ we have

$$\begin{aligned} \varpi(\varpi_1\varpi_2) &= (\varpi\varpi_1)\varpi_2 \\ &= (\varpi_1\varpi)\varpi_2 \\ &= \varpi_1(\varpi\varpi_2) \\ &= \varpi_1(\varpi_2\varpi) \\ &= (\varpi_1\varpi_2)\varpi. \end{aligned}$$

(ii) Since $\qquad \varpi\varpi_1 = \varpi_1\varpi,$

therefore $\qquad \varpi_1^{-1}(\varpi\varpi_1)\varpi_1^{-1} = \varpi_1^{-1}(\varpi_1\varpi)\varpi_1^{-1},$

i.e. $\qquad (\varpi_1^{-1}\varpi)(\varpi_1\varpi_1^{-1}) = (\varpi_1^{-1}\varpi_1)(\varpi\varpi_1^{-1}),$

i.e. $\qquad \varpi_1^{-1}\varpi = \varpi\varpi_1^{-1}.$

COROLLARY 1. *If ϖ commutes with ϖ_i $(i = 1, 2,..., k)$ then ϖ commutes with $\varpi_1\varpi_2 \ldots \varpi_k$.*

COROLLARY 2. *If ϖ commutes with ϖ_1, every positive or negative power of ϖ commutes with every positive or negative power of ϖ_1.*

§ 12. AFFINE GEOMETRY OF ONE DIMENSION

The affine line may be derived from the real projective line $S_1(R)$ as follows. We select a point I of $S_1(R)$, to be called the *point at infinity*, and restrict the class of allowable representations to those representations $\mathscr{R}_A$ of $S_1(R)$ in which the point I has coordinates

(0, 1). This point I is then regarded not as a point actually belonging to the affine line, but as an 'ideal' point adjoined to it. Other ideal points are also to be adjoined to the line, namely a set of complex points, one for each properly complex ratio $x_0:x_1$ in any given representation $\mathscr{R}_A$. Since any two affine representations $\mathscr{R}_A$ are connected by a *real* linear transformation, the distinction between real and complex points is invariant over change of representation.

In this way we obtain a real line, whose points may be represented without exception by a single real non-homogeneous coordinate $X = x_1/x_0$. Adjoined to the line is the ideal point I, with coordinate ∞ in every allowable representation, and the system of ideal complex points, corresponding to complex values of X. Those properties of the affine line which have the same expression in every allowable representation $\mathscr{R}_A$ are the affine properties of the line, and their totality constitutes one-dimensional affine geometry.

The reader may compare this abstract treatment of affine geometry with what was said about the same subject in Chapter II.

THEOREM 20. *If $\mathscr{R}_A$ is any one allowable representation of the affine line, then the whole class $(\mathscr{R}_A)$ of allowable representations consists of all those representations which can be derived from $\mathscr{R}_A$ by applying a transformation of the form*

$$X' = bX+c,$$

where b, c are arbitrary real numbers, with $b \neq 0$.

Proof. Let $\mathscr{R}'_A$ be any allowable representation. Then, in terms of homogeneous coordinates,

$$x'_i = \sum_{k=0}^{1} a_{ik}x_k \quad (i = 0, 1),$$

where $|a_{rs}| \neq 0$.

Since $\mathscr{R}'_A$ is allowable, $x'_0 = 0$ whenever $x_0 = 0$ and $x_1 \neq 0$. Thus $a_{01} = 0$, and the equations of transformation reduce to

$$x'_0 = a_{00}x_0,$$
$$x'_1 = a_{10}x_0+a_{11}x_1,$$

with $a_{00} \neq 0$ and $a_{11} \neq 0$.

Dividing the second equation by the first, we have

$$X' = \frac{a_{10}}{a_{00}}+\frac{a_{11}}{a_{00}}X,$$

and this is an equation of the required form.

Since a_{00}, a_{10}, a_{11} are subject only to the conditions $a_{00} \neq 0$ and $a_{11} \neq 0$, b and c can take any real values provided that $b \neq 0$.

Remark. The set of all transformations of the form

$$X' = bX+c \quad (b \neq 0),$$

is a group, the *affine group* in one dimension.

If A and B are any two actual points of the affine line, the *mid-point* of the segment AB is the harmonic conjugate of I with respect to A and B. Clearly the coordinate X of this mid-point is the mean of the coordinates of A and B.

If A, B, C are three points, with coordinates X_1, X_2, X_3 respectively, the ratio $\dfrac{X_2-X_1}{X_3-X_2}$ is an affine invariant. It is called the position ratio of B with respect to A and C. More generally, any ratio of differences of coordinates $\dfrac{X_2-X_1}{X_4-X_3}$ is an affine invariant.

§ 13. Euclidean Geometry of One Dimension

If we restrict the class of allowable representation still further by selecting two actual points A, B of the affine line, and regarding as allowable only those representations $\mathscr{R}_A$ in which the 'length' $|X_2-X_1|$ of the segment AB is unity, we obtain euclidean geometry of one dimension.

THEOREM 21. *If $\mathscr{R}_E$ is any allowable representation of the euclidean line, then the whole class $(\mathscr{R}_E)$ of allowable representations consists of all those representations which can be derived from $\mathscr{R}_E$ by applying a transformation of the form*

$$X' = \epsilon X+a,$$

where a is an arbitrary real number and $\epsilon = \pm 1$.

Proof. Let $\mathscr{R}'_E$ be any allowable representation, and suppose

$$X' = bX+c \quad (b \neq 0).$$

Then $\quad X'_2-X'_1 = (bX_2+c)-(bX_1+c) = b(X_2-X_1),$

and hence $b = \pm 1$.

Remarks

(i) The set of all transformations of the form

$$X' = \epsilon X+a$$

is a group, the *euclidean group* in one dimension.

(ii) A euclidean transformation leaves every 'length' invariant, not merely that of the chosen unit segment.

(iii) If we require the quantity $X_2 - X_1$ to be equal to unity in sign as well as in magnitude, we obtain the *oriented* euclidean line. Any two allowable representations are now connected by a *proper* euclidean transformation

$$X' = X + a.$$

A concrete realization of the abstract scheme that has just been discussed is provided by the straight line of elementary geometry with an ordinary cartesian coordinate X. This coordinate gives an allowable representation $\mathscr{R}_E$, the reference points being the origin and the point at infinity.

Every homography on the line can be represented by an equation

$$X' = \frac{aX+b}{cX+d},$$

in which a, b, c, d are real numbers such that $ad - bc \neq 0$. The homography is said to be *hyperbolic*, *elliptic*, or *parabolic* according as its united points are real and distinct, conjugate complex (ideal), or coincident. This distinction only arises in the case of the real line.

If the united points M, N are distinct, the homography has a real centre O, the mid-point of MN. If this centre is taken as origin, the equation of the homography reduces to

$$XX' + a(X - X') = k.$$

In this equation k is positive or negative according as the homography is hyperbolic or elliptic. The homography is an involution if and only if $a = 0$.

EXERCISE. If ϖ is a homography with united points M and N, prove that:

(i) if M is finite and N is at infinity, ϖ is a dilatation about M, possibly combined with a reflection in M;

(ii) if N is at infinity and ϖ is involutory, ϖ is a reflection in M;

(iii) if M and N are both at infinity, ϖ is a translation.

EXERCISES ON CHAPTER III

1. If four points A, B, C, D on a line have projective parameters θ_0, θ_1, θ_2, θ_3, prove that A, C separate B, D harmonically if and only if $\theta_1 - \theta_0$, $\theta_2 - \theta_0$, $\theta_3 - \theta_0$ are in harmonic progression.

2. Three points A_0, A_1, B on a line being given, a sequence of points $A_0, A_1, A_2,\ldots$ is constructed by taking as A_n, for each integer n greater than 1, the harmonic conjugate of A_{n-2} with respect to A_{n-1} and B. Show that a projective parameter θ can be defined in such a way that the value of θ at A_i is i $(i = 0, 1,\ldots)$.

3. If A, B, P, Q, U, V are six points on a line, prove that

(i) $\{A, B;\, P, Q\}.\{A, B;\, Q, U\} = \{A, B;\, P, U\}$;
(ii) if $\{A, B;\, P, Q\} = \{A, B;\, U, V\}$ then $\{A, B;\, P, U\} = \{A, B;\, Q, V\}$.

4. If θ is a given projective parameter on a line, find a new parameter $\phi = (a\theta+b)/(c\theta+d)$ which takes the values 0, 1, ∞ at the points for which θ has the values 1, 2, 3. Find the transformed equation, in terms of the new parameter, of the homography given by

$$\theta\theta'-\theta-5\theta'+9 = 0.$$

5. Show that the equation $\theta' = (a\theta+b)/(c\theta+d)$ of any homography with distinct united points $\theta = m$ and $\theta = n$ can be written in the form

$$\frac{\theta'-m}{\theta'-n} = \frac{dm+an}{dn+am}\,\frac{\theta-m}{\theta-n},$$

with appropriate modifications if m or n is infinite, or $m+n = 0$.

If the homography has coincident united points $\theta = m$, show that its equation may be written in the form

$$\frac{1}{\theta'-m} = \frac{1}{\theta-m}+\frac{2c}{a+d}.$$

6. Reduce each of the following equations to one or other of the forms suggested in Exercise 5:

(i) $3\theta\theta'-\theta+5\theta'-7 = 0$;
(ii) $\theta\theta'-2\theta+4 = 0$;
(iii) $4\theta\theta'+13\theta-\theta'+9 = 0$;
(iv) $\theta\theta' = k \quad (k \neq 0)$.

7. If ϖ and σ are the homographies whose equations are respectively $\theta' = (4\theta+6)/(3\theta+2)$ and $\theta' = (4\theta+2)/(\theta+2)$, obtain the equations of the homographies $\varpi^{-1}\sigma$ and $\sigma\varpi^{-1}$; and find the common pairs of ϖ and σ.

8. If the homography ϖ, given by $\theta' = (a\theta+b)/(c\theta+d)$, has distinct united points, show that the two alternative values k and $1/k$ of its modulus are the roots of the following quadratic equation in x:

$$(ad-bc)(x+1)^2-(a+d)^2x = 0.$$

9. Prove that the condition for the homography ϖ of Exercise 8 to be of period 3 is $a^2+d^2+ad+bc = 0$, and the condition for it to be of period 4 (but not of period 2) is $a^2+d^2+2bc = 0$.

10. A homographic transformation carries three points A, B, C of a line into points A', B', C' of the same line. Prove that it transforms the united points of the homography which permutes A, B, C cyclically into the united points of the homography which permutes A', B', C' cyclically.

11. Discuss, for all real values of λ, the character of the (real) homography

$$\theta\theta'+(3+\lambda)\theta+\lambda\theta'+\lambda+8 = 0.$$

12. Find all the homographies that commute with the involution

$$\theta\theta'+1 = 0.$$

[*Answer.* Those whose equations are of one or other of the forms

$$\theta\theta'+b(\theta+\theta')-1 = 0, \qquad \theta\theta'+b(\theta-\theta')+1 = 0.]$$

13. Find a transformation $\theta = (\alpha\phi+\beta)/(\gamma\phi+\delta)$ of the projective parameter θ which reduces the two involutions $\theta\theta'-2(\theta+\theta')+7 = 0$ and $\theta\theta'+3(\theta+\theta')-3 = 0$ simultaneously to the forms $\phi\phi' = m$ and $\phi\phi' = n$.

14. A variable homography ϖ on a line has two assigned pairs of corresponding points. Show that, by choosing the projective parameter suitably, it is possible to reduce the equation of ϖ to the form

$$\theta\theta'+(\lambda+\alpha)\theta+(\lambda-\alpha)\theta'+1-2\alpha = 0,$$

where α is constant and λ is variable. Discuss the character of ϖ for all possible values of λ.

15. If ϖ_1 and ϖ_2 are two given homographic correspondences on a line, show that there exist involutions τ, τ_1, τ_2, in general uniquely determined, such that $\varpi_1 = \tau_1\tau$ and $\varpi_2 = \tau\tau_2$.

16. If ϖ is a homography on a line, with distinct united points M, N, prove that there are exactly two homographies σ_1 and σ_2 which satisfy the condition $\sigma^2 = \varpi$, and show that $\sigma_2\sigma_1^{-1}$ is the involution whose united points are M and N. How is this result modified if ϖ has coincident united points? If ϖ is a real homography, find conditions for σ_1 and σ_2 to be real also.

17. Two pairs of numbers are defined by the quadratic equations

$$ax^2+bx+c = 0 \quad \text{and} \quad a'x^2+b'x+c' = 0$$

respectively. Show that the two possible values of the cross ratio of the numbers of the one pair with respect to those of the other pair are the roots of the quadratic $\lambda^2-2m\lambda+1 = 0$, where

$$m = \frac{(bb'-2ac'-2a'c)^2+(b^2-4ac)(b'^2-4a'c')}{(bb'-2ac'-2a'c)^2-(b^2-4ac)(b'^2-4a'c')}.$$

18. If σ_2 is the sum of all the products, two at a time, of the six cross ratios of four general points, prove that

$$\sigma_2 = 6-\frac{(\lambda^2-\lambda+1)^3}{\lambda^2(\lambda-1)^2},$$

where λ denotes any one of the six cross ratios.

Prove that two unordered sets of four distinct points can be related homographically, when they are paired suitably, if and only if the invariant σ_2 has the same value for each set.

19. Any projective parameter $\theta = X+iY$ for the complex line defines a one–one mapping of this line on the Argand plane, the complex point θ of the line being represented by the real point (X, Y) of the plane; and the general homographic transformation $\theta' = (a\theta+b)/(c\theta+d)$ of the line into

itself then goes over into a corresponding self-transformation of the Argand plane. Show that this latter transformation has the following properties:

(i) every member of the complete family of circles (which includes the straight lines of the plane) is transformed into a member of the same family;

(ii) the angle at which two curves intersect and the angle at which the transformed curves intersect are equal in magnitude and in sense;

(iii) the point given by $Z \equiv X+iY = -d/c$ transforms into the unique *point at infinity* of the Argand plane, and all circles through the point $-d/c$ are transformed into straight lines;

(iv) the transformation has two united points (X_1, Y_1), (X_2, Y_2), and, if they are distinct, the coaxal systems for which they are (*a*) the common points, and (*b*) the limiting points, have the property that every circle of either system is transformed into a circle of the same system.

20. Prove that two pairs of complex numbers (Z_1, Z_2) and (Z_3, Z_4) are harmonic if and only if the points Z_3 and Z_4 in the Argand plane are the intersections of a circle through Z_1 and Z_2 with a circle of the coaxal system for which the points Z_1 and Z_2 are limiting points.

Deduce from this the geemetrical properties, in the Argand plane, of the involution whose equation is

$$ZZ'+p(Z+Z')+q = 0.$$

21. A one-dimensional non-euclidean geometry on a real line l may be defined as follows. Two finite points A, B of l are selected, and only the points of the segment AB are regarded as existent, the interior points of the segments being actual and the end-points A and B ideal. A non-euclidean distance $\overrightarrow{PQ}$ is now defined, for every ordered pair (P, Q) of existent points, by the formula

$$\overrightarrow{PQ} = \tfrac{1}{2}\log\{A, B;\, P, Q\}.$$

With this definition of distance, show that

(i) A and B are both infinitely distant from every actual point P;

(ii) if P, Q, R are any three actual points, then $\overrightarrow{PQ}+\overrightarrow{QR}+\overrightarrow{RP} = 0$;

(iii) the segment PQ determined by any two actual points has a unique mid-point, namely the unique existent united point of the involution on l determined by the two pairs (A, B) and (P, Q).

22. With the notation of Exercise 21, show that, if τ is the involution on l whose united points are A and B, there are two types of homography on l which commute with τ, and these give rise respectively to the non-euclidean translations and non-euclidean reflections for the non-euclidean line AB.

23. Work out the general theory referred to on p. 55 by proving the following theorems on the repetition of a homographic transformation.

(i) If ϖ is a given homography and $\{\varpi^r\}$ denotes the sequence of homographies $\{\ldots\varpi^{-1}, \varpi^0, \varpi, \ldots\}$ then either (*a*) all the powers ϖ^r are distinct, or (*b*) the sequence $\{\varpi^r\}$ is formed by cyclic repetition of a finite set

$$(\varpi^0, \varpi, \ldots, \varpi^{m-1})$$

of distinct powers of ϖ. In case (*b*) $\varpi^r = \varpi^s$ if and only if $r \equiv s \pmod m$.

(ii) If ϖ is a given homography and P_0 is a given point, and the sequence of points $\{...P_{-1}, P_0, P_1,...\}$ is defined by the relation $P_i = \varpi^i P_0$, then either (*a*) all the points P_i are distinct, or (*b*) the sequence $\{P_i\}$ is formed by cyclic repetition of a finite set $(P_0, P_1,..., P_{k-1})$ of distinct points. In case (*b*) $P_i = P_j$ if and only if $i \equiv j \pmod k$.

(iii) If the homography ϖ is cyclic of period m, then every point P_0, other than a united point, generates a cyclic sequence $\{P_i\}$ of period m.

CHAPTER IV

PROJECTIVE GEOMETRY OF TWO DIMENSIONS

§ 1. Definitions

A *two-dimensional projective domain* $S_2(K)$ is a class of entities that admits of certain allowable representations $\mathscr{R}$ by means of triads of homogeneous coordinates (x_0, x_1, x_2), drawn from a given ground field K. This means that in any representation $\mathscr{R}$, the elements of $S_2(K)$ are represented by vectors $\mathbf{x}$ belonging to $V_3(K)$, and that two vectors represent the same element of $S_2(K)$ if and only if they differ at most by a scalar factor.

The class $(\mathscr{R})$ of all allowable representations has to satisfy the following condition. If $\mathscr{R}$ is any one such representation, then

(i) every other allowable representation $\mathscr{R}'$ is related to $\mathscr{R}$ by a transformation of the form

$$x'_i = \sum_{k=0}^{2} a_{ik} x_k \quad (i = 0, 1, 2)$$

i.e.
$$\mathbf{x}' = \mathbf{A}\mathbf{x},$$

where
$$|a_{rs}| \equiv |\mathbf{A}| \neq 0;$$

(ii) every representation $\mathscr{R}'$ which is related to $\mathscr{R}$ by a transformation of this form is allowable.

The allowable representations $\mathscr{R}$ of $S_2(K)$ are thus connected by a group of non-singular linear transformations, the projective group $PGL(2; K)$ in two dimensions, and the projective properties of $S_2(K)$ are invariant for all transformations of this group.

We shall continue to omit explicit reference to the ground field K whenever, as is usually the case, we wish to take the complex field as ground field. We shall also fix our attention from now on upon one particular S_2, supposed fixed once and for all, which we call the *projective plane*; and the elements of this system will be called the *points* of the plane. Our task is to study the projective properties of this projective plane S_2, and to introduce formal equivalents of the familiar notions that find a place in elementary projective geometry. Much of the theory can be derived from the single notion of linear dependence of points, which we now define.

§ 2. Linear Dependence of Points

DEFINITION. A point P of S_2 is said to be *linearly dependent* on a set of points $P_1, P_2,..., P_m$ if a coordinate vector $\mathbf{x}$ which represents P in some allowable representation $\mathscr{R}$ is linearly dependent on coordinate vectors $\mathbf{x}^{(1)}$, $\mathbf{x}^{(2)},..., \mathbf{x}^{(m)}$, which represent $P_1, P_2,..., P_m$ in $\mathscr{R}$; i.e., if there exist scalars $\lambda_1, \lambda_2,..., \lambda_m$ such that

$$\mathbf{x} = \lambda_1 \mathbf{x}^{(1)} + \lambda_2 \mathbf{x}^{(2)} + ... + \lambda_m \mathbf{x}^{(m)}. \tag{1}$$

We have to show that this definition is legitimate, that is to say that it does in fact define an intrinsic relation between points.

It is clear, first of all, that if we take different vectors to represent the points in the same representation $\mathscr{R}$, so that P is now represented by $\mathbf{y} = \alpha \mathbf{x}$ and so on, then a corresponding relation of linear dependence necessarily holds between the new vectors. We have, therefore, only to show that if a relation (1) holds in one representation $\mathscr{R}$ then a similar relation holds in every representation. Suppose, then, that a new representation $\bar{\mathscr{R}}$ is introduced by means of the transformation $\bar{\mathbf{x}} = \mathbf{A}\mathbf{x}$. Multiplying equation (1) by the matrix $\mathbf{A}$, we have

$$\mathbf{A}\mathbf{x} = \lambda_1 \mathbf{A}\mathbf{x}^{(1)} + \lambda_2 \mathbf{A}\mathbf{x}^{(2)} + ... + \lambda_m \mathbf{A}\mathbf{x}^{(m)},$$

or

$$\bar{\mathbf{x}} = \lambda_1 \bar{\mathbf{x}}^{(1)} + \lambda_2 \bar{\mathbf{x}}^{(2)} + ... + \lambda_m \bar{\mathbf{x}}^{(m)},$$

which is the result required.

We can now say that a set of points in S_2 is a linearly independent set if none of the points is linearly dependent on the other $m-1$. It follows from simple theorems in linear algebra that the maximum number of points in S_2 that can be linearly independent is three, and that three points represented by the coordinates $(x_0^{(i)}, x_1^{(i)}, x_2^{(i)})$ $(i = 1, 2, 3)$ are linearly independent if and only if

$$\begin{vmatrix} x_0^{(1)} & x_1^{(1)} & x_2^{(1)} \\ x_0^{(2)} & x_1^{(2)} & x_2^{(2)} \\ x_0^{(3)} & x_1^{(3)} & x_2^{(3)} \end{vmatrix} \neq 0.$$

THEOREM 1. *If X_0, X_1, X_2, E are four points of S_2, every three of which are linearly independent, then there exists a unique allowable representation $\mathscr{R}$ in which the four points are represented respectively by the vectors* $(1, 0, 0)$, $(0, 1, 0)$, $(0, 0, 1)$, $(1, 1, 1)$.

Proof. This theorem, like Theorem 1 of Chapter III, follows immediately from Theorem 1 of the Appendix.

X_0, X_1, X_2 are called the *reference points* of $\mathscr{R}$, and E the *unit point*. Specifying these four points is usually the most convenient way of characterizing a given representation $\mathscr{R}$.

EXERCISE. Determine the equations of transformation from $\mathscr{R}$ to $\mathscr{R}'$ if the reference points and unit point of $\mathscr{R}'$ are represented in $\mathscr{R}$ by the coordinate vectors $(4, 5, 1)$, $(3, -1, \frac{1}{2})$, $(6, 16, 2)$, $(5, 1, 1)$ respectively.

§ 3. Lines and Collinearity

Suppose Q and R are fixed points, with coordinate vectors $\mathbf{y}$ and $\mathbf{z}$ in some representation $\mathscr{R}$. If the points are distinct they are clearly linearly independent, and their coordinates then satisfy one, and essentially only one, linear equation

$$u_0 x_0 + u_1 x_1 + u_2 x_2 = 0.$$

The equation may, in fact, be written in determinantal form as

$$\begin{vmatrix} x_0 & x_1 & x_2 \\ y_0 & y_1 & y_2 \\ z_0 & z_1 & z_2 \end{vmatrix} = 0,$$

and it follows that a point P is linearly dependent on Q and R if and only if its coordinates (x_0, x_1, x_2) satisfy this equation.

DEFINITION. The set of all points linearly dependent on two given (distinct) points Q, R is called the *line* determined by Q and R, or simply the line QR.

EXERCISES

(i) Show that a line is determined by any two of its points.

(ii) Show that the set of all points whose coordinates satisfy a given (homogeneous) linear equation is a line in the sense of the above definition.

The line QR admits of two kinds of algebraic representation. Since a general point of the line is linearly dependent on Q and R, it may be represented by a coordinate vector of the form

$$\mathbf{x} = \lambda \mathbf{y} + \mu \mathbf{z};$$

and the ratio $\lambda : \mu$ is uniquely determined. In this way we obtain a representation of the points of the line by the homogeneous pair of parameters (λ, μ). Alternatively, we may use the non-homogeneous representation $\mathbf{x} = \mathbf{y} + \theta \mathbf{z}$, if we allow θ to take the value ∞ at R.

The other mode of representation of the line is by its equation $\sum_{i=0}^{2} u_i x_i = 0$. If we take the coefficients u_0, u_1, u_2 in this equation as components of a column-vector $\mathbf{u}$, we may write the equation in matrix form $\mathbf{u}^T\mathbf{x} = 0$. The left-hand side is just the inner product $(\mathbf{u}, \mathbf{x})$ of the vectors $\mathbf{u}$ and $\mathbf{x}$, and so the equation may also be written $(\mathbf{u}, \mathbf{x}) = 0$.

Since the line is defined in terms of the projectively invariant notion of linear dependence of points it is a projective entity. This may also be seen directly as follows. Let the representation $\mathscr{R}$ be changed by the transformation $\mathbf{x}' = \mathbf{A}\mathbf{x}$. Then the equation

$$\mathbf{u}^T\mathbf{x} = 0$$

gives

$$\mathbf{u}^T\mathbf{A}^{-1}\mathbf{x}' = 0,$$

i.e.

$$(\mathbf{A}^{-1T}\mathbf{u})^T\mathbf{x}' = 0,$$

or

$$\mathbf{u}'^T\mathbf{x}' = 0,$$

where $\mathbf{u}' = \mathbf{A}^{-1T}\mathbf{u}$; and the equation $\mathbf{u}'^T\mathbf{x}' = 0$ is of the same form as the original equation $\mathbf{u}^T\mathbf{x} = 0$.

THEOREM 2. *The lines of the projective plane form a two-dimensional projective domain.*

Proof. Let us select, arbitrarily, an allowable representation $\mathscr{R}$ of the plane. Then every line has a linear equation $(\mathbf{u}, \mathbf{x}) = 0$, and this equation may be specified by the vector $\mathbf{u}$. Furthermore, two vectors $\mathbf{u}$, $\mathbf{v}$ give rise to the same equation, and therefore the same line, if and only if they differ at most by a scalar factor. Thus the vector $\mathbf{u}$ furnishes a set of three homogeneous coordinates (u_0, u_1, u_2) of the line.

We now have a class $(\overline{\mathscr{R}})$ of representations of the lines of the plane, one for each representation $\mathscr{R}$ of the points, and it only remains for us to show that $(\overline{\mathscr{R}})$ is a class of allowable representations in the sense previously defined. Let $\overline{\mathscr{R}}$, $\overline{\mathscr{R}}'$ denote the representations of lines arising from two given allowable representations $\mathscr{R}$, $\mathscr{R}'$ of points. If $\mathscr{R}$ and $\mathscr{R}'$ are connected by the transformation $\mathbf{x}' = \mathbf{A}\mathbf{x}$, $\overline{\mathscr{R}}$ and $\overline{\mathscr{R}}'$ are connected, as has already been shown, by the transformation $\mathbf{u}' = \mathbf{A}^{-1T}\mathbf{u}$. This is a non-singular linear transformation; and, by choosing $\mathbf{A}$ suitably, we can make it coincide with any assigned non-singular linear transformation. The representations $\overline{\mathscr{R}}$ of the lines of the plane therefore form a class of allowable representations, and this proves the theorem.

Since the algebra of line-geometry now runs parallel to the algebra of point-geometry, the projective geometry of the plane may be said to exhibit a twofold character. This is something to which we shall return below, on p. 78. For the present we merely formulate the following theorem, which, in spite of its apparent triviality, is of some importance because it brings out the reciprocal dependence of points and lines on each other.

THEOREM 3. *Any two distinct points of the plane determine a unique line, which contains them both. Any two distinct lines of the plane determine a unique point, which they both contain.*

§ 4. Subordinate Projective Geometries

The one-dimensional projective geometry discussed in Chapter III and the two-dimensional projective geometry that forms the subject of the present chapter are not merely two analogous but otherwise quite independent systems. There is a much closer connexion between them than mere analogy, and this is brought out in the next theorem.

THEOREM 4. *The two-dimensional geometry of the projective plane induces a subordinate one-dimensional projective geometry on every line of the plane.*

Proof. Let l be a given line in the plane, and let two points Q, R be taken on it. When a definite representation $\mathscr{R}$ of the plane is assigned, and definite vectors $\mathbf{y}$, $\mathbf{z}$ are chosen to represent Q, R in this representation, every point P of l has a unique representation by a vector $\mathbf{x}$ of the form $\lambda\mathbf{y}+\mu\mathbf{z}$; and we may take (λ, μ) as homogeneous coordinates of P on l. If the underlying representation $\mathscr{R}$ is changed, all coordinate vectors are subjected to the same linear transformation, and the values of the parameters λ, μ remain unchanged. We may accordingly confine ourselves to the one representation $\mathscr{R}$ of the plane.

Now suppose a new parametric representation of l is defined by means of a fresh pair of points Q', R'. If Q', R' have coordinate vectors $\mathbf{y}'$, $\mathbf{z}'$, we may write

$$\mathbf{y} = \alpha\mathbf{y}'+\beta\mathbf{z}',$$
$$\mathbf{z} = \gamma\mathbf{y}'+\delta\mathbf{z}',$$

where, since $\mathbf{y}$ and $\mathbf{z}$ are linearly independent, $\alpha\delta-\beta\gamma \neq 0$. Then

$$\begin{aligned}\mathbf{x} &= \lambda\mathbf{y}+\mu\mathbf{z}\\ &= (\alpha\lambda+\gamma\mu)\mathbf{y}'+(\beta\lambda+\delta\mu)\mathbf{z}'\\ &= \lambda'\mathbf{y}'+\mu'\mathbf{z}', \text{ say.}\end{aligned}$$

Thus when new coordinates λ', μ' on l are introduced by means of a fresh choice of the reference points Q and R on l, they are connected with λ, μ by the non-singular linear transformation

$$\lambda' = \alpha\lambda+\gamma\mu,$$
$$\mu' = \beta\lambda+\delta\mu.$$

By taking into account all possible choices of Q and R, we obtain a class of allowable (one-dimensional) representations of l, and l therefore has the geometrical structure of an S_1.

COROLLARY 1. *All projective entities on any line in the plane—such as, for instance, cross ratio, the harmonic relation, and homographic correspondences—are projective entities in the plane also.*

COROLLARY 2. *If four points P_i of the line QR are represented by vectors $\mathbf{x}^{(i)} = \lambda_i \mathbf{y}+\mu_i \mathbf{z}$ $(i = 1, 2, 3, 4)$ then*

$$\{P_1, P_2;\ P_3, P_4\} = \{\theta_1, \theta_2;\ \theta_3, \theta_4\},$$

where $\theta_i = \mu_i/\lambda_i$.

COROLLARY 3. *In particular, the harmonic conjugate with respect to Q and R of the point represented by $\mathbf{y}+\alpha\mathbf{z}$ is represented by $\mathbf{y}-\alpha\mathbf{z}$.*

Let us now consider once again the reference points X_0, X_1, X_2 and the unit point E, which serve to determine a given representation $\mathscr{R}$ of the plane. Any three of the four points X_0, X_1, X_2, E are linearly independent, and therefore non-collinear. X_0, X_1, X_2 are often called the vertices of the *triangle of reference*, and the lines X_1X_2, X_2X_0, X_0X_1 are the opposite sides of this triangle. The equations of these sides are respectively $x_0 = 0$, $x_1 = 0$, $x_2 = 0$. X_0E is the line $x_1-x_2 = 0$, and it meets X_1X_2 in the point E_0 with coordinates $(0, 1, 1)$. This point, and the similar points E_1, E_2 on X_2X_0, X_0X_1 respectively, are called the *subordinate unit points* on the sides of the triangle of reference.

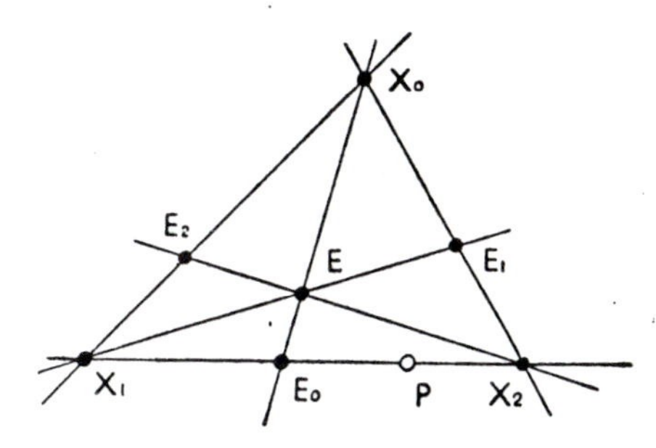

A generic point P of X_1X_2 has coordinates of the form $(0, \lambda, \mu)$; and if the coordinate vectors of X_1, X_2, E_0 are taken as $\mathbf{x}^{(1)}$, $\mathbf{x}^{(2)}$, $\mathbf{e}^{(0)}$ respectively, where $\mathbf{x}^{(1)} = (0, 1, 0)$, $\mathbf{x}^{(2)} = (0, 0, 1)$, $\mathbf{e}^{(0)} = (0, 1, 1)$,† then $\mathbf{e}^{(0)} = \mathbf{x}^{(1)}+\mathbf{x}^{(2)}$ and P has the coordinate vector $\lambda\mathbf{x}^{(1)}+\mu\mathbf{x}^{(2)}$. Thus *when the coordinates of P, regarded as a point of S_2, are written*

† Strictly, of course, these are column-vectors.

as $(0, \lambda, \mu)$, (λ, μ) *are coordinates of* P *in the one-dimensional geometry induced on* $X_1 X_2$, *the reference points and unit point being respectively* X_1, X_2, and E_0.

EXERCISE. With the notation used in the proof of Theorem 4, show that an allowable representation of l may be uniquely specified either (*a*) by giving the particular coordinate vectors $\mathbf{y}$ and $\mathbf{z}$ which are to represent Q and R, or (*b*) by giving the points Q and R and also the point which is to be the unit point of the representation.

The results just obtained make it easy to establish the connexion between *harmonic pole and polar* with respect to a triangle.

THEOREM 5. *If* D, E, F *are the points of intersection of the sides* BC, CA, AB *of a triangle with the lines joining* A, B, C *to a point* P *which does not lie on any of the sides, and if* D', E', F' *are the harmonic conjugates of* D, E, F *with respect to the point-pairs* (B, C), (C, A), (A, B) *respectively, then* D', E', F' *are collinear.*

Proof. Take ABC as triangle of reference and P as unit point. Then B, C, D are respectively $(0, 1, 0)$, $(0, 0, 1)$, and $(0, 1, 1)$, and hence D' is $(0, 1, -1)$. The coordinates of D' therefore satisfy the equation

$$x_0 + x_1 + x_2 = 0;$$

and since this equation is symmetric, D', E', F' all lie on the line p which it represents.

Remarks

(i) The point P and the line p are said to be harmonic pole and polar with respect to the triangle ABC.

(ii) The harmonic polar with respect to the triangle of reference of the unit point is, as has just been shown, the unit line $(1, 1, 1)$. It follows that the coordinate representation $\mathscr{R}$ is determined equally by the vertices of the triangle of reference and the unit point, and by the sides of the triangle of reference and the unit line.

EXERCISES

(i) Prove the converse of Theorem 5.

(ii) Show that the harmonic polar of the point (a, b, c) with respect to the triangle of reference is the line $(1/a, 1/b, 1/c)$.

(iii) In ordinary euclidean geometry, what is the harmonic polar of the centroid of a triangle with respect to the triangle?

§ 5. The Principle of Duality

We have already seen that the lines of the projective plane, as well as its points, are elements of a two-dimensional projective domain, so that there is a projective geometry of lines as well as of points. It is now necessary to examine the relation between the two geometries in greater detail.

A line can be represented either by a linear equation $\sum u_i x_i = 0$ in point-coordinates or by a triad (u_0, u_1, u_2) of line-coordinates; and similarly a point can be represented either by its point-coordinates (x_0, x_1, x_2) or by a linear equation in line-coordinates. The equation $\sum u_i x_i = 0$ expresses, in fact, the symmetrical relation of *incidence* between the point (x_0, x_1, x_2) and the line (u_0, u_1, u_2). If u_0, u_1, u_2 are thought of as constant, the equation means that the variable point (x_0, x_1, x_2) describes the fixed line (u_0, u_1, u_2) as its locus; and if x_0, x_1, x_2 are thought of as constant, it means that the variable line (u_0, u_1, u_2) describes the fixed point (x_0, x_1, x_2) as its envelope.

The class of all the points of the plane and the class of all the lines of the plane are symmetrically related to each other, and to every property of lines in the (original) geometry of points there corresponds a property of points in the geometry of lines. In view of this parallelism, we do not need to discuss both sets of properties in detail. Having once proved a theorem in the geometry of points, we can immediately write down the corresponding theorem about lines, simply by changing the wording suitably, and we do not need to repeat the details of the proof. This is the essence of the important *Principle of Duality*.

If T is any theorem that is valid in the projective geometry of the plane, and T' is the theorem that is obtained from T by changing the word 'point' into the word 'line' and vice versa throughout the enunciation and making the appropriate linguistic adjustments, then T' is also valid in the same geometry.

Remarks

(i) Typical 'linguistic adjustments' are the replacement of 'intersection' by 'join' and 'collinear' by 'concurrent'.

(ii) Two theorems T, T' that are related in the manner described are said to be *dual* to each other.

Among the important theorems that can be obtained by dualizing results already established, the following is especially worthy of mention.

THEOREM 6. *If q, r are two distinct lines, represented by coordinate vectors* $\mathbf{v}$, $\mathbf{w}$, *a general line through their point of intersection is given by the vector*

$$\mathbf{u} = \lambda\mathbf{v} + \mu\mathbf{w}.$$

If four lines p_i of the system are given by

$$\mathbf{u}^{(i)} = \lambda_i \mathbf{v} + \mu_i \mathbf{w} \quad (i = 1, 2, 3, 4),$$

then the cross ratio $\{\theta_1, \theta_2; \theta_3, \theta_4\}$, where $\theta_i = \mu_i/\lambda_i$, depends only on the four lines, and not on their mode of representation.

The cross ratio referred to in Theorem 6 is called the cross ratio of the ordered pairs (p_1, p_2) and (p_3, p_4) of lines in the pencil determined by q and r, and it is denoted by $\{p_1, p_2; p_3, p_4\}$.

§ 6. Fundamental Incidence Theorems

Now that we have seen how the points and lines of the projective plane may be represented by suitably chosen coordinates, we are in a position to prove a number of theorems of a more geometrical character. We shall consider first a group of theorems which embody the so-called incidence properties of the projective plane, theorems that are of fundamental importance in any development of plane projective geometry.

THEOREM 7 (*Desargues's Theorem*). *If two triangles correspond to each other in such a way that the joins of their corresponding vertices are concurrent, then the intersections of their corresponding sides are collinear, and conversely.*

Proof. Let the triangles be ABC, $A'B'C'$, and let AA', BB', CC' be concurrent in P. Let the points $BC.B'C'$, $CA.C'A'$, $AB.A'B'$ be respectively L, M, N.

For simplicity we shall adopt the convention that, in some chosen representation $\mathscr{R}$, each point is represented by a coordinate vector denoted by the corresponding letter, A being represented by $\mathbf{a}$ and so on.

Then

$$\mathbf{a}' = \mathbf{p} + \lambda \mathbf{a},$$

$$\mathbf{b}' = \mathbf{p} + \mu \mathbf{b},$$

$$\mathbf{c}' = \mathbf{p} + \nu \mathbf{c},$$

and therefore

$$\mathbf{b}' - \mathbf{c}' = \mu \mathbf{b} - \nu \mathbf{c}.$$

The vector $\mathbf{b}' - \mathbf{c}'$ consequently represents a point lying on $B'C'$ and also on BC, i.e. the point L. We may accordingly put $\mathbf{l} = \mathbf{b}' - \mathbf{c}'$, and similarly $\mathbf{m} = \mathbf{c}' - \mathbf{a}'$ and $\mathbf{n} = \mathbf{a}' - \mathbf{b}'$, so that we have

$$\mathbf{l} + \mathbf{m} + \mathbf{n} = \mathbf{0}.$$

Thus the points L, M, N are collinear, which proves the direct theorem. The converse follows at once by duality.

Remark. Two triangles related as in Theorem 7 are said to be *in perspective.* P is called the vertex of perspective and the line LMN the axis of perspective.

THEOREM 8 (*Pappus's Theorem*). *If* (A_1, A_2, A_3), (B_1, B_2, B_3) *are two triads of collinear points, not on the same line, then the three cross intersections* $A_i B_j . A_j B_i$ *are collinear.*

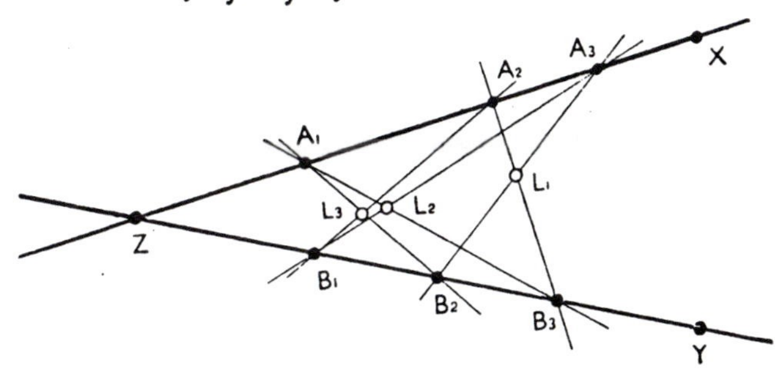

Proof. Let the two given lines be XZ and YZ, and let

$$A_2 B_3 . A_3 B_2, \quad A_3 B_1 . A_1 B_3, \quad A_1 B_2 . A_2 B_1$$

be denoted by L_1, L_2, L_3. We may write

$$\mathbf{a}_i = \mathbf{x} + \lambda_i \mathbf{z} \quad (i = 1, 2, 3),$$

and

$$\mathbf{b}_j = \mathbf{y} + \mu_j \mathbf{z} \quad (j = 1, 2, 3).$$

Then L_1 can be represented both by a vector

$$p(\mathbf{x} + \lambda_2 \mathbf{z}) + q(\mathbf{y} + \mu_3 \mathbf{z})$$

and also by a vector

$$p'(\mathbf{x} + \lambda_3 \mathbf{z}) + q'(\mathbf{y} + \mu_2 \mathbf{z});$$

and therefore

$$\frac{p}{p'} = \frac{q}{q'} = \frac{p\lambda_2 + q\mu_3}{p'\lambda_3 + q'\mu_2}.$$

Then

$$p\lambda_3 + q\mu_2 = p\lambda_2 + q\mu_3,$$

i.e.

$$p(\lambda_2 - \lambda_3) = q(\mu_2 - \mu_3),$$

and we may accordingly write

$$\begin{aligned} \mathbf{l}_1 &= (\mu_2 - \mu_3)(\mathbf{x} + \lambda_2 \mathbf{z}) + (\lambda_2 - \lambda_3)(\mathbf{y} + \mu_3 \mathbf{z}) \\ &= (\mu_2 - \mu_3)\mathbf{x} + (\lambda_2 - \lambda_3)\mathbf{y} + (\lambda_2 \mu_2 - \lambda_3 \mu_3)\mathbf{z}, \end{aligned}$$

with similar expressions for $\mathbf{l}_2$ and $\mathbf{l}_3$. Then, by addition,

$$\mathbf{l}_1 + \mathbf{l}_2 + \mathbf{l}_3 = \mathbf{0},$$

and the points L_1, L_2, L_3 are therefore collinear.

THEOREM 9 (*The harmonic construction*). *Let A, B, C be three (distinct) collinear points, and let three arbitrary lines AXO, BYO, CYX be drawn, one through each of the points, to meet in pairs in Y, X, O. Then, if AY and BX meet in T, the line OT meets the line ABC in the harmonic conjugate D of C with respect to A and B.*

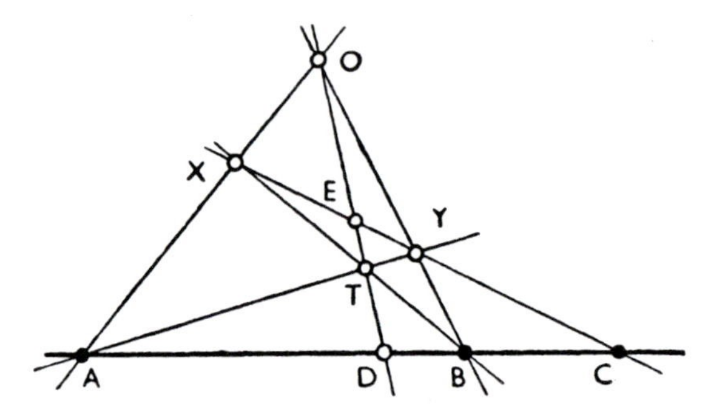

Proof. Take the representation $\mathscr{R}$ for which A, B, O are reference points and T is unit point. Then the coordinates of Y, X, D are respectively $(0, 1, 1)$, $(1, 0, 1)$, and $(1, 1, 0)$. XY is therefore the line $x_0+x_1-x_2 = 0$, and C is the point $(1, -1, 0)$. Thus $\mathbf{d} = \mathbf{a}+\mathbf{b}$ and $\mathbf{c} = \mathbf{a}-\mathbf{b}$, and hence $\{C, D; A, B\} = -1$.

Remarks

(i) Theorem 9 furnishes a construction, using only relations of incidence, for the harmonic conjugate of a given point with respect to two other given points collinear with it. In real projective geometry the construction can be carried out with the straight-edge alone.

(ii) The construction may be thought of as based upon the harmonic property of the quadrangle $ABYX$ (Chapter II, Theorem 2).

EXERCISE. If XY meets OD in E, prove that E is the harmonic conjugate of C with respect to X and Y.

§ 7. HOMOGRAPHIC RANGES AND PENCILS

In Chapter III we defined the notion of homographic correspondence between two variable points P and Q of S_1. If, now, we select any line a in S_2, the points of a form a one-dimensional projective domain, and we may accordingly consider homographies on a. We are no longer confined, however, to a single line, but have the whole doubly infinite system of lines in S_2 at our disposal; and we can now introduce the idea of a homographic correspondence between a variable point P which moves on one fixed line a and another variable point Q which moves on a second fixed line b. Correspondences of this kind are especially interesting because, as we shall see, they are equivalent to certain simple incidence constructions in the plane, and this gives them a specifically geometrical significance. We shall now give a formal definition of homography in the new extended sense, and then go on to establish the main properties of such homographies.

The set of all points of a fixed line a will be called a *range of points*, having a as its axis; and dually, the set of all lines through a fixed point A will be called a *pencil of lines*, having A as its vertex. Both these systems are one-dimensional projective domains, as we have already seen, and the equations

$$\mathbf{x} = \mathbf{y}+\theta\mathbf{z} \quad \text{and} \quad \mathbf{u} = \mathbf{v}+\theta\mathbf{w}$$

give rise, in the two cases, to allowable representations by a parameter θ. Whenever we refer to a parameter of a point of a range or of a line of a pencil we shall mean an allowable parameter defined in this way.

DEFINITION. A (1, 1) correspondence between (a) two variable points P, Q, lying on fixed lines a, b, or (b) two variable lines p, q, passing through fixed points A, B, or (c) a variable point P, lying on a fixed line a, and a variable line q, passing through a fixed point B, is said to be *homographic* if the parameters of corresponding elements of the ranges or pencils concerned, referred in each case to an arbitrary allowable representation, satisfy a fixed equation of the form

$$\theta' = \frac{\alpha\theta+\beta}{\gamma\theta+\delta} \quad (\alpha\delta-\beta\gamma \neq 0).$$

Since cross ratio is invariant over bilinear transformation, this definition is in harmony with the less formal definition of homography already given in Chapter II (p. 28). We need, however, to establish its legitimacy in the formal system, i.e. to show that the concept of homography is independent of the particular choice of representations. This may be shown by a mode of argument that is already familiar, and it will be sufficient here to indicate the result. In terms of homogeneous parameters (λ, μ) and (λ', μ'), the equation of the homography may be written

$$\begin{pmatrix}\lambda'\\ \mu'\end{pmatrix} = \begin{pmatrix}\alpha & \beta\\ \gamma & \delta\end{pmatrix}\begin{pmatrix}\lambda\\ \mu\end{pmatrix},$$

or

$$\boldsymbol{\lambda}' = \mathbf{A}\boldsymbol{\lambda}, \text{ say.}$$

If we change the representation of the two ranges or pencils according to the scheme
$$\bar{\boldsymbol{\lambda}} = \mathbf{P}\boldsymbol{\lambda}, \qquad \bar{\boldsymbol{\lambda}}' = \mathbf{Q}\boldsymbol{\lambda}',$$
we then have
$$\bar{\boldsymbol{\lambda}}' = \bar{\mathbf{A}}\bar{\boldsymbol{\lambda}},$$
where $\bar{\mathbf{A}}$ is the non-singular matrix $\mathbf{QAP}^{-1}$.

Since ranges and pencils enter into our system in exactly the same way, it is often convenient to have a neutral term meaning 'range or pencil', and the term usually adopted is *one-dimensional form*, or simply *form* if there is no risk of ambiguity. The above definition then allows us to speak of a homographic correspondence between any two forms. When such a correspondence is defined we say that the forms are homographic, projective, or related.

If P and Q are variable corresponding points of two homographic ranges, we often write
$$(P) \barwedge (Q);$$
and a similar notation is used to indicate that any two one-dimensional forms are homographically related.

THEOREM 10. *A homographic correspondence between two given one-dimensional forms is uniquely determined by three corresponding pairs.*

This theorem, which is an immediate consequence of Theorem 7, Chapter III, is of fundamental importance in projective geometry—like many other theorems which specify the number of degrees of freedom of a geometrical entity. It plays a central role in the non-algebraic axiomatic development of projective geometry, where it appears as the Projective Axiom or Correspondence Axiom.

Homographies between ranges and pencils can all be defined, as we have already mentioned, by suitable incidence constructions. The reason for this is that the basic operations of projection and section, from which all incidence constructions are built up, transform ranges and pencils always into homographically related pencils and ranges. Thus every correspondence set up by means of an incidence construction is homographic; and we have only to show that the general incidence construction has the right amount of freedom in order to be able to infer, from Theorem 10, that every homographic correspondence is obtainable in this way. The reduction of homographies to incidence constructions, then, depends ultimately on the basic connexion established by the following theorem.

THEOREM 11. *If a is a fixed line and A is a fixed point that does not lie on a, then the correspondence between a variable point P of a and the line p which joins P to A is homographic: i.e. $(P) \barwedge (p)$.*

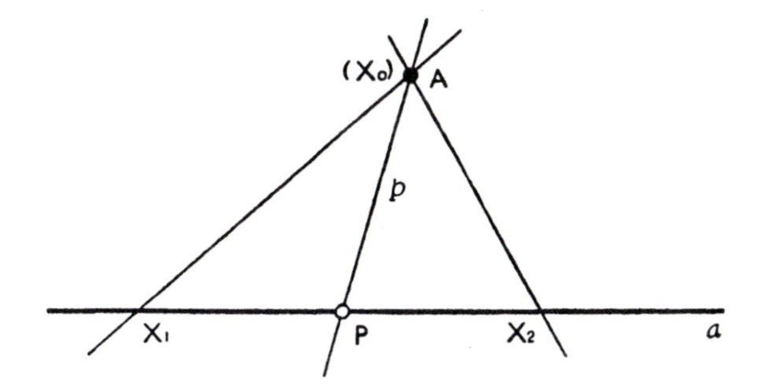

Proof. Take the triangle of reference as shown, and let P be the point $(0, 1, \theta)$. Then p is the line $\theta x_1 - x_2 = 0$, i.e. the line $(0, \theta, -1)$. Thus the parameters of P, referred to X_1, X_2 as reference points, and of p, referred to AX_2, AX_1 as reference lines, are respectively θ and $-1/\theta$; and these are connected by a non-singular bilinear equation.

COROLLARY. *The rays of a pencil cut homographic ranges on any pair of transversals.*

Proof. If the variable ray p cuts the transversals a and b in points P and Q respectively, then

$$(P) \barwedge (p) \barwedge (Q).$$

Note. In this and similar arguments where we construct a chain of homographies, we use the transitivity of the relation $\barwedge$; i.e. the fact that if $(P) \barwedge (Q)$ and $(Q) \barwedge (R)$, then $(P) \barwedge (R)$. The relation is necessarily transitive because the set of all non-singular bilinear transformations $\theta' = \dfrac{\alpha\theta+\beta}{\gamma\theta+\delta}$ is a group.

Theorem 11 describes the simplest possible homographic correspondence between a range and a pencil. There also exist homographic correspondences between two forms of the same kind which are almost as simple.

THEOREM 12. *If variable points P, Q, describing ranges with distinct axes a, b, correspond in such a way that the line PQ always passes through a fixed point V, not lying on a or b, then the correspondence is homographic. Dually, if variable lines p, q, describing pencils with distinct vertices A, B, correspond in such a way that the point pq*

always lies on a fixed line v, not passing through A or B, then the correspondence is homographic.

Proof. The first part of the theorem is equivalent to the corollary to Theorem 11, and the second part follows from the first by duality.

DEFINITION. Two ranges related as in Theorem 12 are said to be in perspective from the vertex V; and two pencils related as in Theorem 12 are said to be in perspective from the axis v.

A range and a pencil related as in Theorem 11 are also sometimes said to be in perspective.

The correspondence between two forms in perspective is called a *perspectivity*.

Now that we have discovered these very simple kinds of homography between two forms, the question naturally arises whether perspectivities are the only kinds of homography that exist, or whether there are other kinds. We shall see that there are more general homographies, but that they can all be expressed as products of perspectivities.

THEOREM 13. *Two homographic ranges (P), (Q), with axes a, b respectively, are in perspective if and only if the point ab is self-corresponding.*

Proof. Let O be the point ab. If the ranges are in perspective, then O is clearly self-corresponding.

Suppose, conversely, that $(P) \barwedge (Q)$ and O is self-corresponding. Take two fixed corresponding pairs (P_1, Q_1), (P_2, Q_2) and let $P_1 Q_1$, $P_2 Q_2$ meet at V. Then the perspectivity with vertex V is a homography which has three pairs (O, O), (P_1, Q_1), (P_2, Q_2) in common with the given homography, and the given homography therefore coincides with this perspectivity.

COROLLARY 1. *Dually, two homographic pencils (p), (q), with vertices A, B respectively, are in perspective if and only if the line AB is self-corresponding.*

COROLLARY 2. *If A, B, C, D are collinear and A, B', C', D' are collinear, and if $\{A, B; C, D\} = \{A, B'; C', D'\}$, then BB', CC', DD' are concurrent. Dually, if a, b, c, d are concurrent and a, b', c', d' are concurrent, and if $\{a, b; c, d\} = \{a, b'; c', d'\}$, then bb', cc', dd' are collinear.*

It follows from Theorem 13 that not every homographic correspondence between two ranges is a perspectivity. For suppose we take two lines a, b, meeting in O, and choose points P_2, P_3 on a and Q_1, Q_2, Q_3 on b, all distinct from O. Then there is a homography connecting the ranges on a and b for which (O, Q_1), (P_2, Q_2), (P_3, Q_3) are pairs, and this homography is certainly not a perspectivity since O is not self-corresponding.

We shall now show that every homography between two distinct lines which is not already a perspectivity can be resolved into a product of two perspectivities.

THEOREM 14. *If ϖ is a homography, but not a perspectivity, between two ranges (P), (Q) with distinct axes a, b, there exists a third related range (R), with a third line c as axis, such that, for any point P and the corresponding points Q and R, the line PR passes through a fixed point V and RQ through a fixed point W.*

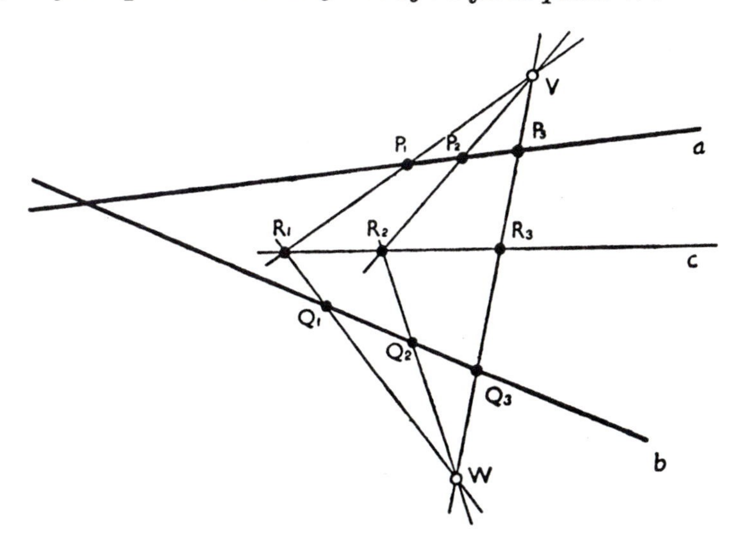

Proof. Choose three pairs of ϖ: (P_1, Q_1), (P_2, Q_2), (P_3, Q_3). On the line $P_3 Q_3$ take two points V, W, distinct from each other and from P_3 and Q_3. Let $VP_1 . WQ_1$, $VP_2 . WQ_2$ be denoted by R_1, R_2, and let the line $R_1 R_2$ be denoted by c. Then, clearly, c, V, and W satisfy the conditions of the theorem; for the product of the perspectivities defined by V, a, c and W, c, b is a homography which has three pairs (P_i, Q_i) $(i = 1, 2, 3)$ in common with ϖ.

COROLLARY. *Any homography that relates variable points P, Q of the same line can be resolved into a product of not more than three perspectivities.*

Harmonic ranges and pencils

By making use of Theorem 11 we can now give a simple alternative proof of Theorem 9 on the harmonic construction; and we give this proof here because the method used is of wide applicability.

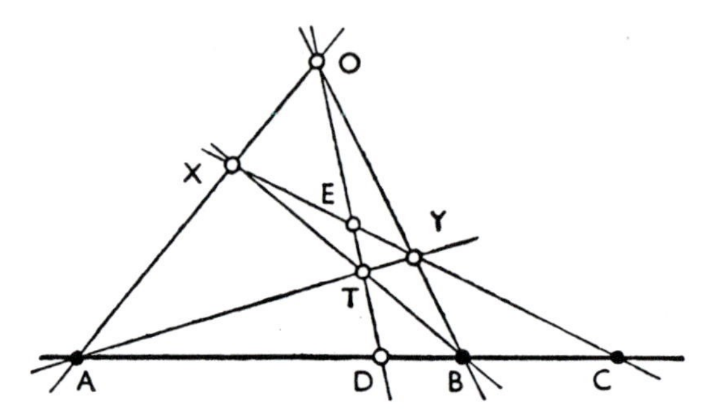

By Theorem 11,

$$\begin{aligned}(A,B,C,D) &\barwedge T(A,B,C,D)\\ &\barwedge (Y,X,C,E)\\ &\barwedge O(Y,X,C,E)\\ &\barwedge (B,A,C,D).\end{aligned}$$

Therefore $\{A,B;\,C,D\} = \{B,A;\,C,D\}$, and the pairs (A,B) and (C,D) are harmonic.

Four collinear points A, B, C, D such that $\{A,B;\,C,D\} = -1$ are said to form a harmonic range; and dually, four concurrent lines a, b, c, d such that $\{a,b;\,c,d\} = -1$ are said to form a harmonic pencil. By Theorem 11, any transversal cuts the rays of a harmonic pencil in the points of a harmonic range, and the rays joining the points of a harmonic range to a common vertex form a harmonic pencil.

Exercise. A figure made up of points and lines is constructed as follows: A, B, C, D are four points, no three of which are collinear, and they are joined in pairs by six lines; these lines meet in three further points X, Y, Z, which are also joined by lines. Show that the three pencils of four lines which occur in the figure are harmonic.

The figure just described is called a *complete quadrangle.* A, B, C, D are the vertices; (AB, CD), (AC, BD), (AD, BC) are the pairs of opposite sides; and X, Y, Z are the diagonal points or vertices of the diagonal triangle. The dual figure is called a *complete quadrilateral* (cf. p. 19).

THEOREM 15 (*The Cross Axis Theorem*). *If P, Q describe homographic ranges on distinct lines a, b, then the point of intersection $P_1 Q_2 . P_2 Q_1$ of the cross joins of any two corresponding pairs (P_1, Q_1), (P_2, Q_2) lies on a fixed line.*

Proof. Let a and b meet at O, and let $P_1 Q_2 . P_2 Q_1$ be the point T. We have to show that the locus of T is a fixed line, and in order to do this we take separately the cases in which the correspondence between P and Q is and is not a perspectivity.

(i) Suppose the ranges (P), (Q) are in perspective from a vertex V.

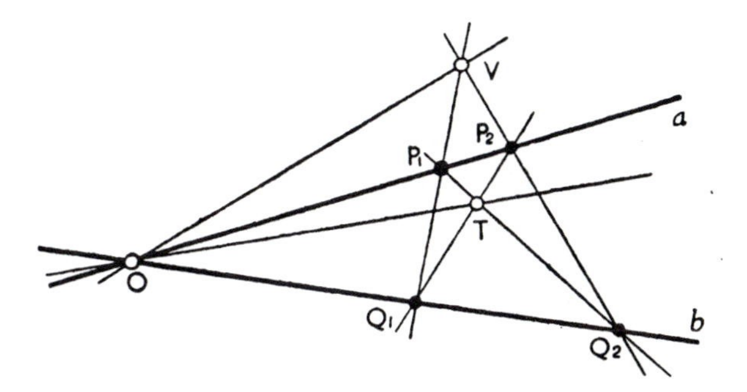

Since OVT is the diagonal triangle of the quadrangle $P_1 P_2 Q_1 Q_2$, the lines OT, OV are harmonic with respect to a, b. But OV, a, b are all fixed lines, and therefore OT is also a fixed line. Thus T always lies on a fixed line through O.

(ii) Suppose the ranges (P), (Q) are not in perspective. Then if to O on a corresponds V on b, and to O on b corresponds U on a, U and V are both distinct from O.

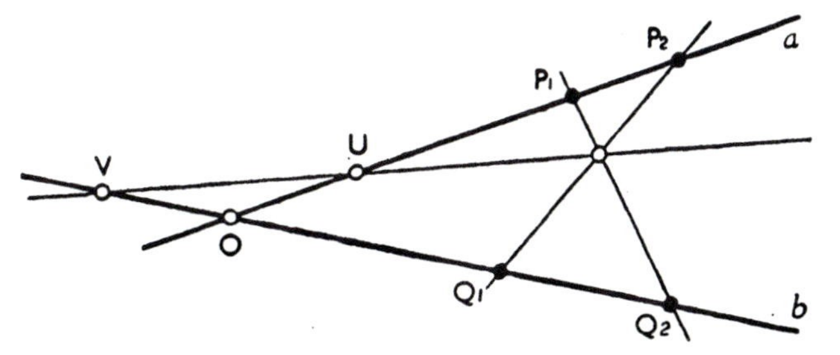

Since (U, O), (O, V), (P_1, Q_1), (P_2, Q_2) are all pairs of the given homography,

$$\{U, O; P_1, P_2\} = \{O, V; Q_1, Q_2\} = \{V, O; Q_2, Q_1\},$$

and therefore (U, V), (O, O), (P_1, Q_2), (P_2, Q_1) are four pairs of a homography. This homography, by Theorem 13, is a perspectivity, and UV, $P_1 Q_2$, $P_2 Q_1$ are therefore concurrent. Thus $P_1 Q_2$ and $P_2 Q_1$ meet on the fixed line UV.

Remark. The line which contains all points of intersection of pairs of cross joins of a homographic correspondence between two ranges is called the *cross axis* of the correspondence. In the general case, when the homography is not a perspectivity, this line is the line joining the points of the two ranges which correspond, in each direction, to the common point of the axes of the ranges.

The Cross Axis Theorem is a key theorem in plane projective geometry, for it enables us actually to construct, by means of the straight-edge, the homographic correspondence between two given lines that is determined by three assigned pairs. Suppose P_1, P_2, P_3 are three given points of a line a and Q_1, Q_2, Q_3 are three given points of a second line b, and P is an arbitrary point of a. If we wish to determine the point Q of b that corresponds to P in the unique homography determined by the three pairs (P_1, Q_1), (P_2, Q_2), (P_3, Q_3), we find the cross axis of the homography by constructing the points $P_1 Q_2 . P_2 Q_1$ and $P_2 Q_3 . P_3 Q_2$. Then Q may be obtained by joining P_1 to the point where the cross axis is met by PQ_1, and marking where this join meets b.

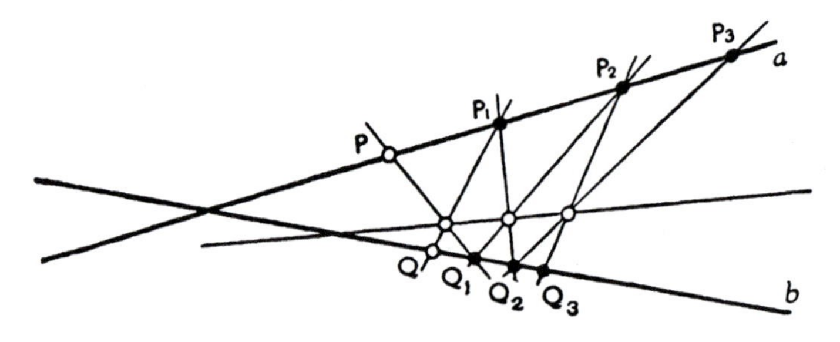

EXERCISES

1. Show how to construct the homography on a single line a that is determined by three given pairs (P_1, Q_1), (P_2, Q_2), (P_3, Q_3). [*Hint.* Project Q_1, Q_2, Q_3 on to another line b.]

2. Show how to construct the homography on a that is determined by (i) one of its united points, M, and two pairs (P_1, Q_1), (P_2, Q_2), and (ii) its two united points M, N and one pair (P_1, Q_1).

3. Deduce Pappus's Theorem (Theorem 8) from Theorem 15.

4. State the dual of Theorem 15 (the Cross Vertex Theorem) and also the dual of Pappus's Theorem.

§ 8. Specializations of Two-dimensional Projective Geometry

We have already seen, in Chapter III, how the affine line and the euclidean line can be obtained from the projective line by suitable specialization, and we now wish to see how to carry out analogous specializations in two dimensions. Since additional complications arise in this more general case, we shall go into rather more detail here than in Chapter III.

All points in the projective plane S_2 are indistinguishable within the system of projective geometry, and all lines are similarly indistinguishable. If, therefore, we now allow certain points or lines to have recognizable individuality, we shall obtain a geometrical system with a more elaborate structure. By singling out special points and lines suitably we can in fact construct abstract models of affine and euclidean geometry, and this is our present purpose. The idea may easily be translated into algebraic language. A point will possess recognizable individuality if its representation in every allowable coordinate system is the same; and we may therefore take special points into account by reducing the class ($\mathscr{R}$) of allowable representations to a subclass of itself—this subclass being defined by the invariance of the representation of the points concerned.

Affine geometry involves an invariant line (invariant as a whole, though not necessarily point by point), while euclidean geometry involves an invariant point-pair. This means that the affine specialization is linear, whereas the euclidean specialization is quadratic; and although we can consider affine geometry over a general field K, we encounter difficulties with euclidean geometry if the ground field is not the real field. We require the 'distance' $\sqrt{\{(X_1-X_2)^2+(Y_1-Y_2)^2\}}$ to be real, for example. This being so, we shall presuppose throughout this section that the ground field is the real field R.

Affine geometry of two dimensions

We begin, then, with the real projective plane $S_2(R)$, and its allowable representations $\mathscr{R}$, which are connected by the group $PGL(2; R)$. If we select one particular representation $\mathscr{R}$, which we agree to regard as fixed for the time being, we may embed the real plane in a complex plane by adjoining to it points which correspond to sets of coordinates (x_0, x_1, x_2) which are essentially complex—i.e. which cannot all be made real by multiplication by a common complex factor. If we now change the representation to $\mathscr{R}'$, by a real non-singular linear transformation, the coordinates of the points are in general changed; but, since the coefficients in the equations of transformation are real, the coordinates of the original points are still real and the coordinates of the adjoined points essentially complex. The distinction between the two kinds of points is thus invariant over the group $PGL(2; R)$, and we may legitimately refer to the real projective plane with complex points adjoined to it. We shall call the original points real or actual and the adjoined points complex or ideal. The real projective plane, extended in this way by ideal complex points, will be denoted by $\hat{S}_2$.

Our next step is to pass from $\hat{S}_2$ to the real affine plane; and to this end we single out a real line i of $\hat{S}_2$, to be called the *line at infinity*. This line is to be removed from the plane, in the sense that its points will be treated not as actual points of the plane but as ideal points. We now have two kinds of ideal points, namely complex points and points at infinity (both real and complex). In one sense they do not form part of the affine plane, for every figure that we consider as an actual figure in affine geometry consists wholly of actual points; but in another sense they do form part of the plane, and we often argue about actual and ideal points at the same time, using the properties of ideal points in order to deduce properties of actual points. The reader may recall the remarks about ideal points made in Part I.

Having once introduced the line at infinity, we can easily define various affine concepts. Thus two lines of the affine plane are said to be parallel if and only if they meet in a point of i. Again, if A and B are two points which do not lie on i, and if AB meets i at N, then the harmonic conjugate of N with respect to A and B is called the mid-point of the segment AB.

In order to handle affine geometry algebraically, we have to characterize the line i by an invariant equation, and we shall suppose that this equation is $x_0 = 0$. Since the points of i are now regarded as ideal points, no point with $x_0 = 0$ is actual, and this means that we can represent the actual points of the affine plane by pairs of non-homogeneous coordinates (X_1, X_2), where

$$X_1 = x_1/x_0, \qquad X_2 = x_2/x_0.$$

The allowable representations $\mathscr{R}_A$ of the affine plane are those representations $\mathscr{R}$ of $\hat{S}_2$ in which the line i has the equation $x_0 = 0$; and this leads at once to the following theorem.

THEOREM 16. *If $\mathscr{R}_A$ is any one allowable representation of the affine plane, then the whole class $(\mathscr{R}_A)$ of allowable representations consists of all those representations which can be derived from $\mathscr{R}_A$ by applying a transformation of the form*

$$X_1' = b_{11}X_1 + b_{12}X_2 + c_1,$$

$$X_2' = b_{21}X_1 + b_{22}X_2 + c_2,$$

where the coefficients are arbitrary real numbers, subject to the condition $|b_{rs}| \neq 0$.

Proof. As Theorem 20, Chapter III.

Remark. The set of all transformations of the type specified in Theorem 16 is a group, the affine group for the plane. It is the subgroup of $PGL(2; R)$ which is picked out by the invariance of the equation $x_0 = 0$.

Euclidean geometry of two dimensions

The euclidean plane is obtained from the (real) affine plane by further specialization. Two conjugate complex points I, J on the line at infinity i are singled out as *absolute points*, and they are required to be recognizable (either individually or as a pair). Since these points are already ideal, the actual points of the euclidean plane are the same as the actual points of the affine plane, but we can now distinguish additional relations between them.

The lines which join any point P to I and J are called the *isotropic lines* through P. If P is real they are, of course, conjugate complex lines. Two lines a and b are said to be perpendicular if they are harmonic with respect to the isotropic lines through their point of intersection.

Let a, b be any two lines, both distinct from i, and let j, j' be the isotropic lines through the point ab. Putting

$$m(a,b) = \{a,b;j,j'\},$$

we call $m(a,b)$ the *angle modulus* of the ordered pair of lines (a,b). It is, of course, presupposed that a definite order is assigned once for all to I and J; otherwise the angle modulus can have either of two reciprocal values. The angle modulus has the following properties:

(i) $m(a,b).m(b,a) = 1$;

(ii) if a and b are perpendicular, $m(a,b) = m(b,a) = -1$, and conversely;

(iii) if a, b, c are concurrent, $m(a,b).m(b,c) = m(a,c)$;

(iv) $m(a,b) = 1$ if and only if a and b coincide.

If a, b and c, d are two pairs of lines such that

$$m(a,b) = m(c,d),$$

we say that the angle between a and b (in that order) is equal to the angle between c and d, and we write $\widehat{ab} = \widehat{cd}$. It will be noted that we have defined equality of angle without first defining an angular measure.

If, now, a and b are neither of them isotropic lines, and j, j' are the isotropic lines through ab, there is a unique pair of lines x, y, harmonic with respect to both the pairs a, b and j, j', namely the common pair of rays of the two involutions whose self-corresponding rays are respectively a, b and j, j'. Then, since

$$\{x,y;\, a,b\} = \{x,y;\, j,j'\} = -1,$$

x and y are the self-corresponding rays of the involution determined by the two pairs (a,b) and (j,j'); and hence

$$\{a,x;\, j,j'\} = \{b,x;\, j',j\} = \{x,b;\, j,j'\}.$$

Thus $m(a,x) = m(x,b)$, or $\widehat{ax} = \widehat{xb}$; and similarly $\widehat{ay} = \widehat{yb}$. We may therefore call x and y the bisectors of the angles between a and b.

By means of the angle modulus $m(a,b)$ we can handle all the properties of angles that interest us in the present book. If, however, we wish to have a formal counterpart of the familiar angular measure of euclidean geometry, this can easily be introduced by means of Laguerre's formula. Let us adopt for the moment the point of view of elementary coordinate geometry, working with a pair of rectangular axes Ox, Oy. If a and b are two lines through the origin we may write their equations as $y = m_1 x$ and $y = m_2 x$, where m_1 and m_2 are the gradients of the lines. Since the coordinates of the circular points are $(1, \pm i, 0)$ (cf. p. 33) the equations of the isotropic lines j and j' through O are $y = ix$ and $y = -ix$.

We have, therefore, $m(a,b) = \{a,b;j,j'\} = \{m_1,m_2;i,-i\}$; and since the angle α between a and b is given by

$$\tan\alpha = (m_1-m_2)/(1+m_1 m_2)$$

we can express α in terms of the angle modulus. In this way we obtain *Laguerre's formula*

$$\alpha = \frac{1}{2i}\log\{a,b;j,j'\}.$$

If, therefore, we wish to introduce angular measure into the formal system, we need only include the further definition

$$\widehat{ab} = \frac{1}{2i}\log m(a,b).$$

In order to handle euclidean geometry algebraically, we must assign invariant coordinates to I and J. It does not matter what coordinates we choose, as long as we do not violate the condition $x_0 = 0$; but in order that our algebra may agree with that familiar from ordinary coordinate geometry we naturally choose the coordinates $(0,1,i)$ and $(0,1,-i)$. This choice determines the class $(\mathscr{R}_E)$ of allowable coordinate representations of the euclidean plane.

THEOREM 17. *If $\mathscr{R}_E$ is any one allowable representation of the euclidean plane, then the whole class $(\mathscr{R}_E)$ of allowable representations consists of all those representations which can be derived from $\mathscr{R}_E$ by applying a transformation of the form*

$$X_1' = c(X_1\cos\alpha+X_2\sin\alpha)+a,$$
$$X_2' = c(-X_1\sin\alpha+X_2\cos\alpha)+b,$$

where a, b, c, α are arbitrary real numbers, with $0 \leqslant \alpha < 2\pi$ and $c > 0$.

Proof. Let (X_1,X_2) and (X_1',X_2') be non-homogeneous coordinates in $\mathscr{R}_E$ and a second allowable representation $\mathscr{R}_E'$. Then, by Theorem 16,

$$X_1' = b_{11}X_1+b_{12}X_2+c_1,$$
$$X_2' = b_{21}X_1+b_{22}X_2+c_2,$$

where $b_{11}b_{22}-b_{12}b_{21} \neq 0$.

Since $(0,1,i)$ and $(0,1,-i)$ are invariant homogeneous coordinates,

$$\frac{i}{1} = \frac{b_{21}.1+b_{22}.i+c_2.0}{b_{11}.1+b_{12}.i+c_1.0},$$

and

$$\frac{-i}{1} = \frac{b_{21}.1+b_{22}.(-i)+c_2.0}{b_{11}.1+b_{12}.(-i)+c_1.0};$$

i.e. $$(b_{11}-b_{22})i-(b_{12}+b_{21}) = 0,$$

and $$-(b_{11}-b_{22})i-(b_{12}+b_{21}) = 0.$$

Therefore $b_{11}-b_{22} = b_{12}+b_{21} = 0$, and the equations of transformation reduce to

$$X_1' = b_{11}X_1+b_{12}X_2+c_1,$$
$$X_2' = -b_{12}X_1+b_{11}X_2+c_2;$$

and these equations may be written in the form given in the enunciation of the theorem. The values of a, b, c, α may be assigned freely except that, if each transformation is to be obtained once only, α must be confined to an interval of length 2π and c taken positively.

Remarks

(i) If we apply the transformation just obtained to the expression

$$\rho^2 = (X_1-Y_1)^2+(X_2-Y_2)^2$$

we obtain the equation $\rho'^2 = c^2\rho^2.$

The function $|\rho|$ is called the distance between the points (X_1, X_2) and (Y_1, Y_2). It is not a euclidean invariant, but any ratio of distances is invariant. For this reason the euclidean plane, as we have defined it, is sometimes referred to as the similarity euclidean plane.

(ii) If we require I and J to be recognizable only as a point-pair, and not individually, the class of allowable representations is correspondingly widened. We now have the improper euclidean transformations

$$X_1' = c(X_1\cos\alpha+X_2\sin\alpha)+a,$$
$$X_2' = -c(-X_1\sin\alpha+X_2\cos\alpha)+b,$$

as well as the proper transformations with $+c$ in the second equation.

(iii) If we start with a system of rectangular coordinates (X_1, X_2) in the plane, a transformation of the type given in the theorem represents a rotation through an angle α, a magnification in the ratio $1:c$, a reflection in the axis of X_1 (for the improper transformation only), and a change of origin to $(-a, -b)$. Distinguishing between I and J is thus equivalent to distinguishing right-handed from left-handed axes. (Compare the remarks made above on the angle modulus.) The constant c corresponds to the arbitrariness of the unit of length—there is no 'natural' unit of length in euclidean geometry, although there is a natural unit of angle.

(iv) If we go farther, and fix the length of a given segment as unit of length, the transformation from any allowable representation to any other is an orthogonal transformation, with $c = 1$ —or $c = \pm 1$ if improper transformations are permitted. We shall not make any such restriction in our system.

(v) The reader will now see that the distinction between affine and euclidean geometry does not fully reveal itself until we have two dimensions at our disposal. Thus in one dimension there is no distinction between the affine line and the similarity euclidean line; and in one dimension the distance between two points is a linear function of the coordinates, namely $X-Y$, whereas in two dimensions it is an irrational expression

$$\sqrt{\{(X_1-Y_1)^2+(X_2-Y_2)^2\}}.$$

Non-euclidean geometry of two dimensions

We mention non-euclidean geometry here for completeness although, as we have not yet introduced conics, it is logically out of place. We shall not take up the study of non-euclidean geometry systematically in this book, but it is interesting to see how it is related to projective geometry.

The absolute point-pair (I, J) may be looked upon as a degenerate conic envelope, and this leads us to ask what happens if we assign a similar role to a proper conic envelope. If we do this we obtain not euclidean geometry but non-euclidean geometry.

Let a proper conic Ω in the real projective plane be singled out as the *absolute conic*. Two lines which meet in a point of Ω will then be said to be parallel, and two lines that are conjugate for Ω will be said to be perpendicular. If A, B are two points, not lying on Ω, and if AB meets Ω in M, N, a distance modulus of A and B is defined as $m(A, B) = \{A, B; M, N\}$. Dually, if a, b are two lines, not touching Ω, and if the tangents to Ω from ab are m, n, an angle modulus of a and b is defined as $m(a, b) = \{a, b; m, n\}$.

It will be seen from this that euclidean geometry may be looked upon as a degenerate case of non-euclidean geometry. The duality, for example, which is so prominent a feature of projective geometry may still be traced out in non-euclidean geometry, but in euclidean geometry it no longer exists.

EXERCISES ON CHAPTER IV

[In these exercises, and throughout the rest of the book, we shall adopt the convention that cyclic order is always to be assumed in the absence of any explicit statement to the contrary. Thus, for example, the statement that XYZ and $X'Y'Z'$ are triangles in perspective implies that (X, X'), (Y, Y'), (Z, Z') are the pairs of vertices which correspond, and the statement that the triangle $X'Y'Z'$ is inscribed in the triangle XYZ means that YZ, ZX, XY pass respectively through X', Y', Z'.]

1. The coordinates of a general point are (x, y, z) in a coordinate system for which A, B, C are the reference points and D is the unit point, and they are (x', y', z') in the system for which D, B, C are the reference points and A is the unit point. Find equations which express the ratios of x', y', z' in terms of x, y, z.

Find also the equations of transformation from the first coordinate system to a system in which A, B, C, D have coordinates $(-1, 1, 1)$, $(1, -1, 1)$, $(1, 1, -1)$, $(1, 1, 1)$ respectively.

2. The equations of transformation from one allowable coordinate representation to another are

$$x' : y' : z' = 8x+3y+2z : 3x+4y : 2x+2z.$$

Find the three points whose coordinates are the same in both representations.

3. Show that the triangle $X'Y'Z'$ whose vertices are $(1, p, p')$, $(q', 1, q)$, $(r, r', 1)$ is in perspective with the triangle of reference if and only if

$$pqr = p'q'r'.$$

If XYZ is in perspective with $X'Y'Z'$ and with $Y'Z'X'$, show that it is also in perspective with $Z'X'Y'$.

4. Triangles $X'Y'Z'$ and $X''Y''Z''$ are respectively circumscribed to and inscribed in a given triangle XYZ. If each of them is also in perspective with XYZ, prove that they are in perspective with each other.

5. Two triangles XYZ and $X'Y'Z'$ are such that XX', YY', ZZ' meet YZ, ZX, XY in collinear points. Show that the three lines which join X', Y', Z' to the intersections of pairs of corresponding sides of the triangles are concurrent.

6. The line joining the points U, V, whose coordinates are $(1, 1, 1)$ and (a, b, c), meets the sides YZ, ZX, XY of the triangle of reference in L, M, N, and the harmonic conjugates of these points with respect to U, V are L', M', N'. Prove that XL', YM', ZN' are concurrent.

7. What is the geometrical relationship between a point and its harmonic polar line with respect to the triangle formed by a pair of rectangular axes OX, OY and the line at infinity?

8. The vertices of a triangle in the extended euclidean plane are an actual point O and the two absolute points I, J. Show that the harmonic polar line of any point A with respect to this triangle lies along the base of the equilateral triangle whose apex is A and whose centre is O.

Deduce, or prove otherwise, that if XYZ is a given triangle in the projective plane, any general point P is one vertex of a unique triangle PQR

of which each side is the harmonic polar of the opposite vertex with respect to XYZ.

9. Four points A, B, C, D and four lines a, b, c, d in the plane are such that the points bc, ca, ab, ad, bd lie respectively on the lines AD, BD, CD, BC, CA. Prove that cd lies on AB.

10. If three triangles are in perspective, two by two, with a common vertex of perspective, prove that the three axes of perspective are concurrent.

State the dual of this result, and interpret it in the case when the common axis of perspective is the line at infinity.

11. If (P) and (P') are homographic ranges on two lines which meet in O, prove that the cross axis of the ranges passes through O if and only if the ranges are in perspective.

12. A homographic correspondence between two ranges (P) and (P') on the same line l is determined by three pairs (U, U), (A, A'), (B, B'). Show that there exists, on any other line m through U, an intermediate range which is in perspective with (P) from a vertex X and with (P') from a vertex Y.

Show that XY meets l in the second united point of the correspondence, and that the correspondence is an elation if XY passes through U.

13. If l and l' are two lines which meet in a point O, and a range (P) on l is projected from two different vertices V_1, V_2 into ranges (P_1), (P_2) on l', show that a necessary and sufficient condition for the correspondence between P_1 and P_2 to be an involution is that OV_1 and OV_2 should be harmonically conjugate with respect to l and l'.

14. Desargues's Theorem for two triangles in perspective fails when one of the triangles is replaced by a triad of concurrent lines. If ABC is a proper triangle and a, b, c are three lines which are concurrent in a point O, prove that a, b, c meet BC, CA, AB in collinear points if and only if the pairs of rays (a, OA), (b, OB), (c, OC) are in involution. [*Hint*. Let a, b, c meet BC, CA, AB in L, M, N, and let AO meet BC in X. Then L, M, N are collinear if and only if $O(L, M, N, A) \barwedge A(L, M, N, O)$; and the result follows at once when we observe that

$$A(L, M, N, O) \barwedge (L, C, B, X) \barwedge O(L, C, B, A) \barwedge O(A, B, C, L).]$$

15. If ABC is a proper triangle and A', B', C' are three points lying on a line l, prove that AA', BB', CC' are concurrent if and only if l meets BC, CA, AB in points U, V, W, such that (A', U), (B', V), (C', W) are pairs of an involution.

Deduce the theorem on the concurrence of the altitudes of a triangle.

16. If a fixed line l meets the sides of two triangles ABC and $A'B'C'$ in the triads of points L, M, N and L', M', N', prove that the lines AL', BM', CN' are concurrent if and only if the lines $A'L$, $B'M$, $C'N$ are concurrent.

What does this theorem become when l is the line at infinity?

17. A fixed line l is met by the pairs of opposite sides of a quadrangle $ABCD$ in the pairs of points (L, L'), (M, M'), (N, N'). Prove that the three pairs belong to an involution on l.

18. If XYZ and $X'Y'Z'$ are two given triangles, prove that there exist in general two solutions to the problem of constructing a triangle PQR which is both inscribed in XYZ and circumscribed to $X'Y'Z'$.

Show also that if $X'Y'Z'$ is itself circumscribed to XYZ the problem has an infinity of solutions, each point of YZ being the vertex P of a triangle PQR which fulfils the required conditions.

19. XYZ is the triangle of reference, and $X'Y'Z'$ is another triangle whose vertices have coordinates (x_i, y_i, z_i) $(i = 1, 2, 3)$. If there exists only one triangle which is inscribed in XYZ and circumscribed to $X'Y'Z'$, prove that

$$x_1^2 X_1^2 + y_2^2 Y_2^2 + z_3^2 Z_3^2 - 2y_2 Y_2 z_3 Z_3 - 2z_3 Z_3 x_1 X_1 - 2x_1 X_1 y_2 Y_2 + 4X_1 Y_2 Z_3 = 0,$$

where

$$X_1 = y_2 z_3 - y_3 z_2, \qquad Y_2 = z_3 x_1 - z_1 x_3, \qquad Z_3 = x_1 y_2 - x_2 y_1.$$

20. A range (P) on a line l_1 is projected into a range (Q) on a line l_2 from a vertex V_{12}, and (Q) is projetcted into a range (R) on a line l_3 from a vertex V_{23}. The lines joining P and R to two further points V_{41} and V_{34} respectively meet in S. Show that the locus of S as P varies on l_1 is a line if and only if $V_{41}V_{34}$ meets l_1 and l_3 respectively in points whose joins to V_{12} and V_{23} meet on l_2.

If this condition is satisfied, and the locus of S is l_4, describe the relation of the quadrangle $V_{12}V_{23}V_{34}V_{41}$ to the quadrilateral $l_1 l_2 l_3 l_4$.

21. $ABCD$ is a square of side a and $A'B'C'D'$ is a similarly situated concentric square of side b. If $a = (1+\sqrt{2})b$, prove that any general point P of AB is one vertex of a quadrangle $PQRS$ such that P, Q, R, S lie respectively on AB, BC, CD, DA and PQ, QR, RS, SP pass respectively through B', C', D', A'.

22. If Δ denotes the expression

$$\frac{1}{2}\begin{vmatrix} X_1 & Y_1 & 1 \\ X_2 & Y_2 & 1 \\ X_3 & Y_3 & 1 \end{vmatrix},$$

prove that Δ is a *relative* affine invariant, i.e. that any affine transformation of the coordinates X, Y multiplies it by a constant which depends only on the transformation.

Show that the affine transformations of determinant $+1$ form a group, and that area, in the ordinary sense, is an *absolute* invariant for this group.

23. In a given allowable (projective) coordinate system, the absolute points I, J have coordinates $(0, 7, -3)$ and $(4, 5, -5)$ respectively. Find the modulus of the angle between the lines $y+4z = 0$ and $2x-3y = 0$.

Find also the equations of the bisectors of the angles between these lines, and the moduli of the angles which the bisectors make with the line

$$y+4z = 0.$$

24. Show that the transformations of the form

$$X_1' = c(X_1 \cos\alpha + X_2 \sin\alpha) + a,$$

$$X_2' = \pm c(-X_1 \sin\alpha + X_2 \cos\alpha) + b,$$

are the only affine transformations which leave the distance function $(X_1-Y_1)^2+(X_2-Y_2)^2$ relatively invariant, i.e. which multiply it by the same constant factor for all pairs of points (X_1, X_2), (Y_1, Y_2).

25. A real line l lies in a real plane π, and ϖ is a homographic correspondence between variable points P, P' of l. If ϖ has conjugate complex united points, show that there are two possible choices for a point A of π such that the pairs P, P' are cut on l by the arms of a variable angle, fixed in magnitude and sense, whose vertex is A.

CHAPTER V

CONIC LOCI AND CONIC ENVELOPES

§ 1. ALGEBRAIC LOCI AND ENVELOPES

IN this chapter we begin a systematic study of the conic locus and its dual figure the conic envelope. It is convenient to give, first of all, general definitions of the concepts of algebraic locus and algebraic envelope, of which conic locus and conic envelope are particular cases.

DEFINITION. An algebraic curve or locus of order n in S_2 is the totality of points whose coordinates in some assigned allowable representation $\mathscr{R}$ satisfy a fixed homogeneous equation of the nth degree:

$$f(x_0, x_1, x_2) = 0.$$

Dually, an algebraic envelope of class n in S_2 is the totality of lines whose coordinates in some assigned allowable representation $\mathscr{R}$ satisfy a fixed homogeneous equation of the nth degree:

$$\phi(u_0, u_1, u_2) = 0.$$

Since the change from one allowable representation to another is equivalent to a non-singular homogeneous linear transformation of coordinates, the order of a locus and the class of an envelope are projective properties.

We have already seen, in Chapter IV, that a locus of order 1 is a line and an envelope of class 1 is a point.

THEOREM 1. *A locus of order n meets a general line of the plane in n points, distinct or coincident. An envelope of class n has exactly n lines, distinct or coincident, through a general point of the plane.*

Proof. Consider a locus of order n, whose equation is

$$f(x_0, x_1, x_2) = 0.$$

A general point of the line joining the fixed points represented by $\mathbf{y}$ and $\mathbf{z}$ is given by $\mathbf{y}+\theta\mathbf{z}$, and this point belongs to the locus if and only if

$$f(y_0+\theta z_0, y_1+\theta z_1, y_2+\theta z_2) = 0.$$

Since this is an equation of degree n in θ there are exactly n points of intersection. The only possible exception is when the equation for θ is an identity, and in this case the line forms part of the locus. The second part of the theorem follows by duality.

DEFINITION. The locus $f(x_0, x_1, x_2) = 0$ is said to be *irreducible* if the polynomial $f(x_0, x_1, x_2)$ is irreducible over the ground field K, i.e. if it has no non-trivial factor with coefficients in K. An irreducible envelope is defined dually.

§ 2. CONIC LOCI AND CONIC ENVELOPES

DEFINITION. An algebraic locus of the second order is called a *conic locus*; and it is said to be *proper* or *degenerate* according as it is or is not irreducible. Dually, an algebraic envelope of the second class is called a *conic envelope*; and it is said to be *proper* or *degenerate* according as it is or is not irreducible.

There are three distinct kinds of conic locus:

(i) the proper conic locus, given by an irreducible quadratic equation;
(ii) the pair of distinct lines, given by a quadratic equation with distinct linear factors;
(iii) the repeated line, given by a quadratic equation whose left-hand side is the square of a linear form.

Dually, there are three distinct kinds of conic envelope: (i) the proper conic envelope; (ii) the pair of distinct points; (iii) the repeated point.

A conic locus is represented, in terms of any chosen representation $\mathscr{R}$ of S_2, by an equation of the second degree in x_0, x_1, x_2, and for different purposes it is convenient to write the general equation in different forms.

(i) Using suffix notation, we may write it as

$$S \equiv \sum_{i=0}^{2} \sum_{k=0}^{2} a_{ik} x_i x_k = 0.$$

It is no restriction to impose the symmetry condition $a_{ki} = a_{ik}$ $(i, k = 0, 1, 2)$ and we shall always suppose this done; the coefficients in the quadratic form on the left-hand side are then uniquely determined by the form itself. The quadratic form will always be denoted by S or, if we need to be more explicit, by $S(x_0, x_1, x_2)$ or $S(x_i)$.

(ii) Using matrix notation, we may write the same equation as

$$S(\mathbf{x}) \equiv \mathbf{x}^T \mathbf{A} \mathbf{x} = 0.$$

In this equation, $\mathbf{A}$ is the symmetric 3×3 matrix (a_{rs}) and $\mathbf{x}$ stands as usual for the column-vector with components (x_0, x_1, x_2).

(iii) In solving specific problems it is often an advantage to revert to the more elementary usage, in which the coordinates are denoted by x, y, z. Then the equation of the conic may be written as

$$S(x,y,z) \equiv ax^2+by^2+cz^2+2fyz+2gzx+2hxy = 0,$$

or, in Cayley's abridged notation,

$$(a,b,c,f,g,h \between x,y,z)^2 = 0.$$

The general quadratic form in line-coordinates will regularly be denoted by Σ, and we have the following ways of writing the equation of the general conic envelope:

(i) $$\Sigma(u_i) \equiv \sum_{i=0}^{2} \sum_{k=0}^{2} A_{ik} u_i u_k = 0,$$

with $A_{ki} = A_{ik}$ $(i,k = 0,1,2)$;

(ii) $$\Sigma(\mathbf{u}) \equiv \mathbf{u}^T \mathfrak{A} \mathbf{u} = 0;$$

(iii) $$\Sigma(u,v,w) \equiv Au^2+Bv^2+Cw^2+2Fvw+2Gwu+2Huv = 0,$$

or $$(A,B,C,F,G,H \between u,v,w)^2 = 0.$$

The equation of a given conic locus, referred to any allowable representation whatever, is necessarily of the form

$$\sum_i \sum_k a_{ik} x_i x_k = 0,$$

but if the representation is specially related to the conic the equation may be correspondingly simplified. We have a number of standard equations in projective geometry, arising from special ways of choosing the triangle of reference, which may be compared with the standard cartesian equations used in the elementary coordinate geometry of conics.

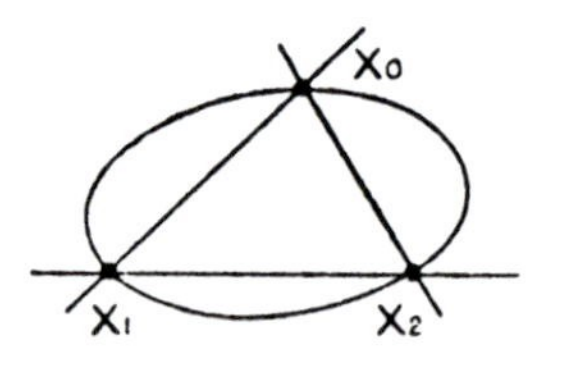

Suppose, for instance, three points of a given conic locus S are chosen as vertices of the triangle of reference. Then, since $(1,0,0)$ is a point of S, we have $a_{00} = 0$. Similarly $a_{11} = a_{22} = 0$, and the equation reduces to

$$a_{12}x_1x_2+a_{20}x_2x_0+a_{01}x_0x_1 = 0.$$

This is the general equation of a conic which circumscribes the triangle of reference. We shall come across other special forms of the equation of the conic later on (see Chapter VI, § 1).

§ 3. The Freedom of the Conic

The general equation $S = 0$ contains six coefficients. If we require a conic S to pass through a given point this imposes one linear condition upon the coefficients; and if five independent conditions of this kind are imposed, the ratios of the coefficients are uniquely determined. This suggests the following fundamental theorem:

There is a unique conic locus that contains five general points; and dually, there is a unique conic envelope that contains five general lines.

When, in enunciations such as this, we refer to a 'general' point or a 'general' figure, we mean one that is chosen so as to avoid any special configurations to which the theorem does not apply. In any particular case the reader can discover what the postulated generality amounts to by looking into the details of the algebra used to prove the theorem and seeing in what circumstances the argument breaks down. In the present instance it is sufficient to exclude cases in which two of the points coincide or four of them are collinear. There is an additional complication, however, in that the linear condition imposed upon the a_{ik} by making the conic pass through a fixed point is not the most general linear condition on these six quantities, and it is not quite clear that even if the five points are 'general' the five linear conditions will be linearly independent. In view of the basic character of the theorem, therefore, we shall restate it more precisely and give a rigorous proof of it in the new form.

THEOREM 2. *There is a unique conic locus that contains five given distinct points, no four of which are collinear. Dually, there is a unique conic envelope that contains five given distinct lines, no four of which are concurrent.*

Proof. Since no four of the points are collinear, we may select three which are the vertices of a proper triangle, and take this triangle as triangle of reference. Then the equation of any conic through the three points is of the form

$$a_{12}x_1x_2 + a_{20}x_2x_0 + a_{01}x_0x_1 = 0,$$

and we have to show that requiring the conic to pass through two further points, which do not both lie on the same side of the triangle of reference, determines the ratios $a_{12}:a_{20}:a_{01}$ uniquely. Let the points be (y_0, y_1, y_2) and (z_0, z_1, z_2). Then

$$a_{12}y_1y_2+a_{20}y_2y_0+a_{01}y_0y_1 = 0$$

and

$$a_{12}z_1z_2+a_{20}z_2z_0+a_{01}z_0z_1 = 0.$$

If the point $\mathbf{y}$ does not lie on a side of the triangle of reference, these conditions on a_{12}, a_{20}, a_{01} are certainly independent. If, however, $y_0 = 0$, then $y_1 \neq 0$ and $y_2 \neq 0$, and the first equation becomes $a_{12}y_1y_2 = 0$, which gives simply $a_{12} = 0$. The second equation cannot also reduce to $a_{12} = 0$, since $z_0 \neq 0$ by hypothesis. Thus the ratios $a_{12}:a_{20}:a_{01}$ are uniquely determined in every case, and this proves the theorem.

The conclusion of Theorem 2 can be expressed in slightly different terms by saying that both the conic locus and the conic envelope are geometrical entities with five degrees of freedom. We also say sometimes that the plane contains ∞^5 conic loci and ∞^5 conic envelopes.

Since a conic locus is an entity with five degrees of freedom we can specify it, if we so choose, by means of a set of five non-homogeneous coordinates or a set of six homogeneous coordinates. The simplest way of doing this is to take as coordinates of the conic (as we have already taken as coordinates of a line) the coefficients in its equation. In terms of a fixed allowable representation $\mathscr{R}$ of the plane, every conic locus is then represented by its six coordinates $(a_{00}, a_{11}, a_{22}, a_{12}, a_{20}, a_{01})$.

The number of degrees of freedom of a conic locus in S_2 is the same as the number of degrees of freedom of a point in S_5, and we may therefore represent the system of all conics in S_2 by the system of all points in S_5. If we choose arbitrary allowable representations of S_2 and S_5, the equations $x_0' = a_{00}$, $x_1' = a_{11}$, $x_2' = a_{22}$, $x_3' = a_{12}$, $x_4' = a_{20}$, $x_5' = a_{01}$, define a one–one mapping of the conics in S_2 on the points of S_5.

§ 4. General Properties of the Conic Locus

We come now to a number of fundamental properties of the conic locus which serve as a foundation for the whole theory of the conic. We shall at first consider only the 'general' case, which in fact means the case of a proper conic locus. It will become clear in due course in what way the assumption of generality comes in, and which of the results remain valid, either as they stand or with suitable modification, when the conic breaks up.

Consider a fixed conic locus S, whose equation in the chosen representation $\mathscr{R}$ is

$$S(\mathbf{x}) \equiv \sum_i \sum_k a_{ik} x_i x_k = 0 \quad (a_{ki} = a_{ik}).$$

If Q, R are two points of the plane, represented by $\mathbf{y}$, $\mathbf{z}$ respectively, a variable point of the line QR is given by $\mathbf{y}+\theta\mathbf{z}$, and this point lies on S if and only if

$$S(\mathbf{y}+\theta\mathbf{z}) = 0. \tag{1}$$

This equation, which is known as the *Joachimsthal equation* for Q and R, is quadratic in θ; and its roots θ_1 and θ_2 are the parameters of the points A_1 and A_2 in which QR is met by the conic.

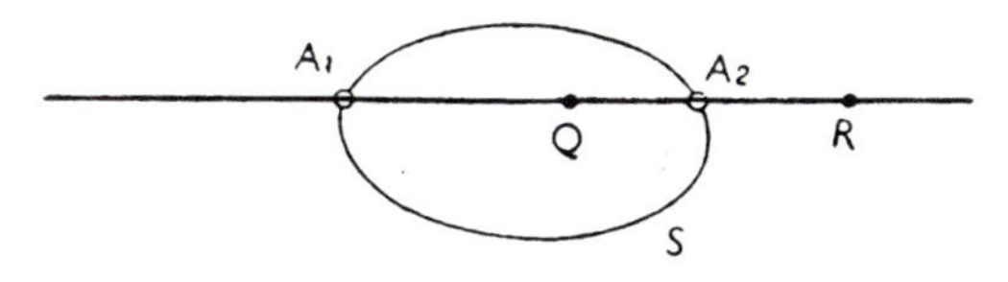

Equation (1) may be written:

$$\sum_i \sum_k a_{ik}(y_i+\theta z_i)(y_k+\theta z_k) = 0,$$

i.e.

$$\sum_i \sum_k a_{ik} y_i y_k + \theta \sum_i \sum_k a_{ik}(y_i z_k + z_i y_k) + \theta^2 \sum_i \sum_k a_{ik} z_i z_k = 0,$$

i.e.

$$S_{yy} + 2\theta S_{yz} + \theta^2 S_{zz} = 0,$$

where

$$S_{yy} = S(\mathbf{y}), \qquad S_{zz} = S(\mathbf{z}),$$

and

$$\begin{aligned} S_{yz} &= \tfrac{1}{2}\Big(\sum_i \sum_k a_{ik} y_i z_k + \sum_i \sum_k a_{ik} y_k z_i\Big) \\ &= \tfrac{1}{2}\Big(\sum_i \sum_k a_{ik} y_i z_k + \sum_i \sum_k a_{ki} y_i z_k\Big) \\ &= \sum_i \sum_k a_{ik} y_i z_k, \end{aligned}$$

since

$$a_{ki} = a_{ik}.$$

S_{yy} and S_{zz} are simply the quadratic form S, evaluated for the vectors $\mathbf{y}$ and $\mathbf{z}$ respectively, while S_{yz} is an intermediate expression. S_{yz} is a symmetrical bilinear form in $\mathbf{y}$ and $\mathbf{z}$ which reduces to S_{yy} when $\mathbf{z}$ is put equal to $\mathbf{y}$, and these characteristics define it uniquely. It is called the *polarized form* of the quadratic form S_{yy}. We shall have much to do with polarized forms in this and subsequent chapters.

S_{yz} has a neat expression in terms of partial derivatives, which may be obtained as follows:

$$\begin{aligned}\frac{\partial S}{\partial x_p} &= \frac{\partial}{\partial x_p}\sum_i\sum_k a_{ik}x_i x_k \\ &= \sum_i\sum_k a_{ik}\left(x_i\frac{\partial x_k}{\partial x_p}+x_k\frac{\partial x_i}{\partial x_p}\right) \\ &= \sum_i\sum_k a_{ik}(x_i\,\delta_{kp}+x_k\,\delta_{ip}) \\ &= \sum_i a_{ip}x_i+\sum_k a_{pk}x_k \\ &= 2\sum_i a_{ip}x_i,\end{aligned}$$

and hence

$$\begin{aligned}\sum_k z_k\frac{\partial S_{yy}}{\partial y_k} &= 2\sum_k z_k\sum_i a_{ik}y_i \\ &= 2\sum_i\sum_k a_{ik}y_i z_k \\ &= 2S_{yz}.\end{aligned}$$

We have, therefore,

$$S_{yz} = \frac{1}{2}\sum_k z_k\frac{\partial S_{yy}}{\partial y_k} = \frac{1}{2}\sum_k y_k\frac{\partial S_{zz}}{\partial z_k}.$$

DEFINITION. Q and R are said to be *conjugate points* with respect to the conic locus S if they are harmonic with respect to the points A_1 and A_2 in which QR meets S.

Since A_i is represented by $\mathbf{y}+\theta_i\mathbf{z}$, A_1 and A_2 are harmonically separated by Q and R if and only if $\theta_1+\theta_2 = 0$. But θ_1 and θ_2 are the roots of the quadratic equation

$$S_{yy}+2\theta S_{yz}+\theta^2 S_{zz} = 0,$$

and so we have as the condition for conjugacy

$$S_{yz} = 0.$$

Since this equation is linear in z_0, z_1, z_2, the locus of a variable point that is conjugate to a fixed point Q is a line. This line is called the *polar line* or *polar* of Q, and its equation may be written

$$S_y \equiv \frac{1}{2}\sum_i x_i\frac{\partial S_{yy}}{\partial y_i} = 0.$$

The conic locus S thus defines a transformation of the points of the plane into the lines of the plane, which is given algebraically by

$$u'_i = \frac{1}{2}\frac{\partial S_{xx}}{\partial x_i} \equiv \sum_k a_{ik}x_k,$$

or, in matrix notation,

$$\mathbf{x} \to \mathbf{u}' = \mathbf{A}\mathbf{x}.$$

This transformation is known as the *polarity* defined by S. If $|a_{rs}| \neq 0$ (which means, as we shall show on p. 112, that S is a proper conic locus) each line is the polar of a unique point.

THEOREM 3. *If the polar of Q passes through R, then the polar of R passes through Q.*

Proof. The condition for conjugacy, $S_{yz} = 0$, is symmetrical in $\mathbf{y}$ and $\mathbf{z}$.

EXERCISE. The derivation of the Joachimsthal equation (1) is only significant when Q and R are distinct points. What becomes of the condition for conjugacy when Q and R coincide?

If Q_1 is a general point of the plane, all points conjugate to Q_1 with respect to S lie on the polar line q_1 of Q_1. If, now, Q_2 is any point of q_1, the polar q_2 of Q_2 passes through Q_1 (Theorem 3). If q_2 meets q_1 in Q_3, then Q_3 is conjugate to both Q_1 and Q_2, and the three points Q_1, Q_2, Q_3 are such that the polar of each is the line joining the other two. Three such points are said to form a *self-polar triangle* for S. Since Q_1 has two degrees of freedom in the plane and Q_2 then has one degree of freedom on the line q_1, the conic locus S has ∞^3 self-polar triangles.

THEOREM 4. *If A, B, C, D are four points of the conic locus S, the diagonal triangle of the quadrangle $ABCD$ is self-polar for S.*

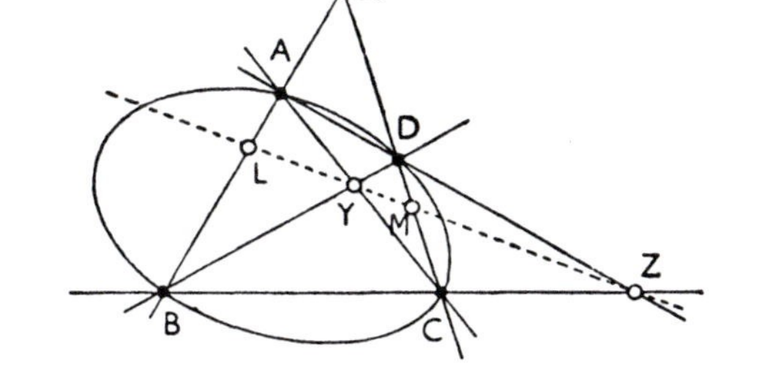

Proof. Let XYZ be the diagonal triangle, and let YZ meet AB, CD in L, M respectively. Then, by Theorem 9 of Chapter IV,

$$Z\{X, Y; A, B\} = -1,$$

and hence $\{X, L; A, B\} = \{X, M; D, C\} = -1.$

The pairs of points X, L and X, M are both conjugate for S, and the polar of X is therefore LM, i.e. YZ. The theorem now follows by symmetry.

DEFINITION: A line t is said to be a tangent to the conic locus S at a point Q of S if the two points in which t meets S both coincide with Q.

THEOREM 5. *If Q is a point of the conic locus S, there is one and only one tangent to S at Q, namely the polar of Q.*

Proof. Let l be a line through Q, and let R be any point of l other than Q. The points in which l is met by S are then given by the Joachimsthal equation

$$S_{yy} + 2\theta S_{yz} + \theta^2 S_{zz} = 0.$$

But, since Q lies on S, $S_{yy} = 0$, and one root of the equation is zero. The roots are therefore coincident if and only if the second root is also zero, i.e. if $S_{yz} = 0$. Thus l is a tangent at Q if and only if the coordinates (z_0, z_1, z_2) of every point on it satisfy the equation $S_{yz} = 0$; in other words, there is a unique tangent to S at Q, and its equation is $S_y = 0$.

When the points A_1, A_2 in which QR meets S both coincide with Q, the parameters of the pairs (A_1, A_2) and (Q, R) are $(0, 0)$ and $(0, \infty)$, and these satisfy the relation

$$(\theta_1 + \theta_2)(\theta_3 + \theta_4) = 2(\theta_1 \theta_2 + \theta_3 \theta_4).$$

Thus the two pairs of points are harmonic, and R is conjugate to Q with respect to S. It follows from this that the tangent at Q is the polar of Q, which completes the proof of the theorem.

THEOREM 6. *Through any point Q of the plane there pass exactly two tangents (distinct or coincident) to a given conic locus S, and their points of contact are the points in which S is met by the polar of Q.*

Proof. The tangent to S at a point A is the polar of A, and it therefore passes through Q if and only if A is conjugate to Q, i.e. if A is one of the two points in which S is met by the polar of Q.

THEOREM 7. *The two tangents from Q to S coincide if and only if Q lies on S.*

Proof. The tangents coincide if and only if the polar of Q meets S in coincident points, at A say. But then the polar of Q is the tangent at A, i.e. the polar of A, and Q therefore coincides with A.

THEOREM 8. *If Q has the coordinate vector* $\mathbf{y}$, *the joint equation of the pair of tangents from Q to S is*

$$S_y^2 - SS_{yy} = 0.$$

Proof. Let R, with coordinate vector $\mathbf{z}$, be any point on either of the tangents from Q. Then QR meets S in coincident points, and therefore

$$S_{yz}^2 - S_{yy} S_{zz} = 0.$$

Replacing $\mathbf{z}$ by a variable vector $\mathbf{x}$, we at once have the equation of the pair of tangents in the form required.

The matrix and rank of a conic locus

If a conic locus S is represented by the equation

$$\sum_i \sum_k a_{ik} x_i x_k = 0 \quad (a_{ki} = a_{ik}),$$

the matrix $\mathbf{A} = (a_{rs})$ is called the matrix of S. When once the coordinate representation $\mathscr{R}$ has been chosen, this matrix is determined apart from an arbitrary scalar factor. If the representation is changed from $\mathscr{R}$ to $\mathscr{R}'$ by the substitution $\mathbf{x} = \mathbf{P}\mathbf{x}'$, $|\mathbf{P}| \neq 0$, the matrix of S is changed in a simple manner. For the equation of the conic may be written both as

$$\mathbf{x}^T \mathbf{A} \mathbf{x} = 0$$

and as

$$(\mathbf{P}\mathbf{x}')^T \mathbf{A} (\mathbf{P}\mathbf{x}') = 0,$$

i.e.

$$\mathbf{x}'^T \mathbf{P}^T \mathbf{A} \mathbf{P} \mathbf{x}' = 0,$$

and the new matrix $\mathbf{A}'$ is therefore $\mathbf{P}^T\mathbf{A}\mathbf{P}$.

This matrix is in general different from $\mathbf{A}$, but $\mathbf{A}$ and $\mathbf{A}'$ have some essential features in common. Since $|\mathbf{A}'| = |\mathbf{P}|^2 . |\mathbf{A}|$, $|\mathbf{A}|$ and $|\mathbf{A}'|$ are both zero or both non-zero. The determinant $|\mathbf{A}|$ is called the *discriminant* of S in the original representation. Its vanishing is a projective property of the conic locus, whose geometrical interpretation will be obtained shortly.

Not only is the vanishing or non-vanishing of $|\mathbf{A}|$ an invariant property, but the rank $\rho[\mathbf{A}]$ of the matrix $\mathbf{A}$ is an invariant number. For, since $\mathbf{A}' = \mathbf{P}^T\mathbf{A}\mathbf{P}$, every r-rowed minor of $|\mathbf{A}'|$ is a linear combination of r-rowed minors of $|\mathbf{A}|$, and hence $\rho[\mathbf{A}'] \leqslant \rho[\mathbf{A}]$. But $\mathbf{A} = \mathbf{P}^{-1T}\mathbf{A}'\mathbf{P}^{-1}$, and therefore by the same argument

$$\rho[\mathbf{A}] \leqslant \rho[\mathbf{A}'].$$

It follows that $\rho[\mathbf{A}'] = \rho[\mathbf{A}]$. The number $\rho[\mathbf{A}]$ is called the *rank of the conic locus* S.

THEOREM 9. *If the rank of the conic locus S is greater than* 1, *then the totality of tangents of S is a conic envelope Σ. If the equation of S is*

$$\sum_i \sum_k a_{ik} x_i x_k = 0,$$

then the equation of Σ, referred to the same representation $\mathscr{R}$, is

$$\sum_i \sum_k A_{ik} u_i u_k = 0,$$

where A_{ik} is the cofactor of a_{ik} in the determinant $|a_{rs}|$.

Proof. Let the line $\mathbf{u}$ be a tangent to S, with point of contact $\mathbf{y}$. Then, since the polar of $\mathbf{y}$ is $\mathbf{u}$,

$$\sum_k a_{ik} y_k = \lambda u_i \quad (i = 0, 1, 2)$$

and further, since $\mathbf{y}$ lies on $\mathbf{u}$,

$$\sum_k u_k y_k = 0.$$

Eliminating y_0, y_1, y_2, λ from these four linear equations, we have

$$\begin{vmatrix} a_{00} & a_{01} & a_{02} & u_0 \\ a_{10} & a_{11} & a_{12} & u_1 \\ a_{20} & a_{21} & a_{22} & u_2 \\ u_0 & u_1 & u_2 & 0 \end{vmatrix} = 0,$$

and this equation may be expressed in the required form

$$\sum_i \sum_k A_{ik} u_i u_k = 0.$$

If $\rho[\mathbf{A}] = 1$, all the A_{ik} are zero, and there is no condition on $\mathbf{u}$. If $\rho[\mathbf{A}] = 2$ or 3, the coordinates u_0, u_1, u_2 of every tangent to S satisfy the equation of a certain conic envelope Σ; and, by the reversibility of the algebra, every line of this envelope is a tangent to S.

THEOREM 10. *A necessary and sufficient condition for the conic locus S to be degenerate is the vanishing of its discriminant $|a_{rs}|$.*

Proof. (i) Let S consist of two lines, distinct or coincident. Then there is at least one point $\mathbf{y}$ that is common to the two lines, and this point is conjugate to every point of the plane. Thus

$$\sum_i \sum_k a_{ik} x_i y_k \equiv 0$$

identically in x_0, x_1, x_2, and therefore

$$\sum_k a_{ik} y_k = 0 \quad (i = 0, 1, 2).$$

Since y_0, y_1, y_2 cannot all be zero, it follows that

$$|a_{rs}| = 0.$$

(ii) Now suppose, conversely, that $|a_{rs}| = 0$. Then there exists a point $\mathbf{y}$ such that

$$\frac{1}{2}\frac{\partial S_{yy}}{\partial y_i} \equiv \sum_k a_{ik} y_k = 0 \quad (i = 0, 1, 2)$$

and hence
$$S_{yy} \equiv \frac{1}{2}\sum_i y_i \frac{\partial S_{yy}}{\partial y_i} = 0.$$

If, now, $\mathbf{z}$ is any point of S, we have

$$S_{zz} = 0,$$

and also
$$S_{yz} \equiv \frac{1}{2}\sum_i z_i \frac{\partial S_{yy}}{\partial y_i} = 0.$$

Thus, for all values of θ,

$$S(\mathbf{y}+\theta\mathbf{z}) \equiv S_{yy}+2\theta S_{yz}+\theta^2 S_{zz} = 0.$$

This means that if $\mathbf{z}$ is any point of S every point of the line joining $\mathbf{y}$ to $\mathbf{z}$ belongs to S; and S is accordingly degenerate.

EXERCISE. Show that if S is a pair of distinct lines its rank is 2, and if it is a repeated line its rank is 1.

§ 5. General Properties of the Conic Envelope

The properties of the conic envelope follow at once by duality from the properties of the conic locus, and we shall merely summarize the main results.

Let Σ be a fixed conic envelope, whose equation in a given coordinate representation $\mathscr{R}$ is

$$\Sigma(u) \equiv \sum_i \sum_k A_{ik} u_i u_k = 0 \quad (A_{ki} = A_{ik}),$$

and let q, r be two lines of the plane, represented by the coordinate vectors $\mathbf{v}$, $\mathbf{w}$. A variable line through the point qr is given by $\mathbf{v}+\theta\mathbf{w}$, and it belongs to the envelope Σ if and only if

$$\Sigma(\mathbf{v}+\theta\mathbf{w}) = 0.$$

This equation is called the Joachimsthal equation for q and r, and its roots θ_1, θ_2 are the parameters of the lines a_1, a_2 through qr which belong to Σ. It may be written

$$\Sigma_{vv}+2\theta\Sigma_{vw}+\theta^2\Sigma_{ww} = 0,$$

where Σ_{vw} is the polarized form of the quadratic form Σ_{vv}.

The lines q and r are said to be conjugate lines with respect to Σ if they are harmonically separated by the two lines of the envelope which pass through their point of intersection; and the algebraic condition for conjugacy is $\Sigma_{vw} = 0$.

The envelope of a variable line which is conjugate to q is the point given by

$$\Sigma_v \equiv \frac{1}{2}\sum_i u_i \frac{\partial \Sigma_{vv}}{\partial v_i} = 0.$$

This point is called the *pole* of q with respect to Σ.

The conic envelope Σ determines in this way a transformation of lines into points, the associated polarity, with equations

$$x_i' = \sum_k A_{ik} u_k.$$

The polarity is non-singular if $|A_{rs}| \neq 0$, and when this is so it has a well-defined inverse. Every point is then the pole of a unique line of the plane.

The dual of the tangent to a locus at one of its points is the point of contact of a line of an envelope. We accordingly say that a point T is a point of the conic envelope Σ if the two lines of Σ which pass through it are coincident. If the repeated line is q, we say that T is the point of contact of the line q with the envelope.

THEOREM 11. *If q is a line of a conic envelope Σ, there is one and only one point of contact of q, namely the pole of q.*

THEOREM 12. *On any line q of the plane there lie two points of a given conic envelope Σ, and they are the points of contact of the lines of Σ which pass through the pole of q.*

THEOREM 13. *The two points of Σ which lie on q coincide if and only if q is a line of Σ.*

If the equation of a conic envelope Σ in a given representation $\mathscr{R}$ is

$$\sum_i \sum_k A_{ik} u_i u_k = 0 \quad (A_{ki} = A_{ik}),$$

the matrix $\mathfrak{A} = (A_{rs})$ is called the matrix of Σ. The determinant $|A_{rs}|$ is called the discriminant of Σ, and the rank $\rho[\mathfrak{A}]$ the rank of Σ. The rank of a conic envelope is a projective invariant, and its value determines the nature of the envelope. Σ is a proper conic envelope, a pair of distinct points, or a repeated point according as the value of $\rho[\mathfrak{A}]$ is 3, 2, or 1.

THEOREM 14. *If the rank of* Σ *is greater than* 1, *the totality of points of* Σ *is a conic locus* S. *If the equation of* Σ *is*

$$\sum_i \sum_k A_{ik} u_i u_k = 0,$$

then the equation of S, *referred to the same representation* $\mathscr{R}$, *is*

$$\sum_i \sum_k a_{ik} x_i x_k = 0,$$

where a_{ik} *is the cofactor of* A_{ik} *in the determinant* $|A_{rs}|$.

§ 6. THE CONIC AS A SELF-DUAL FIGURE

We have seen above that with every conic locus of rank 2 or 3 there is associated a conic envelope, namely the assemblage of its tangents, and with every conic envelope of rank 2 or 3 is associated a conic locus, namely the assemblage of its points. We can now go farther than this and show that a proper conic locus and its associated envelope are, in a natural sense, one and the same figure, so that a proper conic is a self-dual system of points and lines. In the case of degenerate conics, the self-duality is only partial. We now establish the fundamental connexion between the proper conic locus and its envelope.

THEOREM 15. *If* S *is a proper conic locus and* Σ *is the associated envelope, then* Σ *is a proper conic envelope. Furthermore, the conic locus associated with the envelope* Σ *is the original locus* S.

Proof. Let the equation of the proper conic locus S be

$$S \equiv \sum_i \sum_k a_{ik} x_i x_k = 0.$$

Then the associated envelope is represented by

$$\Sigma = \sum_i \sum_k A_{ik} u_i u_k = 0,$$

where A_{ik} is the cofactor of a_{ik} in $|a_{rs}|$. But then

$$|A_{rs}| = |a_{rs}|^2 \neq 0,$$

and Σ is therefore proper. The locus associated with the conic envelope is now given by the equation

$$\bar{S} \equiv \sum_i \sum_k \bar{a}_{ik} x_i x_k = 0,$$

where $\bar{a}_{ik}$ is the cofactor of A_{ik} in $|A_{rs}|$. But, by a known theorem on determinants, $\bar{a}_{ik} = |a_{rs}| a_{ik}$, and therefore $\bar{S} \equiv |a_{rs}| S$. Since $|a_{rs}| \neq 0$, the equations $S = 0$ and $\bar{S} = 0$ represent the same locus; and this completes the proof of the theorem.

Theorem 15 establishes the fact that proper conic loci and proper conic envelopes go together in pairs, being reciprocally associated with each other. The next theorem will show that a proper conic locus S and its associated envelope Σ determine the same (non-singular) polarity of the plane.

THEOREM 16. *If S is a proper conic locus and Σ is the associated conic envelope, then a line p is the polar of a point P with respect to S if and only if P is the pole of p with respect to Σ.*

Proof. If the equation of S is

$$\mathbf{x}^T\mathbf{A}\mathbf{x} = 0 \quad (|\mathbf{A}| \neq 0),$$

the equation of Σ may be written as

$$\mathbf{u}^T\mathbf{A}^{-1}\mathbf{u} = 0.$$

If, now, P and p have coordinate vectors $\mathbf{x}$ and $\mathbf{u}$, the condition for p to be the polar of P for S is

$$\mathbf{u} = \mathbf{A}\mathbf{x},$$

and the condition for P to be the pole of p for Σ is

$$\mathbf{x} = \mathbf{A}^{-1}\mathbf{u};$$

and these are the same condition.

From now on we shall regard a locus and an envelope such as S and Σ above as forming a single figure, the *proper conic* $s = (S, \Sigma)$. This conic has both a point-equation

$$S \equiv \sum_i \sum_k a_{ik} x_i x_k = 0$$

and a line-equation

$$\Sigma \equiv \sum_i \sum_k A_{ik} u_i u_k = 0,$$

and each of these equations determines the other. The conic gives rise to a polarity of the plane, a $(1, 1)$ correspondence between points and lines given by

$$u_i = \sum_i a_{ik} x_k \quad (i = 0, 1, 2),$$

and a point and its polar line are incident if and only if the line is a tangent to s and the point is its point of contact.

Alternatively, we may begin with a polarity—i.e. a non-singular point–line correspondence

$$u_i = \sum_k a_{ik} x_k \quad (i = 0, 1, 2)$$

whose matrix (a_{rs}) is symmetric—and use it to define a conic s. The points of s are the points which lie on their polar lines, and the lines of s are the lines which pass through their poles. This approach to the projective geometry of the conic has been adopted in some standard treatments of the subject. We shall leave it to the reader to show in detail that such a definition of the proper conic is equivalent to our definition.

§ 7. Degenerate Conics

When we try to extend Theorems 15 and 16 to cover degenerate as well as proper conics we find that the situation becomes more complicated, and only partial generalizations are possible.

THEOREM 17. *If S is a pair of distinct lines, the associated envelope Σ is the point of intersection of the lines, taken twice. If S is a repeated line, the associated envelope is indeterminate.*

Dually, if Σ is a pair of distinct points, the associated locus S is the line joining the points, taken twice. If Σ is a repeated point, the associated locus is indeterminate.

Proof. Choose the coordinate representation $\mathscr{R}$ so that the equation of S assumes the form $x_1 x_2 = 0$ or $x_2^2 = 0$, as the case may be. It is then found, on working out the cofactors, that in the first case the equation of Σ is $u_0^2 = 0$ and in the second case every coefficient in the equation of Σ vanishes.

COROLLARY. *A conic locus of rank 2 determines a conic envelope of rank 1, but a conic locus of rank 1 determines no conic envelope at all.*

We have now discovered two kinds of degenerate conic, which are dual to each other:

(*a*) a locus consisting of a pair of distinct lines, with a repeated point as the associated envelope;

(*b*) an envelope consisting of a pair of distinct points, with a repeated line as the associated locus.

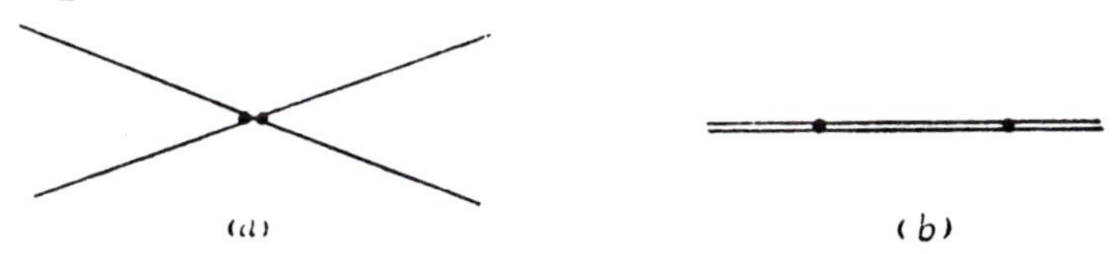

(*a*) (*b*)

In the case of a degenerate conic of type (*a*) the point-rank (rank of S) is 2, while the line-rank is 1. The point-equation determines the line-equation, but not vice versa. The conic gives rise to a linear transformation of points into lines

$$u_i = \sum_k a_{ik} x_k \quad (i = 0, 1, 2),$$

but this transformation is singular ($|a_{rs}| = 0$) and has no inverse. Thus the degenerate conic does not define a proper polarity of the plane.

These results may be dualized for the degenerate conic of type (*b*).

There is yet a third type (*c*) of degenerate conic—the repeated line and repeated point—which may be regarded as a specialization of both (*a*) and (*b*).

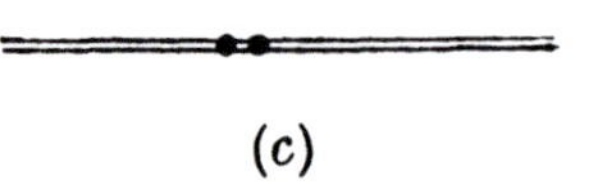

(*c*)

Both the point-rank and the line-rank of such a conic are 1, and neither of the two equations $S = 0$ and $\Sigma = 0$ determines the other.

EXERCISES

(i) If s is a degenerate conic of type (*a*), consisting of two distinct lines meeting in a point V, show that every point P of the plane, other than V, has a unique polar, namely the harmonic conjugate of VP with respect to the lines of s, and that the polar of V is indeterminate. Show further that every line of the plane which does not pass through V has V for its pole, and that the pole of a line through V can be taken to be any point of the harmonic conjugate of this line with respect to the lines of s.

Dualize these results for a degenerate conic of type (*b*), and examine also the third type (*c*) of degenerate conic from this point of view.

(ii) If the equations

$$\Sigma \equiv \sum_i \sum_k A_{ik} u_i u_k = 0, \qquad S' \equiv \sum_i \sum_k a'_{ik} x_i x_k = 0$$

represent respectively a conic envelope of rank not less than 2 and a pair of lines, show that a necessary and sufficient condition for the lines of S' to be conjugate for Σ is

$$\Theta \equiv \sum_i \sum_k A_{ik} a'_{ik} = 0.$$

§ 8. AFFINE GEOMETRY OF THE CONIC

When we turn to the study of the affine and euclidean properties of the conic in the real plane we need to limit the choice of coordinate representation to the 'cartesian' coordinate systems, in which the line at infinity has an invariable equation. For simplicity, we shall now denote the point-coordinates and line-coordinates in the chosen system by (x, y, z) and (u, v, w), and we shall suppose that the equation of the line at infinity i is $z = 0$.

Consider, first of all, a proper conic s, whose point-equation is

$$S \equiv (a,b,c,f,g,h \between x,y,z)^2 = 0,$$

all the coefficients being real numbers.

The curve cuts the line at infinity in two points H, K, its points at infinity, and their coordinates are given by

$$ax^2+2hxy+by^2 = 0 = z.$$

H, K are real and distinct, coincident, or conjugate complex according as h^2-ab is greater than, equal to, or less than zero; and in the three cases s is said to be a *hyperbola*, a *parabola*, or an *ellipse* respectively.

The line at infinity has a pole C with respect to s, and this point is called the *centre* of s. C is conjugate to every point of i and hence, provided it is a finite point, it is the mid-point of every chord that passes through it. We may accordingly refer to it as a centre of symmetry of s. Since both the ellipse and the hyperbola have centres of symmetry, these conics are known as central conics. In the case of the parabola, for which H and K coincide, the line at infinity touches the curve, and its pole C is a point at infinity. The parabola therefore has no finite centre.

The coordinates of the pole of the line $(0, 0, 1)$ with respect to the general conic are (G, F, C), and these are therefore the coordinates of the centre. If the conic is a parabola, $C = ab-h^2 = 0$, and the centre of the general parabola is the point at infinity $(G, F, 0)$.

The tangents to s at H and K are called the *asymptotes* of s, and they are the lines CH, CK. The hyperbola has real asymptotes, whereas the asymptotes of the ellipse are two conjugate complex lines. The asymptotes of the parabola both coincide with the line at infinity.

Exercise. Show that the equation of the pair of asymptotes of s may be written in each of the following forms:

(i) $S(x,y,z)S(G,F,C)-\{S(x,y,z;\ G,F,C)\}^2 = 0$,

where $S(x,y,z;\ G,F,C)$ is a polarized form;

(ii) $a\left(X-\frac{G}{C}\right)^2+2h\left(X-\frac{G}{C}\right)\left(Y-\frac{F}{C}\right)+b\left(Y-\frac{F}{C}\right)^2 = 0;$

(iii) $S+\lambda z^2 = 0$, where λ is given a suitable value.

Any line through the centre of s is called a *diameter* of s. The diameters form a pencil of lines; and since the polars of the points

$(1,0,0)$ and $(0,1,0)$ are both diameters, the equation of a general diameter may be written

$$\frac{\partial S}{\partial x}+\lambda\frac{\partial S}{\partial y} = 0.$$

The diameters of a parabola are all parallel.

If two diameters of s are conjugate lines with respect to s, they are called *conjugate diameters.*

THEOREM 18. *Every diameter of a central conic s bisects all chords parallel to the conjugate diameter.*

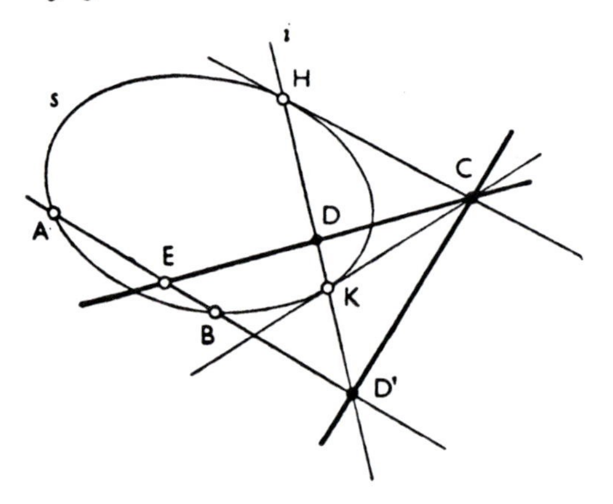

Proof. Let CD, CD' be a pair of conjugate diameters, meeting i in D, D', and let AB be a chord parallel to CD'. Then AB passes through D'. Now D' is conjugate to C, since it lies on the polar of C, and also conjugate to D, since

$$\{D,D';\,H,K\} = C\{D,D';\,H,K\} = -1;$$

and the polar of D' is therefore CD. If CD meets AB in E, E is conjugate to D', and therefore $\{E,D';\,A,B\} = -1$. Thus E is the mid-point of AB.

EXERCISE. Show that the mid-points of all chords of a parabola which are parallel to a fixed line lie on a diameter of the parabola.

If CD, CD' are conjugate diameters of a central conic s, and if we take C, D, D' as vertices of the triangle of reference in an allowable representation $\mathscr{R}_A$ (i.e. if we take CD, CD' as oblique cartesian axes) the equation of the conic may be written non-homogeneously in the form

$$aX^2+bY^2 = 1.$$

For since the origin is a centre of symmetry the equation can contain no linear terms, and since each axis bisects all chords parallel to the other axis there can be no term in XY. The signs of a and b are the same or different according as s is an ellipse or a hyperbola.

EXERCISE. Show that if s is a parabola, and we take as axes of reference a diameter and the tangent to s at the unique finite point where it is met by the diameter, the equation of s assumes the form $Y^2 = 4aX$.

§ 9. EUCLIDEAN GEOMETRY OF THE CONIC

We now restrict the class of allowable representations still farther, by admitting only those affine representations $\mathscr{R}_A$ in which the absolute points I and J (which lie on the line i) have the coordinates $(1, i, 0)$ and $(1, -i, 0)$ respectively. The allowable euclidean representations $\mathscr{R}_E$ which are thus obtained may for convenience be called rectangular cartesian coordinate systems, and we may use the ordinary language of coordinate geometry in order to describe our present abstract system of euclidean geometry.

It is often desirable to treat the absolute point-pair as a degenerate conic Ω, for which the envelope is the pair of points and the locus is the line at infinity taken twice. The line-equation of the absolute conic is then

$$\Omega \equiv u^2+v^2 = 0.$$

Since two lines are perpendicular if they are harmonic with respect to the isotropic lines through their point of intersection, perpendicular lines are simply lines that are conjugate with respect to Ω.

Let us now consider a central conic s. In general s has a unique pair of perpendicular conjugate diameters, namely the common pair of the involution of conjugate diameters (defined by its self-corresponding rays CH, CK) and the orthogonal involution at C (defined by its self-corresponding rays CI, CJ). These special diameters are called the *principal axes* or simply the *axes* of s.

The axes of s may be taken as axes of reference for a rectangular cartesian coordinate system, and s then has an equation of the form

$$aX^2+bY^2 = 1. \tag{1}$$

The axes are uniquely defined except in the one case in which the two involutions at C are identical. In this case any two conjugate diameters of s are perpendicular, so that the axes of s are indeterminate, and we then say that s is a circle. Since the line-equation corresponding to the point-equation (1) is

$$\frac{U^2}{a}+\frac{V^2}{b} = 1,$$

the conic represented by (1) is a circle if and only if the conditions

$$\frac{U_1 U_2}{a}+\frac{V_1 V_2}{b} = 0,$$

and

$$U_1 U_2+V_1 V_2 = 0$$

are equivalent; i.e. if $a = b$. Thus the equation of every circle, referred to two perpendicular diameters as axes, is of the form

$$X^2+Y^2 = k^2.$$

The right-hand side must be positive if the circle is to have real points.

EXERCISE. Prove that (i) all the points of a circle are equidistant from the centre; (ii) a central conic is a circle if and only if it passes through the absolute points.

The conic s represented by the equation

$$aX^2+bY^2 = 1$$

is a circle when the coefficients are equal. When these coefficients are equal but opposite in sign we get another special conic with interesting properties, the rectangular hyperbola.

EXERCISE. Show that (i) a hyperbola is rectangular if and only if its asymptotes are perpendicular; (ii) the general equation

$$(a,b,c,f,g,h \between x,y,z)^2 = 0,$$

referred to an arbitrary rectangular cartesian frame of reference $\mathscr{R}_E$, represents a rectangular hyperbola if and only if $a+b = 0$.

THEOREM 19. *The locus of a variable point from which two perpendicular tangents may be drawn to a fixed central conic s is a circle concentric with s.*

Proof. The theorem may be derived from equation (1) above as in elementary coordinate geometry, but we prefer to give the following proof because it yields a useful form of the equation of the circle.

Let the equation of s, in some representation $\mathscr{R}_E$, be

$$S \equiv (a,b,c,f,g,h \between x,y,z)^2 = 0.$$

The equation of the pair of tangents from (x',y',z') is then

$$S(x,y,z)S(x',y',z')-\{S(x,y,z;\, x',y',z')\}^2 = 0.$$

The condition for the lines represented by this equation to be perpendicular is that the sum of the coefficients of x^2 and y^2 is zero, and this gives

$$(a+b)S(x',y',z')-(ax'+hy'+gz')^2-(hx'+by'+fz')^2 = 0.$$

Introducing the cofactors A, B, C, F, G, H, we may write this equation as

$$C(x'^2+y'^2)-2Gz'x'-2Fy'z'+(A+B)z'^2 = 0,$$

and the locus of the point (x',y',z') is therefore the circle whose non-homogeneous equation is

$$C(X^2+Y^2)-2GX-2FY+A+B = 0.$$

The circle just obtained is known as the *director circle* of s.

DEFINITION. A point F is called a *focus* of a conic s if the isotropic lines FI and FJ are tangents to s. The polar of a focus of s is called the *directrix* associated with that focus.

THEOREM 20. *A central conic has four foci, two of which are real and two conjugate complex. The real foci lie on one axis and the complex foci on the other.*

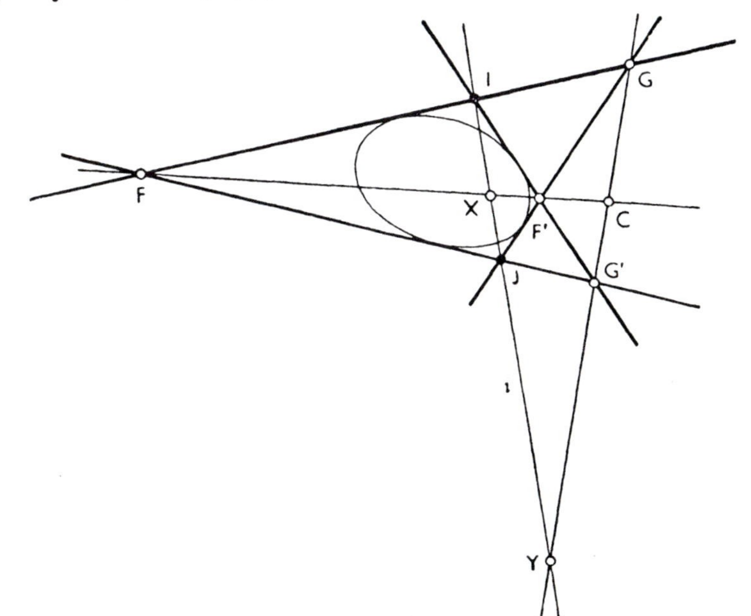

Proof. There are two tangents to s from I and two from J; and, since I, J are conjugate complex points and s has a real equation, the first two lines are the complex conjugates of the second two. The two pairs of lines meet in four points other than I and J —two real points F, F' and two conjugate complex points G, G'.

By the dual of Theorem 4, the diagonal trilateral of the quadrilateral formed by the four tangents is self-polar for s. Thus, in the figure on p. 123, the triangle CXY is self-polar for s. It follows that C is the pole of IJ, i.e. the centre of s, and CX, CY are conjugate diameters. Further, by the harmonic properties of the quadrangle $IJG'G$, $\{X, Y; I, J\} = -1$, and CX, CY are therefore perpendicular. These two lines are consequently the axes of s; and this completes the proof.

The properties of the parabola are unsymmetrical compared with those of the central conics, in consequence of the fact that the parabola touches the line at infinity. The relation between the parabola and the central conics is grasped most easily if we regard the parabola as a limiting case of a variable central conic, in which the points H and K coincide. In saying this, we are not introducing limiting processes into our formal system; for when once we have arrived at properties of the parabola by a continuity argument we can interpret them as algebraic specializations of the more general properties of the central conics. In the limit, then, as H and K tend to coincidence:

(i) C coincides with H and K;

(ii) one real focus F remains finite, the other real focus F' coincides with C, and the complex foci G, G' coincide respectively with I, J;

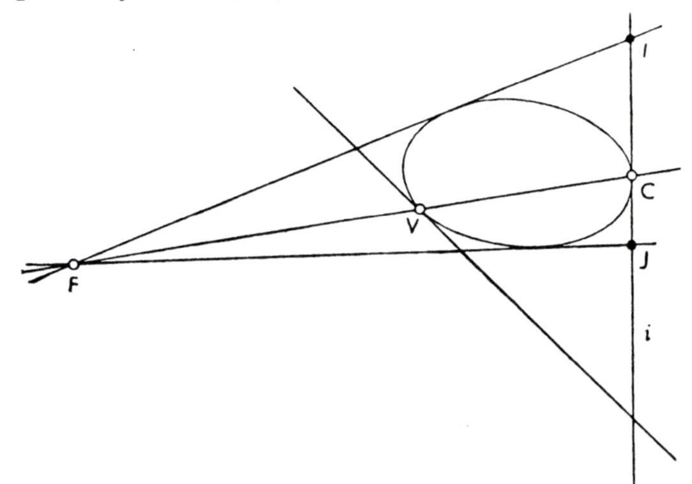

(iii) the axis CX remains finite, but CY coincides with i. Thus the parabola has only one axis, and only one (real) focus F, which lies on this axis. The point at infinity on the axis is the centre C, and the axis cuts the curve in one finite point V, which is called the *vertex*.

EXERCISE. Prove that the axis of the parabola and the tangent at the vertex are perpendicular.

If we take the axis and the tangent at the vertex as axes of reference, the equation of the parabola becomes

$$Y^2 = 4aX.$$

From this equation we may deduce, by familiar reasoning, many standard properties of the parabola. Among these is the useful result that perpendicular tangents to the parabola always meet on the directrix. This is useful because it enables us easily to locate a parabola, in relation to the axes of reference, when its equation is given. If the equation is $(a,b,c,f,g,h \between x,y,z)^2 = 0$, with $C = 0$, the equation of the directrix is, by Theorem 19,

$$2GX+2FY-A-B = 0.$$

The focus is then the pole of this line $(2G, 2F, -A-B)$, and its coordinates may at once be found.

EXERCISES ON CHAPTER V

1. Express each of the following quadratic forms as $\mathbf{x}^T\mathbf{A}\mathbf{x}$, where $\mathbf{A}$ is a symmetric matrix:

$$S_1 \equiv x_0^2+x_1^2+x_2^2-2x_1x_2-2x_2x_0+2x_0x_1;$$
$$S_2 \equiv 4x_0^2+6x_1^2-2x_2^2-x_1x_2-7x_2x_0+14x_0x_1;$$
$$S_3 \equiv x_0^2+3x_1^2-5x_2^2-7x_1x_2-13x_2x_0+8x_0x_1.$$

Show that the ranks of the three forms are 1, 2, 3 respectively.

Exhibit the locus $S_1 = 0$ as a repeated line and the locus $S_2 = 0$ as a line-pair; and find a new allowable coordinate representation in which the conic $S_3 = 0$ has the equation $y_0^2+y_1^2+y_2^2 = 0$.

2. Find the equation of the conic locus determined by the five points $(1,0,1)$, $(2,0,1)$, $(0,1,1)$, $(0,2,1)$, $(\lambda,\lambda,1)$, and the equation of the conic envelope determined by the five lines with the same coordinates. Discuss the cases, arising for particular values of λ, in which the locus is a line-pair or the envelope a point-pair. Draw rough sketches to illustrate your results.

3. Find the combined equation of the pair of tangents from the point $(1,1,1)$ to the conic $x^2+2y^2+z^2+yz+3zx+xy = 0$, and obtain the separate equations of the tangents and the coordinates of their points of contact. Give also the dual interpretation of the algebra in terms of line-coordinates.

4. Show that, for all values of λ, the conic whose line-equation is

$$2\lambda u^2-(2\lambda+\lambda^2)v^2-w^2+2\lambda wu = 0$$

passes through the point $(1,1,1)$ and through three further fixed points. Find the coordinates of these points.

5. With the usual notation, prove that the coordinate vectors of (*a*) the polars of the vertices of the triangle of reference and (*b*) the poles of the sides of this triangle for the general conic are the rows of the matrices

$$\begin{pmatrix} a & h & g \\ h & b & f \\ g & f & c \end{pmatrix} \quad \text{and} \quad \begin{pmatrix} A & H & G \\ H & B & F \\ G & F & C \end{pmatrix}$$

respectively.

Show that the point-equation of the conic can be written as

$$(A, B, C, F, G, H \between x', y', z')^2 = 0,$$

where $x' = ax+hy+gz$, $y' = hx+by+fz$, $z' = gx+fy+cz$.

6. Prove that the lines (u, v, w) and (u', v', w') intersect in a point of the conic $(a, b, c, f, g, h \between x, y, z)^2 = 0$ if and only if

$$\begin{vmatrix} a & h & g & u & u' \\ h & b & f & v & v' \\ g & f & c & w & w' \\ u & v & w & 0 & 0 \\ u' & v' & w' & 0 & 0 \end{vmatrix} = 0.$$

7. Given three linearly independent linear forms $L_i \equiv u_i x+v_i y+w_i z$ $(i = 1, 2, 3)$, prove that any conic locus for which the lines $L_i = 0$ form a self-polar triangle has an equation of the form $\lambda_1 L_1^2+\lambda_2 L_2^2+\lambda_3 L_3^2 = 0$, and, conversely, that any equation of this form (in which none of the λ_i is zero) represents a proper conic for which the lines form a self-polar triangle.

8. Find the equation of the conic envelope for which the points $(1, 0, 1)$, $(0, 1, 1)$, $(1, 1, 1)$ form a self-polar triangle and which touches the lines $x = 2z$ and $y = 2z$.

9. If k is a proper conic and XYZ is a given triangle, show that there exist, in general, two triangles, inscribed in XYZ, which are self-polar for k.

Find a condition that must be satisfied by the coefficients in the line-equation of k (with XYZ as triangle of reference) in order that there shall exist only one such triangle.

10. The lines joining a point P to the vertices of a triangle XYZ meet YZ, ZX, XY in L, M, N; and the harmonic conjugates of P with respect to the pairs of points (X, L), (Y, M), (Z, N) are U, V, W. Show that the triangle UVW is circumscribed to XYZ, and that there exists a conic which touches VW, WU, UV at X, Y, Z respectively.

11. A conic touches the sides YZ, ZX, XY of a triangle XYZ at L, M, N. Show that the three points $MN.YZ$, $NL.ZX$, $LM.XY$ lie on a line. If the lines joining X, Y, Z to any point of this line meet MN, NL, LM at L', M', N', prove that the triangle $L'M'N'$ is self-polar for the conic.

12. The line joining the points Q and R, whose coordinate vectors are $\mathbf{y}$ and $\mathbf{z}$, meets the conic $S = 0$ in A_1 and A_2. If $\{A_1, A_2;\, Q, R\} = k$, prove that

$$4kS_{yz}^2 = (1+k)^2 S_{yy} S_{zz}.$$

If the point Q and the constant k are fixed, show that the locus of R is a conic which touches the given conic at the points of contact of the tangents from Q.

Give a euclidean interpretation of this result when the points of contact are the absolute points I, J.

13. Obtain the projective generalization of the theorem that a circle is transformed by radial expansion about its centre into a concentric circle.

14. If k is the conic envelope whose equation is

$$(A, B, C, F, G, H \between u, v, w)^2 = 0,$$

show that the equation

$$Hz^2 + Cxy - Gyz - Fzx = 0$$

represents the locus of a point whose joins to the reference points X and Y are conjugate lines for k. Under what conditions does this conic break up into a pair of lines?

Give a euclidean interpretation of your results in the case when X, Y are the absolute points I, J.

15. Find the centre, asymptotes, and principal axes of the hyperbola whose equation in rectangular cartesian coordinates is

$$2X^2 - 13XY + 15Y^2 + 5X - 11Y + 7 = 0.$$

16. Find the coordinates of the centre of the conic whose rectangular cartesian equation is $X^2 + 4\lambda XY + 4Y^2 - 2\lambda X + 4Y - 3 = 0$, and find the equation of the locus of the centre as λ varies. Discuss the nature of the conic for all values of λ.

17. If $T = 0$ is the equation of a line-pair, show that T can be written as a homogeneous quadratic expression in two of the linear forms $\partial T/\partial x$, $\partial T/\partial y$, $\partial T/\partial z$.

If $S = 0$ is the equation of a conic in homogeneous cartesian coordinates, find a homogeneous quadratic equation in $\partial S/\partial x$ and $\partial S/\partial y$ which represents the asymptotes of the conic.

18. Show that the chord of the conic $aX^2 + bY^2 = 1$ whose mid-point is (X_0, Y_0) is given by the cartesian equation $aXX_0 + bYY_0 = aX_0^2 + bY_0^2$. Obtain a similar result for the conic $(a, b, c, f, g, h, \between X, Y, 1)^2 = 0$.

19. Show that the chords which join any point of a conic to the ends of a diameter are parallel to conjugate diameters of the conic. Give the modified form of this theorem appropriate to the case when the conic is a parabola.

20. If the coordinates are rectangular cartesian, show that the line-equation of any conic for which the origin O is one focus is of the form

$$A(u^2 + v^2) + 2Fvw + 2Gwu + Cw^2 = 0;$$

and find the equation of the directrix of this conic which corresponds to the focus O.

Obtain also, for the same conic, the equation of the locus of the foot of the perpendicular from O on to a variable tangent (the *auxiliary circle* of the conic). Discuss the case when the conic is a parabola ($C = 0$) showing that the locus in question is then the *tangent at the vertex*.

21. Obtain the equation of the line at infinity in trilinear coordinates and also in areal coordinates.

If the lengths of the sides of the triangle of reference for trilinear coordinates are a, b, c, and the angles of the triangle are A, B, C, show that the equation of the circumcircle of the triangle is $ayz+bzx+cxy = 0$. Deduce that the equation of the point-pair (I, J) in trilinear coordinates is

$$\Omega \equiv u^2+v^2+w^2-2vw\cos A-2wu\cos B-2uv\cos C = 0,$$

and obtain the corresponding equation in areal coordinates.

22. XYZ is the triangle of reference for a system of areal coordinates, and a conic k touches XY, XZ at Y, Z respectively. If the line $ux+vy+wz = 0$ is an asymptote of k, prove that $2vw = wu+uv$; and show that the equation of the other asymptote is $2x/(v+w)+y/v+z/w = 0$. Show that the asymptotes of all conics such as k envelop a parabola which touches the sides of the triangle of reference, and that the point of contact of the parabola with YZ is the mid-point of YZ.

23. If the absolute points I, J are given, in terms of a general system of projective coordinates, by the equations

$$x^2+y^2+z^2 = 0 = px+qy+rz,$$

find the condition for the lines (u, v, w) and (u', v', w') to be perpendicular.

If a rectangular hyperbola passes through the mid-points of the sides of a triangle ABC and meets the sides again in P, Q, R, prove that AP, BQ, CR are concurrent in a point of the circumcircle of the triangle.

CHAPTER VI

FURTHER PROPERTIES OF CONICS

In Chapter V we laid the foundations for a systematic treatment of the projective geometry of the conic, and showed also how the affine and euclidean properties of conics find a natural place in the scheme. In this chapter we shall develop the theory in greater detail, and it will be part of our purpose to prove a number of well-known geometrical properties of conics, such as, for example, Pascal's Theorem.

The reader should by this time be fully conversant with the relationship between the different geometries of the projective hierarchy, and he should be able from time to time to change his point of view without losing sight of the logical structure of the system as a whole. Our principal object of study, he will understand, is projective geometry over the complex field; but by considering only real points as actual and properly complex points as ideal, and by suitably restricting the class of allowable coordinate representations, we can treat affine or euclidean space as a specialization of complex projective space. Up to the present we have been careful to maintain a clear separation between the different kinds of space, but from now on we shall allow ourselves greater freedom, dealing for the most part with projective space, but permitting digressions into more special spaces whenever these seem appropriate or illuminating.

Let us turn back again now to the complex plane, i.e. to a projective space S_2 over the field K of complex numbers.

§ 1. Special Forms of the Equation of a Conic

We saw, on p. 104, how the equation of a conic may be put in a simple form by taking an inscribed triangle as triangle of reference. There are several standard ways in which the representation of the conic may be simplified by special choice of the frame of reference, and it will be convenient to collect these together at this point.

Case 1: when the triangle of reference is an inscribed triangle.

The point-equation of the conic, as we have seen, takes the form

$$a_{12}x_1x_2+a_{20}x_2x_0+a_{01}x_0x_1 = 0.$$

The corresponding line-equation, found by evaluating the various cofactors, is

$$a_{12}^2u_0^2+a_{20}^2u_1^2+a_{01}^2u_2^2-2a_{20}a_{01}u_1u_2-2a_{01}a_{12}u_2u_0-$$
$$-2a_{12}a_{20}u_0u_1 = 0.$$

Case 2: when the triangle of reference is a circumscribed triangle.

Case 2 is dual to Case 1, and the equations are therefore of the forms

$$A_{12}^2x_0^2+A_{20}^2x_1^2+A_{01}^2x_2^2-2A_{20}A_{01}x_1x_2-2A_{01}A_{12}x_2x_0-$$
$$-2A_{12}A_{20}x_0x_1 = 0,$$

and $$A_{12}u_1u_2+A_{20}u_2u_0+A_{01}u_0u_1 = 0.$$

Case 3: when the triangle of reference is self-polar.

The point-equation is easily found to be

$$a_{00}x_0^2+a_{11}x_1^2+a_{22}x_2^2 = 0,$$

and the corresponding line-equation is

$$a_{11}a_{22}u_0^2+a_{22}a_{00}u_1^2+a_{00}a_{11}u_2^2 = 0.$$

If a_{00}, a_{11}, a_{22} are all non-zero, the latter equation may be written

$$\frac{u_0^2}{a_{00}}+\frac{u_1^2}{a_{11}}+\frac{u_2^2}{a_{22}} = 0.$$

The familiar equation $aX^2+bY^2 = 1$ is of this type, and we have already seen directly that the corresponding triangle of reference is self-polar.

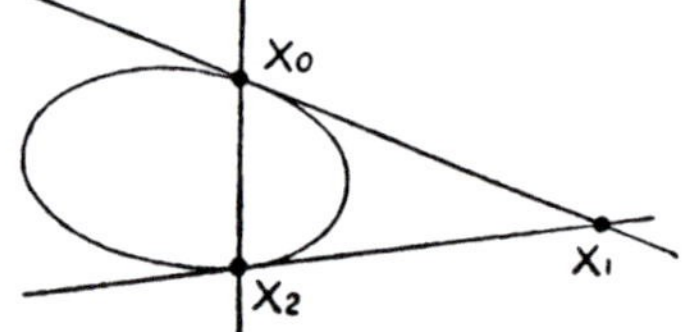

Case 4: when the triangle of reference consists of two tangents and their chord of contact.

The point-equation reduces to the form

$$x_1^2 = kx_2x_0,$$

and the corresponding line-equation is

$$ku_1^2 = 4u_2u_0. \quad \text{(Note the factor 4.)}$$

Special instances of this are the equation of the parabola in the form $Y^2 = 4aX$ and the equation $XY = c^2$ of the hyperbola referred to its asymptotes as cartesian axes.

Case 5: The canonical equation.

Suppose that in addition to choosing the triangle of reference as in Case 4 we also choose some point of the conic as unit point.

Then $k = 1$, and the point-equation assumes the *canonical form*

$$x_1^2 = x_2 x_0.$$

This form gives rise to a most important parametric representation of the conic locus. The equation may be written

$$\left(\frac{x_1}{x_2}\right)^2 = \frac{x_0}{x_2},$$

and hence if $x_1/x_2 = \theta$ then $x_0/x_2 = \theta^2$. Thus

$$x_0 : x_1 : x_2 = \theta^2 : \theta : 1.$$

This is the *canonical parametric representation* of the proper conic, and it is especially easy to handle since it involves only polynomials in the parameter—indeed the simplest possible polynomials that can be linearly independent. The reader may compare the representations $(at^2, 2at)$ for the parabola and $(ct, c/t)$ for the rectangular hyperbola. The real importance of the canonical representation resides, however, not so much in its ease of manipulation as in its theoretical significance. This will become clear very shortly; but before taking up this question we shall obtain one or two useful algebraic results.

Let (θ_1), (θ_2) be two distinct points of the conic. Then the equation of the chord joining them is

$$\begin{vmatrix} x_0 & x_1 & x_2 \\ \theta_1^2 & \theta_1 & 1 \\ \theta_2^2 & \theta_2 & 1 \end{vmatrix} = 0,$$

i.e. $x_0 - (\theta_1 + \theta_2)x_1 + \theta_1 \theta_2 x_2 = 0$, since $\theta_1 - \theta_2 \neq 0$.

Letting θ_2 tend to θ_1, we obtain the equation of the tangent at (θ_1) in the form

$$x_0 - 2\theta_1 x_1 + \theta_1^2 x_2 = 0.$$

This equation is obtained here by a limiting process, but we may now verify at once that the line represented by it does in fact meet the conic in the point (θ_1) taken twice.

In the canonical parametric representation, the coordinates of a general point of the conic are proportional to quadratic polynomials in the parameter θ, and these polynomials are chosen as simply as possible—namely as θ^2, θ, and 1. If, however, we were to take a curve represented by *general* quadratic polynomials in the parameter, we should still obtain only a conic.

Let

$$x_0 = a_{00}\theta^2+a_{01}\theta+a_{02},$$
$$x_1 = a_{10}\theta^2+a_{11}\theta+a_{12},$$
$$x_2 = a_{20}\theta^2+a_{21}\theta+a_{22}$$

be parametric equations of a curve c, the three polynomials being linearly independent. Then $|a_{rs}| \neq 0$, and we may solve the three equations for the ratios $\theta^2:\theta:1$, obtaining the result

$$\theta^2:\theta:1 = (A_{00}x_0+A_{10}x_1+A_{20}x_2):(A_{01}x_0+A_{11}x_1+A_{21}x_2):$$
$$:(A_{02}x_0+A_{12}x_1+A_{22}x_2).$$

Since $|A_{rs}| = |a_{rs}|^2 \neq 0$, the three linear forms on the right are linearly independent, and may be taken as new coordinates x_0', x_1', x_2'. The curve c then has the parametric representation

$$x_0':x_1':x_2' = \theta^2:\theta:1.$$

The canonical representation $(\theta^2, \theta, 1)$ of the proper conic is a *proper* parametric representation of the curve, in the sense that it sets up a $(1, 1)$ correspondence between the points of the curve and the values of the parameter. A representation such as

$$x_0:x_1:x_2 = \phi^4:\phi^2:1,$$

for example, would not satisfy this requirement, since there would be two values of ϕ corresponding to each point. When a curve admits of a proper parametric representation by polynomials with coefficients in the ground field K, it is said to be *rational*; and it has the property that its points can be put into $(1, 1)$ correspondence with the points of a line by means of polynomials (or, if the coordinates are non-homogeneous, by means of rational functions). The proper conic is therefore a rational curve.

§ 2. The Proper Conic as a One-dimensional Projective Domain

We return in this section to the theoretical significance of the canonical parameter θ. As we shall now see, a proper conic locus $x_1^2 = x_2 x_0$ has the structure of a one-dimensional projective domain S_1, and θ is an allowable parameter for this domain. In order to show this, we need to consider certain homographic correspondences that are associated with the conic.

THEOREM 1. *The pencils which project the points of a proper conic s from any two fixed points of s are homographically related.*

Proof. Take the fixed points as X_0 and X_2 in a canonical representation of s, so that $\theta = \infty$ at X_0 and $\theta = 0$ at X_2. Let P be a general point $(\theta^2, \theta, 1)$ of s.

The equations of $X_0 P$ and $X_2 P$ are respectively

$$x_1 - \theta x_2 = 0 \quad \text{and} \quad x_0 - \theta x_1 = 0,$$

and the two lines therefore describe homographic pencils as θ varies.

THEOREM 2 (*Chasles's Theorem*). *If P_1, P_2, P_3, P_4 are four points of a proper conic s, and V is any fifth point of s, then the cross ratio $V\{P_1, P_2; P_3, P_4\}$ is the same for all positions of V.*

Proof. Theorem 2 is an immediate consequence of Theorem 1.

DEFINITION. The cross ratio $V\{P_1, P_2; P_3, P_4\}$ is called the cross ratio of the two pairs of points (P_1, P_2) and (P_3, P_4) *on the conic*, and it is denoted by $\{P_1, P_2; P_3, P_4\}$.

Since θ is a projective parameter for the pencil $X_0(P)$, we have

$$\{P_1, P_2; P_3, P_4\} = X_0\{P_1, P_2; P_3, P_4\} = \{\theta_1, \theta_2; \theta_3, \theta_4\},$$

where θ_i is the parameter of P_i $(i = 1, 2, 3, 4)$. Thus the cross ratio of any four points of s, taken in a definite order, is equal to the cross ratio of their parameters.

THEOREM 3. *Any three distinct points X_0, X_2, E of s define a unique canonical representation $(\theta^2, \theta, 1)$ in which they have the parameters* ∞, 0, 1 respectively.

Proof. If X_1 is the pole of $X_2 X_0$, the four points X, X_1, X_2, E define a unique representation $\mathscr{R}$ of the plane, and in this representation the equation of s has the canonical form $x_1^2 = x_2 x_0$. We thus obtain a parametric representation $(\theta^2, \theta, 1)$ satisfying the stated conditions.

COROLLARY 1. *If θ is the parameter of a variable point P of s, in the representation just defined, then $\theta = \{E, P; X_0, X_2\}$ on s.*

For $\{E, P; X_0, X_2\} = \{1, \theta; \infty, 0\} = \theta$.

COROLLARY 2. *If s is represented parametrically by equations*

$$x_0 : x_1 : x_2 = (a_{00}\theta^2 + a_{01}\theta + a_{02}) : (a_{10}\theta^2 + a_{11}\theta + a_{12}) : (a_{20}\theta^2 + a_{21}\theta + a_{22}),$$

where $|a_{rs}| \neq 0$, and if P_i has parameter θ_i $(i = 1, 2, 3, 4)$ then

$$\{P_1, P_2; P_3, P_4\} = \{\theta_1, \theta_2; \theta_3, \theta_4\}.$$

For, as has been shown on p. 132, θ is a canonical parameter in a suitably chosen representation.

THEOREM 4. *The projective geometry of S_2 induces a subordinate one-dimensional projective geometry on every proper conic locus s.*

Proof. Let X_0, X_2, E be three chosen points of s. These points define a canonical representation of s; i.e. a representation of the points of s in terms of a single non-homogeneous parameter θ or a pair of homogeneous parameters (λ, μ) with $\lambda/\mu = \theta$. Since

$$\theta = \{E, P; X_0, X_2\}$$

on s, the assignment of the values of the parameter is invariant over any change of the underlying allowable representation $\mathscr{R}$ of S_2, and it therefore forms part of the projective geometry of S_2. We have only to show, then, that the class of canonical representations of s is the full class of allowable representations of a one-dimensional projective domain.

Suppose a second canonical representation is defined by the points X_0', X_2', E'. Then, in this representation,

$$\theta' = \{E', P; X_0', X_2'\} = \{\theta_{E'}, \theta; \theta_{X_0'}, \theta_{X_2'}\};$$

and θ' is therefore related to θ by a bilinear transformation. This transformation must be non-singular, since θ' is variable with θ.

Now suppose, conversely, that a new parameter θ' is defined algebraically by the transformation

$$\theta' = \frac{\alpha\theta+\beta}{\gamma\theta+\delta} \quad (\alpha\delta-\beta\gamma \neq 0).$$

Then there are three points X_0', X_2', E' whose new parameters θ' are respectively ∞, 0, 1; and, since cross ratio is invariant over non-singular bilinear transformation,

$$\begin{aligned} \{E', P; X_0', X_2'\} &= \{\theta_{E'}, \theta_P; \theta_{X_0'}, \theta_{X_2'}\} \\ &= \{\theta'_{E'}, \theta'_P; \theta'_{X_0'}, \theta'_{X_2'}\} \\ &= \{1, \theta'_P; \infty, 0\} \\ &= \theta'_P. \end{aligned}$$

Therefore θ' is the parameter in the canonical representation defined by X_0', X_2', E'. This completes the proof of the theorem.

It will be observed that our definition of cross ratio on the conic is related to the canonical parameter θ in such a way that the cross ratio we have defined is the cross ratio that occurs in the subordinate one-dimensional geometry just referred to.

The results which have been obtained for the proper conic locus on the preceding pages may all be dualized, and we have the following theorems for the proper conic envelope.

THEOREM 5. *The ranges which are intercepted by the lines of a proper conic s on any two fixed lines of s are homographically related.*

THEOREM 6. *If p_1, p_2, p_3, p_4 are four lines of the proper conic s, and v is any fifth line of s, then the cross ratio $v\{p_1, p_2; p_3, p_4\}$ is the same for all positions of v.*

DEFINITION. The cross ratio $v\{p_1, p_2; p_3, p_4\}$ is called the cross ratio of the two pairs of lines (p_1, p_2) and (p_3, p_4) on the conic, and it is denoted by $\{p_1, p_2; p_3, p_4\}$.

THEOREM 7. *Any three distinct lines x_0, x_2, e of s define a unique canonical representation $u_0:u_1:u_2 = \theta^2:\theta:1$ of the conic envelope, in which they have the parameters ∞, 0, 1 respectively.*

THEOREM 8. *The projective geometry of S_2 induces a subordinate one-dimensional projective geometry on every proper conic envelope s.*

Every proper conic s gives rise to two subordinate one-dimensional geometries, one for the locus and one for the envelope. As might be expected, these two geometries are linked together; and we now establish the linkage by means of the three theorems which follow.

THEOREM 9. *If (P) is a range of points and p is the polar of P with respect to s, then, as P varies, p generates a pencil of lines, and the range (P) and the pencil (p) are homographically related. Dually, if (p) is a pencil of lines and P is the pole of p with respect to s, then, as p varies, P generates a range of points, and the pencil (p) and the range (P) are homographically related.*

Proof. Let P_1, P_2 have coordinate vectors $\mathbf{x}^{(1)}$, $\mathbf{x}^{(2)}$. Then, by the linearity of the condition for conjugacy of two points, if the polars of P_1 and P_2 are represented by $\mathbf{u}^{(1)}$ and $\mathbf{u}^{(2)}$, the polar of the general point P, given by $\mathbf{x}^{(1)}+\theta\mathbf{x}^{(2)}$, is represented by $\mathbf{u}^{(1)}+\theta\mathbf{u}^{(2)}$; and this is sufficient to prove the theorem.

THEOREM 10. *If P is a variable point of s and p is the tangent at P, then there is a homographic correspondence $(P) \barwedge (p)$ between the locus and the envelope of s.*

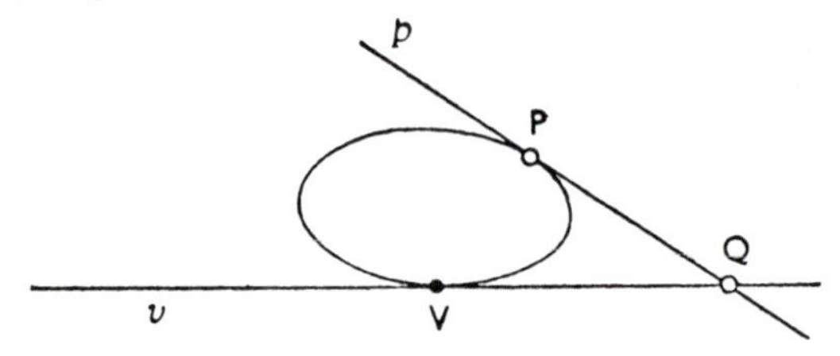

Proof. Take a fixed point V of s, and let v be the tangent at V. If p meets v in the variable point Q, then

$$\begin{aligned}(P) &\barwedge V(P)\\ &\barwedge (Q) \qquad \text{(Theorem 9)}\\ &\barwedge (p).\end{aligned}$$

THEOREM 11. *The cross ratio of any four points of s, taken in a definite order, is equal to the cross ratio of the four tangents at these points, taken in the same order.*

Theorem 11 is an immediate consequence of Theorem 10. Taken in conjunction with Theorem 2 and Theorem 6, it gives a more complete form of Chasles's Theorem.

§ 3. Projective Generation of the Conic

THEOREM 12 (*Steiner's Theorem*). *The locus of the point of intersection P of corresponding rays of two homographically related pencils $A(P)$ and $B(P)$ is a conic s which passes through A and B.*

If the homography is not a perspectivity, s is a proper conic, and its tangent at A is the ray through A which corresponds to the ray BA through B.

If the homography is a perspectivity, s is a degenerate conic, made up of the axis of perspective and the line AB.

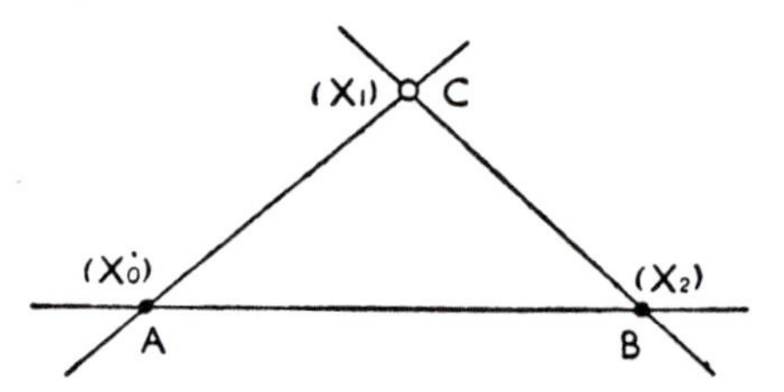

Proof. (i) Suppose the pencils are not in perspective. Then AB does not correspond to BA, and the ray through B which corresponds to AB through A and the ray through A which corresponds to BA through B meet in a point C which does not lie on AB. We may therefore take ABC as triangle of reference. The equations of a general pair of corresponding rays may then be written as

$$x_1 - \theta x_2 = 0 \quad \text{and} \quad x_0 - \theta' x_1 = 0,$$

and, since $(0,0)$ and (∞,∞) are corresponding pairs of parameters, the relation between θ and θ' is of the form $\theta' = k\theta$. Eliminating θ and θ' between the three equations just given, we have the equation of the locus of P in the form

$$kx_1^2 = x_2 x_0.$$

The locus is therefore a proper conic through A and B, which touches AC at A and BC at B.

(ii) Suppose the pencils are in perspective. Then AB is a self-corresponding ray, and every other pair of corresponding rays meet on the axis of perspective. Since every point of AB is a point of intersection of the pair (AB, BA), the complete locus consists of the axis of perspective and the line AB.

EXERCISES

(i) Verify the second part of Theorem 12 algebraically, by taking AB as one side of the triangle of reference and a pair of corresponding rays as the other two sides.

(ii) Deduce Theorem 19 of Chapter V from Theorem 12 by taking I and J as A and B.

(iii) A variable triangle is drawn so that its sides YZ, ZX, XY pass respectively through three fixed points A, B, C. If Y and Z lie on fixed lines, show that the locus of X is, in general, a conic. When does this locus reduce effectively to a line?

THEOREM 13 (*Dual of Theorem* 12). *The envelope of the line p which joins corresponding points of two homographically related ranges $a(p)$ and $b(p)$ is a conic s which touches a and b.*

If the homography is not a perspectivity, s is a proper conic, and its point of contact with a is the point of a which corresponds to the point ba of b.

If the homography is a perspectivity, s is a degenerate conic, made up (as an envelope) of the vertex of perspective and the point ab.

§ 4. Homographic Correspondences on a Conic

The existence of a one-dimensional projective geometry on every proper conic locus and conic envelope makes possible an immediate application to the conic of all the results obtained in Chapter III. In particular, we have a theory of homographic correspondences on the conic. In this section we shall deal only with the conic locus, leaving to the reader the formulation of the dual properties of the conic envelope.

The one-dimensional projective geometry on a conic locus s is, of course, identical (or at least isomorphic) with the one-dimensional geometry of S_1, and if we were considering the two systems in isolation there would be little point in stating the same results over again. The situation is quite different, however, when we think of the conic or the projective line as embedded in the projective plane S_2, with its geometry subordinate to the two-dimensional geometry of S_2. The projective properties of the submanifold are bound up with the projective properties of the whole plane, and this connexion gives them increased significance. We have already seen that there are interesting incidence constructions involving homographies on the projective line, and we shall soon see that there are constructions of a similar nature for homographies on the conic. In many cases the latter constructions turn out to be simpler than the former, so that, in a sense, projective geometry on a conic (in the projective plane) is simpler than projective geometry on a line.

Involutions on a conic

The above remarks are illustrated in a striking manner by the following construction for the general involution on a conic.

THEOREM 14. *If s is a proper conic and V is a point which does not lie on s, the lines of the pencil whose vertex is V cut an involution on s. Conversely, every involution on s is generated by chords which pass through a fixed point.*

Proof. Take a canonical representation of s, and consider a general pair of points P, P', with parameters θ, θ'.

The condition for PP' to pass through a fixed point $V(\alpha_0, \alpha_1, \alpha_2)$, which does not lie on s, is

$$\alpha_2 \theta\theta' - \alpha_1(\theta+\theta') + \alpha_0 = 0 \quad (\alpha_1^2 - \alpha_2 \alpha_0 \neq 0);$$

and the condition for (P, P') to be a pair of a fixed involution τ may be written

$$a\theta\theta' + b(\theta+\theta') + d = 0 \quad (ad - b^2 \neq 0).$$

These conditions are of exactly the same form, and there is therefore a (1, 1) correspondence between points of concurrence V and involutions τ.

Remarks

(i) The point V is called the *vertex* of the involution τ. The united points of τ are the points of contact of the tangents from V to s, i.e. the points in which s is met by the polar of V.

(ii) It follows at once from Theorem 14 that two involutions on s have a unique common pair, the pair of points in which s is cut by the line joining their vertices. Thus the second part of Theorem 16 of Chapter III is trivial for involutions on a conic. This illustrates the way in which homographies on a conic are often simpler to deal with than homographies on a line.

COROLLARY 1. *Two chords of s are conjugate if and only if they meet s in harmonic pairs of points.*

Proof. Let AB and CD be conjugate chords of s. Then the pole of AB lies on CD, and therefore (C, D) is a pair of the involution whose united points are A and B. Hence $\{A, B; C, D\} = -1$ on s. The converse follows by reversing the argument.

COROLLARY 2. *Two involutions on s commute if and only if their vertices are conjugate points.*

Proof. Let τ_i $(i = 1, 2)$ be an involution, with united points M_i, N_i and vertex V_i.

Then, by Theorem 18 of Chapter III, $\tau_1 \tau_2 = \tau_2 \tau_1$ if and only if $\{M_1, N_1; M_2, N_2\} = -1$, i.e. if and only if $M_1 N_1$ and $M_2 N_2$ are conjugate lines. Since $M_1 N_1$ and $M_2 N_2$ are the polars of V_1 and V_2, this is equivalent to the condition that V_1 and V_2 are conjugate points.

COROLLARY 3. *If τ_1, τ_2, τ_3 are three involutions on s, related in the manner described in the corollary to Theorem 18 of Chapter III— i.e. so that the product of any two of them, in either order, is the third— then the triangle formed by their vertices is self-polar for s.*

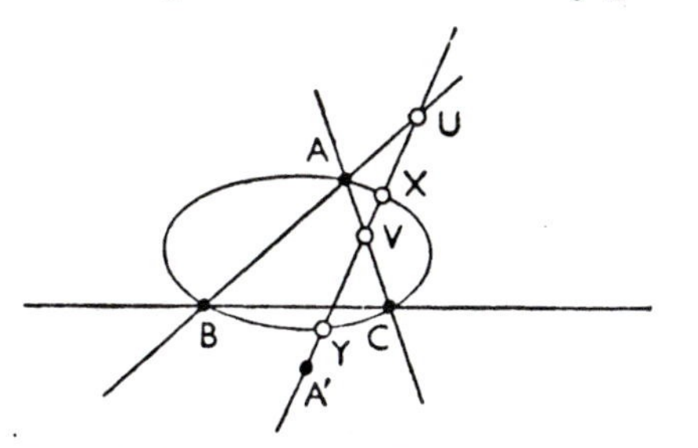

COROLLARY 4. *If ABC is a triangle, inscribed in a proper conic s, then any line through the pole of BC meets AB and AC in points conjugate with respect to s.*

Proof. Let a line through A', the pole of BC, meet AB, AC in U, V respectively and s in X, Y. Since the line is conjugate to BC,

$$\{B, C; X, Y\} = -1 \text{ on } s,$$

and therefore

$$\{U, V; X, Y\} = A\{U, V; X, Y\} = \{B, C; X, Y\} = -1.$$

U and V are thus conjugate points.

EXERCISE. Interpret Corollary 4 as a euclidean theorem by taking B and C as absolute points I and J.

From the general theorems that have now been proved many standard properties of conics may be deduced. In order to illustrate the method we shall establish the following well-known property of the parabola: *Tangents to a parabola, drawn at the ends of a focal chord, meet at right angles on the directrix.*

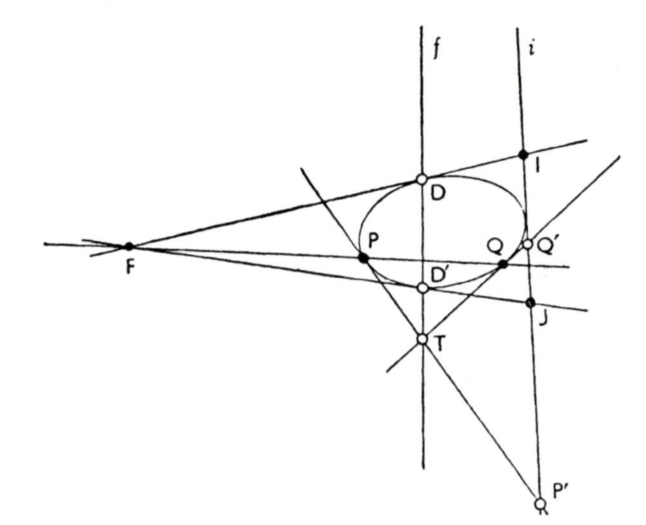

Let k be the given parabola, with focus F and directrix f, and let f meet k in D and D'. Let PQ be a focal chord, and let the tangents at P and Q meet in T. Then since PQ passes through F, the pole of f, PQ and f are conjugate lines; and f therefore passes through the pole of PQ, i.e. T. Thus the tangents at P and Q meet on the directrix f.

Now let the polars of all points be denoted, as usual, by the corresponding small letters, and let the points in which p and q are met by i be denoted by P' and Q'. Then, since PQ and DD' are conjugate lines, $\{P, Q; D, D'\} = -1$ on k (Theorem 14, Corollary 1). It follows that $\{p, q; d, d'\} = -1$ on k (Theorem 11); and hence, taking the section by the tangent i, $\{P', Q'; I, J\} = -1$. We have, therefore, $T\{P', Q'; I, J\} = -1$, and the tangents TP' and TQ' are accordingly at right angles.

The Cross Axis Theorem for the conic

THEOREM 15. *If ϖ is a homography on s, the point of intersection $P_1 P_2'.P_2 P_1'$ of a variable pair of cross joins lies on a fixed line, namely the line joining the united points, distinct or coincident, of ϖ.*

Proof. (i) Suppose ϖ has distinct united points M and N. If we take these as X_0 and X_2 in a canonical representation of s, the equation of ϖ becomes $\theta' = k\theta$. The lines joining the pairs $(\theta_1, k\theta_2)$ and $(\theta_2, k\theta_1)$ are then given by

$$x_0 - (\theta_1 + k\theta_2)x_1 + k\theta_1 \theta_2 x_2 = 0$$

and

$$x_0 - (\theta_2 + k\theta_1)x_1 + k\theta_1 \theta_2 x_2 = 0.$$

Subtracting the first of these equations from the second, we have

$$(\theta_1 - \theta_2)(1-k)x_1 = 0;$$

and the point of intersection of the cross joins thus lies on the line $x_1 = 0$, i.e. MN.

(ii) Suppose ϖ has coincident united points M. If M is taken as X_0 in a canonical representation of s, the equation of ϖ becomes

$$\theta' = \theta + \alpha.$$

The lines joining the pairs $(\theta_1, \theta_2+\alpha)$ and $(\theta_2, \theta_1+\alpha)$ are then given by

$$x_0 - (\theta_1 + \theta_2 + \alpha)x_1 + \theta_1(\theta_2+\alpha)x_2 = 0$$

and

$$x_0 - (\theta_1 + \theta_2 + \alpha)x_1 + \theta_2(\theta_1+\alpha)x_2 = 0.$$

Subtracting, we have $\quad \alpha(\theta_1 - \theta_2)x_2 = 0;$

and the point of intersection of the cross joins thus lies on the line $x_2 = 0$, i.e. the tangent at M.

Remark. In view of the results just obtained, the line joining the united points of ϖ is called the *cross axis* of ϖ.

COROLLARY. *If (A, A') is a fixed pair of corresponding points of ϖ and (P, P') is a variable corresponding pair, then the pencils $A(P')$ and $A'(P)$ are in perspective, with the cross axis of ϖ as axis of perspective.*

Some incidence constructions

Like the Cross Axis Theorem (Theorem 15) of Chapter IV, the theorem that we have just proved gives rise to a number of interesting incidence constructions, and we shall now consider a few typical problems that can be solved with its aid.

PROBLEM 1. Given three pairs of points on a conic s, to find the united points of the homography which they determine.

Using the three given pairs, we have only to put in two pairs of cross joins and join their points of intersection. The line so obtained cuts s in the required united points. (If the construction is actually carried out in the euclidean plane it need not, of course, lead to real points of intersection.)

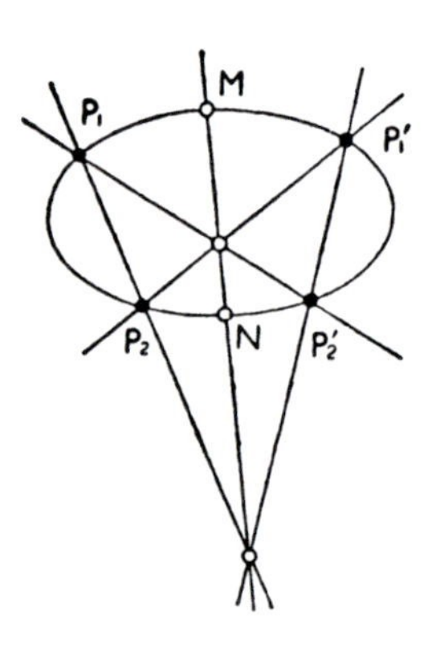

Having constructed the cross axis of the homography determined by the three given pairs, we may now use it to find the point that corresponds to any arbitrary point of the conic.

PROBLEM 2. Given two pairs of points on the conic s, to find the united points of the involution which they determine.

If the given pairs are (P_1, P_1') and (P_2, P_2'), we need only apply the previous construction to the three pairs (P_1, P_1'), (P_2, P_2'), and (P_2', P_2) (see figure).

PROBLEM 3. To construct the polar of a given point V.

The polar of V is the cross axis of the involution whose vertex is V. We accordingly draw two chords $P_1 P_1'$ and $P_2 P_2'$ through V, and then apply the previous construction to the pairs (P_1, P_1') and (P_2, P_2'). It will be seen that this amounts simply to using the quadrangle construction to find two points that are conjugate to V with respect to s.

Remarks

(i) The above constructions involve three geometrical operations only: (*a*) drawing a line through two given points, (*b*) finding the point of intersection of two given lines, (*c*) finding the two points of intersection of a line with a conic that is supposed to be already drawn. In the euclidean plane the constructions can be carried out with the straight-edge alone; but since (*c*) is a quadratic operation, constructions which involve it do not necessarily lead to real points of intersection—i.e. they may fail.

(ii) If we require to find the united points of a homography on a line, we can proceed as follows. We first draw an arbitrary conic (in the euclidean plane a circle will do) and, choosing an arbitrary vertex V on it, we project the line on to the conic from V. After finding the united points of the homography on the conic—a problem of type 1 above—we have only to project back on to the line.

The problem of finding united points is a quadratic problem, and we cannot solve it for the line by means of the straight-edge alone, since this only allows us to perform linear operations. When we are given a conic, however, we have a quadratic locus at our disposal, and the quadratic problem is then soluble by means of the straight-edge, as we have seen.

Pascal's Theorem

THEOREM 16 (*Pascal's Theorem*). *A conic can be drawn to pass through the vertices of a given hexagon if and only if the points of intersection of the opposite sides of the hexagon are collinear.*

Proof. Let A_1, A_2, A_3, A_4, A_5, A_6, in this order, be the vertices of the given hexagon.

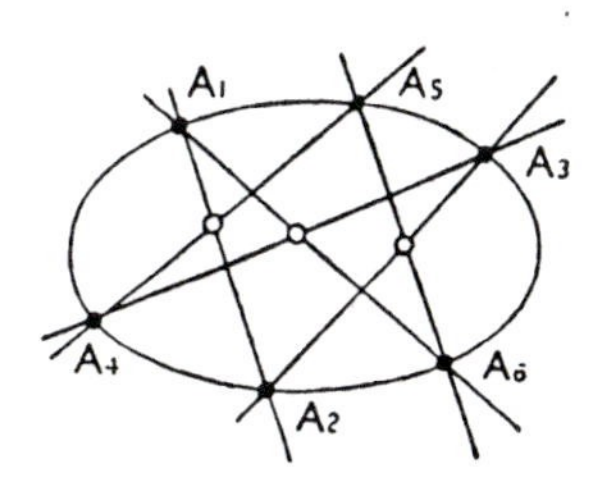

(i) Suppose the hexagon is inscribed in a conic s. Then the three pairs (A_1, A_4), (A_5, A_2), (A_3, A_6) determine a homography ϖ on s; and the three pairs of cross joins $(A_1 A_2, A_4 A_5)$, $(A_5 A_6, A_2 A_3)$, $(A_3 A_4, A_6 A_1)$, which are simply the pairs of opposite sides of the hexagon, meet on the cross axis of ϖ.

(ii) Suppose, conversely, that the three points $A_1 A_2 . A_4 A_5$, $A_5 A_6 . A_2 A_3$, $A_3 A_4 . A_6 A_1$ all lie on a line l. A_1, A_2, A_3, A_4, A_5 determine a unique conic s, which meets $A_6 A_1$ in A'_6, say. Then, by (i), the three points $A_1 A_2 . A_4 A_5$, $A_5 A'_6 . A_2 A_3$, $A_3 A_4 . A'_6 A_1$ lie on a line l'. But the first and the third of these points already lie on l, and l' therefore coincides with l. $A_5 A_6$ and $A_5 A'_6$ then both meet $A_2 A_3$ on l, and hence A'_6 coincides with A_6. The six vertices of the hexagon therefore all lie on the conic s.

COROLLARY (*Brianchon's Theorem*). *A conic can be drawn to touch the sides of a given hexagon if and only if the lines joining the pairs of opposite vertices of the hexagon are concurrent.*

EXERCISES

(i) Investigate what becomes of Pascal's Theorem in the following special cases:

(*a*) when the vertices of the hexagon coincide in pairs;

(*b*) when the circumscribing conic is degenerate.

(ii) If A, B, C, D, E are five general points in the plane and l is a line through A, show how to construct, by use of the straight-edge alone, the second point in which l is met by the conic determined by A, B, C, D, E.

Generation of a homography on a conic

THEOREM 17. *If ϖ is a homography on a conic s, with distinct united points M, N, it may be resolved into a product of two involutions τ_1, τ_2 with vertices V_1, V_2 on MN. V_1 may be chosen arbitrarily on MN, and V_2 is then uniquely determined.*

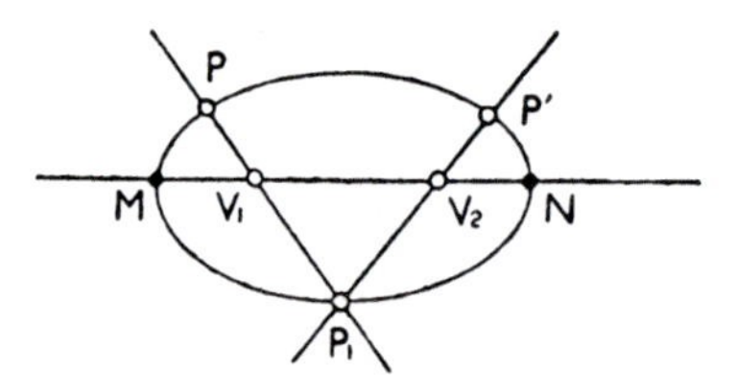

Proof. The theorem follows immediately from Theorem 17 of Chapter III. It may also be proved independently as follows.

Choose some pair (P, P') of ϖ. Then ϖ is determined by the three pairs (M, M), (N, N), (P, P'). Now choose a point V_1 arbitrarily on MN, and let PV_1 meet s a second time in P_1. Let $P_1 P'$ meet MN in V_2. Then, if τ_1, τ_2 are the involutions on s whose vertices are V_1 and V_2, $\tau_2 \tau_1$ is a homography which has (M, M), (N, N), (P, P') as corresponding pairs, i.e. the homography ϖ.

COROLLARY. *With the notation of the theorem, $\{M, N; V_1, V_2\}$ is equal to the modulus of ϖ.*

For
$$\begin{aligned}\{M, N; V_1, V_2\} &= P_1\{M, N; V_1, V_2\} \\ &= \{M, N; P, P'\} \text{ on } s.\end{aligned}$$

THEOREM 18. *If (P, P') is a variable pair of a homography ϖ on a conic s, the envelope of the line PP' is a conic which touches s at the united points of ϖ (or has four-point contact with s if the united points are coincident).*

Proof. (i) Suppose ϖ has distinct united points M, N. If we take these points as X_0 and X_2 in a canonical representation of s, a general pair (P, P') of ϖ is given by $(\theta, k\theta)$; and the coordinates of the line PP' then satisfy the relations

$$\frac{u_0}{1} = \frac{u_1}{-(\theta + k\theta)} = \frac{u_2}{k\theta^2}.$$

Eliminating θ, we have

$$\frac{u_2}{ku_0} = \left\{-\frac{u_1}{(1+k)u_0}\right\}^2,$$

i.e.
$$u_1^2 = \frac{(1+k)^2}{k} u_2 u_0;$$

and this is the condition for PP' to envelop a conic which touches X_0X_1 and X_2X_1 at X_0 and X_2, i.e. a conic which touches s at M and N.

(ii) Suppose ϖ has united points which coincide at M. Then, using the canonical form $\theta' = \theta+\alpha$ for ϖ, we obtain as the line-equation of the envelope of PP'

$$\alpha^2 u_0^2 - u_1^2 + 4u_2 u_0 = 0.$$

The associated point-equation is

$$4x_1^2 + \alpha^2 x_2^2 - 4x_2 x_0 = 0,$$

i.e.
$$x_1^2 - x_2 x_0 + \frac{\alpha^2}{4} x_2^2 = 0;$$

and this represents a conic having four-point contact with s at M (see p. 160).

THEOREM 19. *If s' is a conic, having double contact with s at M and N, there exists a homography ϖ on s, with M, N as united points, such that if a tangent to s' cuts s in P and Q, then either $Q = \varpi P$ or $P = \varpi Q$.*

If s' has four-point contact with s at M, then there exists a homography ϖ on s, with its united points coincident at M, having a similar property.

Proof. We prove the first part of the theorem only, leaving the second part to the reader.

Taking a canonical representation of s, with M, N as X_0, X_2, we can put the point-equation of s and the line-equation of s' in the forms

$$S \equiv x_1^2 - x_2 x_0 = 0, \qquad \Sigma' \equiv u_1^2 - \lambda u_2 u_0 = 0.$$

If θ, θ' are the parameters of the ends of a chord of s which touches s', then

$$(-\theta-\theta')^2 - \lambda.1.\theta\theta' = 0,$$

i.e.
$$\theta'^2 + (2-\lambda)\theta\theta' + \theta^2 = 0,$$

i.e. $$(\theta'-\mu\theta)\left(\theta'-\frac{1}{\mu}\theta\right) = 0,$$

for a suitable value of μ.

Thus either $Q = \varpi P$ or $P = \varpi Q$, where ϖ is the homography on s whose equation is $\theta' = \mu\theta$.

Theorem 18 and Theorem 19 together characterize completely the generation of homographies on a proper conic. The reader should examine how the earlier results concerning involutions (Theorem 14) fit into this more general scheme.

§ 5. Applications of the Theory of Homographic Correspondences on a Conic

The theory of homographies on a conic and their geometrical generation is interesting in itself, and in addition it provides a powerful means of proving well-known theorems in the projective geometry of the conic. We shall now establish some typical results of this kind. In this section the term 'conic' will continue to be used, for the most part, to mean 'proper conic locus'.

THEOREM 20. *If the six vertices of two triangles all lie on a conic, then the six sides of the triangles all touch a conic, and conversely.*

Proof. Let ABC, $A'B'C'$ be two triangles, both inscribed in the same conic s; and let $A'B'$, $A'C'$ meet BC in P, Q, and AB, AC meet $B'C'$ in P', Q'. Then

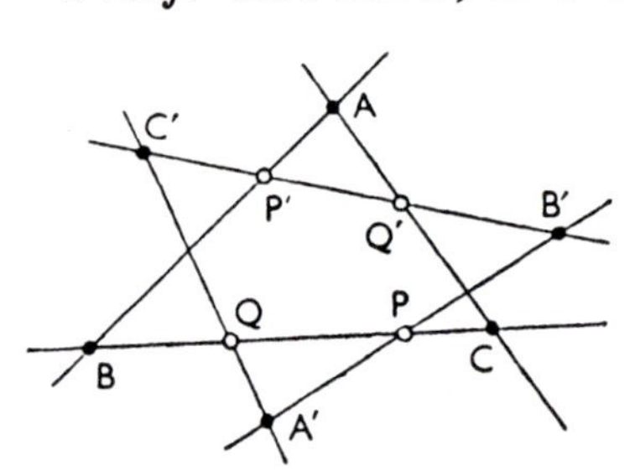

$$\begin{aligned}(P,Q,B,C) &\barwedge A'(P,Q,B,C)\\ &\barwedge A'(B',C',B,C)\\ &\barwedge A(B',C',B,C)\\ &\barwedge (B',C',P',Q'),\end{aligned}$$

and the four joins PB', QC', BP', CQ' therefore all touch a conic which touches $B'C'$ and BC. This proves the direct theorem, and the converse follows by duality.

EXERCISE. Deduce that if a triangle is circumscribed to a parabola its circumcircle passes through the focus of the parabola.

THEOREM 21. *If one triangle exists which is inscribed in a given conic s and circumscribed to a second given conic s', then an infinity of such triangles exists, and one vertex may be chosen arbitrarily on s.*

Proof. Let ABC be the given triangle, inscribed in s and circumscribed to s'. Take a point P arbitrarily on s, and draw the two tangents from P to s', cutting s again in Q and R. Then, since the triangles ABC and PQR are both inscribed in s, their six sides all touch a conic. But five of these sides touch s', which must therefore be the conic so defined. Thus QR also touches s'.

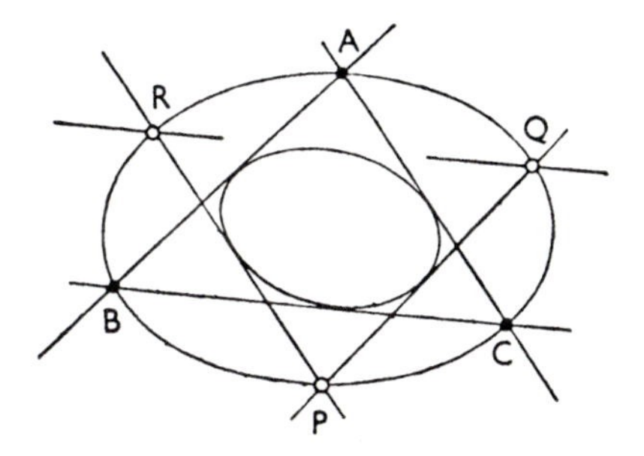

The problem of constructing proper triangles that are both inscribed in one given conic s and circumscribed to another given conic s' has, in general, no solution; but if s and s' happen to be suitably related it has an infinity of solutions. If the conics are related in this special way, s is said to be *triangularly circumscribed* to s', and s' *triangularly inscribed* in s. A problem of this kind which, instead of having a finite number of solutions in every case (as consideration of degrees of freedom would lead us to expect) has either no solution at all or else an infinity of solutions, is said to be *poristic.* Theorem 21 is the simplest case of Poncelet's Porism (Theorem 4 of Chapter VIII).

THEOREM 22. *If two triangles are both self-polar with respect to a given conic, then their six vertices lie on a second conic and their six sides touch a third conic.*

Proof. Let ABC, $A'B'C'$ be two triangles, both self-polar for the same conic s, and let $A'B'$, $A'C'$ meet BC in P, Q, and AB, AC meet $B'C'$ in P', Q'. Then, by Theorem 9,

$$\begin{aligned}(P,Q,B,C) &\barwedge A(C',B',C,B)\\ &\barwedge (C',B',Q',P')\\ &\barwedge (B',C',P',Q'),\end{aligned}$$

since $\qquad \{C',B';\,Q',P'\} = \{B',C';\,P',Q'\}.$

The four joins PB', QC', BP', CQ' therefore all touch a conic which touches $B'C'$ and BC. This proves the first part of the theorem, and the second part follows either by duality or by applying Theorem 20.

EXERCISE. Deduce the following porism from Theorem 22. If one triangle exists which is inscribed in a given conic s and self-polar for a second given conic s', then an infinity of such triangles exists, and one vertex may be chosen arbitrarily on s.

If a conic s has inscribed triangles that are self-polar for a second conic s', s is said to be *harmonically circumscribed* to s'. Dually, if s has circumscribed triangles that are self-polar for s', s is said to be *harmonically inscribed* in s'.

The two relations 's_1 is harmonically circumscribed to s_2' and 's_2 is harmonically inscribed in s_1' are thus only connected by duality, whereas the terminology suggests that they are the same relation looked at from two points of view. In actual fact there is no confusion on this account, since, as we now show, the two relations are equivalent.

LEMMA. *If s and s' are two proper conics of general position, there exists a conic k such that s and s' are transformed into each other by the polarity determined by k, i.e. a conic k for which s and s' are reciprocal.*

Proof. Since s and s' may be represented by quadratic point-equations, they have four common points; and these points are, in general, the vertices of a proper quadrangle. The diagonal triangle of the quadrangle is self-polar for both conics, and if it is taken as triangle of reference, the equations of the conics may be written as

$$S \equiv ax_0^2+bx_1^2+cx_2^2 = 0,$$

$$\Sigma \equiv bcu_0^2+cau_1^2+abu_2^2 = 0;$$

and

$$S' \equiv a'x_0^2+b'x_1^2+c'x_2^2 = 0,$$

$$\Sigma' \equiv b'c'u_0^2+c'a'u_1^2+a'b'u_2^2 = 0.$$

Now if the polarity determined by k transforms s into s', the unique† common self-polar triangle $X_0X_1X_2$ of the conics must be transformed into itself; i.e. it must also be self-polar for k. The point-equation of k may accordingly be written

$$K \equiv px_0^2+qx_1^2+rx_2^2 = 0.$$

Since the polar of the point (y_0, y_1, y_2) with respect to k then has line-coordinates (py_0, qy_1, ry_2), the following equations must be equivalent:

$$ay_0^2+by_1^2+cy_2^2 = 0,$$

and

$$b'c'(py_0)^2+c'a'(qy_1)^2+a'b'(ry_2)^2 = 0;$$

† See the exercise on p. 158, below.

and it follows that

$$\frac{b'c'p^2}{a} = \frac{c'a'q^2}{b} = \frac{a'b'r^2}{c},$$

i.e. $$p^2:q^2:r^2 = aa':bb':cc'.$$

There are thus four possible conics k, given by

$$\sqrt{(aa')}\,x_0^2 \pm \sqrt{(bb')}\,x_1^2 \pm \sqrt{(cc')}\,x_2^2 = 0,$$

each of which defines a polarity which transforms s and s' into each other.

THEOREM 23. *If s_1 and s_2 are two proper conics which meet in four distinct points, s_1 is harmonically circumscribed to s_2 if and only if s_2 is harmonically inscribed in s_1.*

Proof. If k is a conic defining a polarity which transforms s_1 and s_2 into each other, the polarity transforms any triangle inscribed in s_1 and self-polar for s_2 into a triangle circumscribed to s_2 and self-polar for s_1, and vice versa.

THEOREM 24. *If two pairs of opposite vertices of a quadrilateral are each conjugate for a conic s, then the third pair is also conjugate for s.*

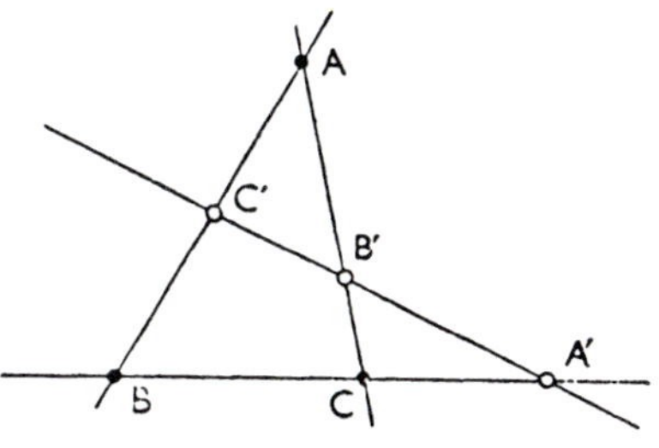

Proof. If (A, A'), (B, B'), (C, C') are the pairs of opposite vertices of the quadrilateral, as shown, we may take ABC as triangle of reference and $A'B'C'$ as unit line. Then A is the point $(1, 0, 0)$ and A' the point $(0, 1, -1)$; and the condition for these points to be conjugate for the conic

$$\sum_i \sum_k a_{ik} x_i x_k = 0$$

is $a_{20} - a_{01} = 0$. But, of the three conditions $a_{20} = a_{01}$, $a_{01} = a_{12}$, and $a_{12} = a_{20}$, any two entail the third, and this proves the theorem.

COROLLARY. *If two pairs of opposite sides of a quadrangle are conjugate for a conic s, then the third pair is also conjugate for s.*

THEOREM 25. *A triangle and its polar triangle with respect to a conic s are in perspective. Conversely, if two triangles are in perspective, there exists a conic with respect to which they are polar.*

Proof. (i) Let ABC be the given triangle, and let $A'B'C'$ be its polar triangle, the sides of $A'B'C'$ being respectively the polars of A, B, and C with respect to s. Let BC, $B'C'$ meet in P and BB', CC' in O. Then the pairs of points (B, C') and (B', C) are each conjugate for s, and therefore, by Theorem 24, O and P are con-

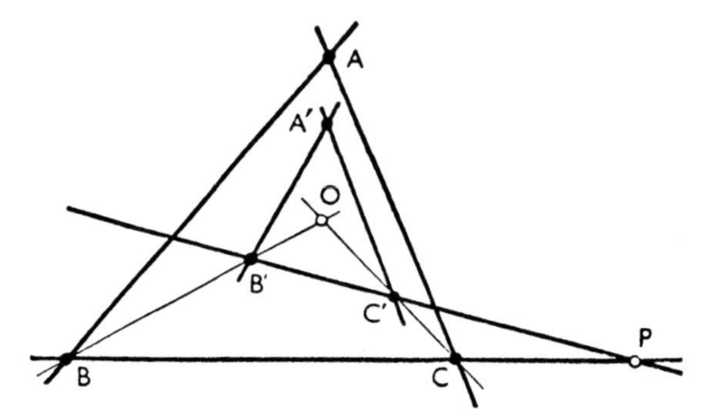

jugate for s. Thus the polar AA' of P passes through O; i.e. the triangles are in perspective from O.

(ii) The algebraic conditions imposed on the coefficients in the equation of a conic s by making (a) a given point to lie on s, and (b) two given points to be conjugate for s, are of the same general form; and it follows from Theorem 2 of Chapter V that there is a unique conic for which five general specified pairs of points are conjugate. This enables us to deduce the converse of the first part of the theorem by *reductio ad absurdum* argument.

Remark. Although Theorem 24 and the first part of Theorem 25 look so different, they are essentially the same proposition, usually referred to as *Hesse's Theorem*. In the above treatment we first proved Theorem 24 algebraically and then deduced Theorem 25 from it; but we could equally well have begun with a direct algebraic proof of Theorem 25 and then inferred Theorem 24 as a corollary. [*Exercise.* Do this.] It follows that the two theorems are equivalent, in the sense that each is deducible from the other.

THEOREM 26. *From a general point A in the plane of a central conic s, fcur normals can be drawn to s. Their feet lie on a rectangular hyperbola h, which passes through A and through the centre C of s, and has its asymptotes parallel to the axes of s.*

Proof.† Let (d, d') be a variable pair of conjugate diameters of s, and let the perpendicular AN from A to d' meet d in P. Then

$$C(P) \equiv (d) \barwedge (d') \barwedge A(N) \equiv A(P),$$

and the locus of P is therefore a conic h through C and A. When d

† Chasles, *Sections coniques* (Paris, 1865), § 219.

is an axis of s, d' is the other axis, and AN is parallel to d. P is then at infinity on the axis d. The conic h therefore passes through the points at infinity on the axes of s, i.e. its asymptotes are parallel to the axes of s.

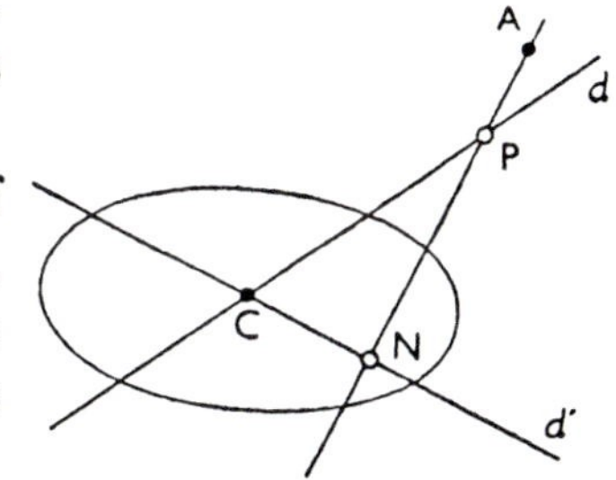

It is easily verified that P lies on s if and only if AP is the normal to s at P. Since h cuts s in four points, there are four normals from A to s, and their feet lie on h.

The hyperbola h is called the *hyperbola of Apollonius* for the point A.

EXERCISES ON CHAPTER VI

1. If ϕ is the eccentric angle of a variable point P on the ellipse
$$X^2/a^2+Y^2/b^2 = 1,$$
show that $t = \tan\frac{1}{2}\phi$ is a projective parameter of P on the curve.

Another projective parameter of P is the gradient m of the line joining it to the point $(a, 0)$. Find the fixed homographic relation which connects the two parameters t and m.

2. The variable line $x+z = \theta(y-3z)$ passes through the fixed point $(-1, 3, 1)$ of the conic k whose equation is $3x^2+y^2-5z^2-2yz+zx = 0$, and its other intersection with k is the variable point P. Show that θ may be taken as a projective parameter of P on k, and that the coordinates of P are $(2\theta^2-4\theta-1,\ 9\theta^2+5\theta-1,\ 3\theta^2+1)$.

Four points A, B, C, D of k have coordinates $(-1, -1, 1)$, $(-3, 13, 4)$, $(5, 3, 4)$, $(6, -5, 7)$. Show that the cross ratio of C, D with respect to A, B on k has the value $3/2$.

3. Verify by direct substitution that the transformation of coordinates given by
$$x : y : z = 2x'-4y'-z' : 9x'+5y'-z' : 3x'+z'$$
reduces the equation of the conic k in the preceding exercise to the form
$$y'^2 = z'x'.$$

4. Being given a system of cartesian coordinates (X, Y) in the real plane, prove that any hyperbola admits of a real parametric representation of the form
$$X = a_1\theta+b_1\theta^{-1}+c_1, \qquad Y = a_2\theta+b_2\theta^{-1}+c_2.$$
Find the centre and asymptotes of the conic with this representation, and show that its line-equation is
$$(c_1u+c_2v+w)^2-4(a_1u+a_2v)(b_1u+b_2v) = 0.$$

5. Prove that the equation of the tangent at the general point (x_0, y_0, z_0) to the conic whose equation is $fyz+gzx+hxy = 0$ may be written in the form $fx/x_0^2+gy/y_0^2+hz/z_0^2 = 0$.

Show also that the equations of any projective parametric representation of the same conic can be written in the form

$$x = a/(\theta-\alpha), \qquad y = b/(\theta-\beta), \qquad z = c/(\theta-\gamma),$$

where $$a(\beta-\gamma) : b(\gamma-\alpha) : c(\alpha-\beta) = f : g : h.$$

6. The sides AB, BC, CD, DA of a quadrilateral $ABCD$ touch a parabola k, and lines through B and D, parallel to DC and BC respectively, meet in T. Prove that AT is parallel to the axis of k.

7. Two triangles XYZ and $X'Y'Z'$ are inscribed in a conic. Two further triangles are formed, the first having XX', YY', ZZ' as its sides, and the second having the points $(YZ.Y'Z')$, $(ZX.Z'X')$, $(XY.X'Y')$ as its vertices. Prove that these two triangles are in perspective. [*Hint.* Apply Pascal's Theorem to the hexagon $XY'ZX'YZ'$.]

8. ABC is a triangle inscribed in a conic s. Prove that an infinite number of triangles can be found which are inscribed in ABC and self-polar for s, and that all these triangles are circumscribed to the triangle formed by the tangents to s at A, B, C.

9. Being given a conic s and a triangle ABC whose vertices are not on s, prove that there exist in general two triangles inscribed in s which are circumscribed to ABC. Show how to construct these triangles.

Show also that if ABC is self-polar for s, then there exist infinitely many triangles inscribed in s which are circumscribed to ABC.

10. If two triangles ABC and $A'B'C'$ are such that ABC is in perspective with the polar triangle of $A'B'C'$ for a conic s, prove that $A'B'C'$ is in perspective with the polar triangle of ABC for s. Discuss the limiting case in which s degenerates as an envelope into the absolute point-pair (I, J).

11. Two conics s, s' have four-point contact at a point A. The ends of a variable chord of s which touches s' are projected from A on to a line parallel to the common tangent of the conics at A. Prove that the distance between the projections is constant.

12. Two conics s and s' pass through a point A, and chords XY and $X'Y'$ of the curves are such that XX' and YY' both pass through A. If XY passes through a fixed point P, show that $X'Y'$ passes through a fixed point P'. What is the locus of P' if P moves on a given line?

State the dual theorems.

13. A variable tangent to a parabola k meets the sides of a fixed circumscribed triangle in points L, M, N. Prove that the ratio $LM : MN$ is constant.

14. Investigate the validity of Theorem 20 in the special case in which one of the given triangles has two of its sides coincident.

Prove that if a triangle is circumscribed to a parabola there exists a rectangular hyperbola which passes through the vertices of the triangle and has the tangent at the vertex of the parabola as one asymptote.

15. Prove that, if a circle is triangularly circumscribed to a conic and has its centre on the conic, then it touches the director circle of the conic.

Show that any circle through the focus of a parabola is triangularly circumscribed to the parabola, and hence deduce the focus–directrix property of the parabola as a special case of the preceding result.

16. Show that chords of a conic s which subtend a right angle at a fixed point A of s all pass through a fixed point F on the normal to s at A. [F is called the *Frégier point* of A with respect to s.]

If s is a rectangular hyperbola, show that F is the point at infinity on the normal at A.

17. Prove that through any general point there pass three normals to a parabola k.

Through any point P of k there are drawn the two lines, other than the normal at P, which are normal to k. Prove that the line joining the feet of these normals passes through a fixed point on the axis of k.

18. Prove that a general point in the plane of a conic k, given by the equation $aX^2+bY^2 = 1$, lies on four normals to k; and that, if the line $uX+vY+w = 0$ joins the feet of two of these normals, the feet of the other two lie on the line $aX/u+bY/v-1/w = 0$.

If Q, R, S are the feet of the other three normals to k from a variable point T on the normal at a fixed point P, prove that the sides of the triangle QRS are tangents to a fixed parabola, which touches the axes of k and the lines joining the vertices of k on each axis to the image of P in that axis.

19. Find the equation of the homography on the conic $x:y:z = \theta^2:\theta:1$ which has the line $ux+vy+wz = 0$ as its cross axis and which carries the point $\theta = \alpha$ into the point $\theta' = \beta$.

Two homographies on a conic k have cross axes a_1, a_2 respectively and a given common pair of corresponding points X, Y. Show that they have no further common pair of corresponding points if and only if a_1 and a_2 intersect on XY (it being assumed that one at least of the homographies is not an involution).

20. Two cobasal involutions τ_1, τ_2 have united points A, B and C, D respectively, and the mate E of A in τ_2 is also the mate of C in τ_1. Prove that the homography $\varpi \equiv \tau_2\tau_1$ is cyclic of period three, and hence show that the mate of B in τ_2 is the mate of D in τ_1.

A triangle ACE is inscribed in a conic k, the tangents at A, C meet CE, AE respectively at T, U, and the polars of T, U meet k again at B, D respectively. Prove that TD, UB intersect in a point F of k, and that the tangent at F meets BD in a point of TU.

21. If ϖ, σ are homographies on a line, show that $\varpi\sigma$ is an involution if $\sigma\varpi$ is an involution. If σ is an involution, prove that $\varpi\sigma$ is an involution if and only if the united points in ϖ correspond in σ.

A variable triangle PQR is inscribed in a conic s, and QR touches a conic s' which has double contact with s. If PQ meets the chord of contact of s and s' at a fixed point, prove that PR passes through one or other of two fixed points on the same chord of contact.

22. If (A, A'), (B, B') are two given pairs of points on a conic k, prove that there exist in general two elations on k which transform A, B into A', B', and that the united points in these elations are mates in the involution on k in which (A, A') and (B, B') are pairs.

Show how to find a point P on k such that the chords AA', BB' are projected from P into segments of equal length, with the same sense, on a given line.

23. If two homographies on a conic k have no common united point, prove that there exist involutions τ, τ_1, τ_2 such that the first of the homographies is $\tau_1 \tau$ and the second is $\tau \tau_2$.

Two elations ϖ_1, ϖ_2 on k have united points M and N respectively; and N_1 is the transform of N by ϖ_1, while M_1 is the point which is transformed into M by ϖ_2. Prove that $\varpi_1 \varpi_2$ is an elation if and only if the harmonic conjugate of M with respect to N, N_1 is the same as that of N with respect to M, M_1.

24. The lines joining a variable point P of a conic k to three fixed points A, B, C meet k again at Q, R, S respectively. Prove that, in general, the envelopes of the chords RS, SQ, QR are three different conics, each of which touches k at its intersections with one side of the triangle ABC.

If A, B, C all lie on a line, show that any general point A' of this line is one member of another fixed triad of points A', B', C' of AB such that $A'Q$, $B'R$, $C'S$ concur at a point of k for every triad QRS.

25. A circle c is rotated about its centre through a given angle in a given sense. If a variable point P of c is displaced in this way to P', prove that the correspondence between P and P' is homographic, with the absolute points I, J as its united points. Show, conversely, that any real homographic correspondence on c, with I and J as united points, is given by a rotation of the circle about its centre.

26. A circle c is given and also a fixed point O, not the centre of c. Through any point P of c a line p is drawn, making a fixed angle α, in the positive sense, with OP. Prove that, as P varies, p envelops a conic, with O as focus, which has double contact with c.

What special relation does c have to the conic when $\alpha = \frac{1}{2}\pi$?

[*Hint.* If p and PO meet c again in Q and R respectively, then $(Q) \barwedge (R)$ and $(P) \barwedge (R)$ on c, so that $(P) \barwedge (Q)$ on c.]

CHAPTER VII

LINEAR SYSTEMS OF CONICS

WE have already seen (p. 79) that if $l_1 = 0$ and $l_2 = 0$ are the equations of two distinct lines of the plane, the equation

$$l_1 + \lambda l_2 = 0$$

represents a pencil (∞^1 linear system) of lines, the set of all lines which pass through the unique common point of l_1 and l_2.

If we were to take three non-concurrent lines we could use them to define an ∞^2 linear system

$$l_1 + \lambda l_2 + \mu l_3 = 0,$$

but to do this would be of little interest. Since, in fact, the linear forms l_1, l_2, l_3 are linearly independent, and a line is uniquely determined by two of its points, the ∞^2 system is simply the totality of all the lines in the plane.

If, now, we consider conics instead of lines, the situation is somewhat different. The totality of all conics in the plane is an ∞^5 linear system, and it contains proper ∞^r linear subsystems for every value of r from 1 to 4. We shall discuss in some detail the ∞^1 system or *pencil* of conics, and then we shall refer more briefly to the ∞^2 system or *net*.

When we say that the conics (or more precisely the conic loci) of the plane form an ∞^5 linear system we mean simply that the equation of the general conic, referred to a fixed representation $\mathscr{R}$, is a homogeneous linear equation in the six coefficients a_{00}, a_{11}, a_{22}, a_{12}, a_{20}, a_{01}. As we have seen on p. 106, these coefficients may be taken as homogeneous coordinates of the conic, and this representation of conics by coordinates is analogous to the representation of lines by line-coordinates (u_0, u_1, u_2). We say that the condition of passing through a fixed point or of having two fixed points as conjugate points is a linear condition on the conic locus because it imposes a linear condition on its coordinates a_{ik}. Having a fixed line (v_0, v_1, v_2) as a tangent, on the other hand, is a quadratic condition on the conic locus, since it leads to an algebraic relation $\sum_i \sum_k A_{ik} v_i v_k = 0$, in which the A_{ik} are all quadratic in the coordinates of the conic locus.

Dually, a conic envelope may be represented by its coordinates $(A_{00}, A_{11}, A_{22}, A_{12}, A_{20}, A_{01})$. Touching a given line or having two given lines as a conjugate pair are linear conditions on the conic envelope, but passing through a given point is a quadratic condition. It will be seen from this that we cannot refer significantly to a linear condition on a conic unless we specify whether the curve is to be treated as a locus or as an envelope.

§ 1. Pencils and Ranges of Conics Elementary Properties

DEFINITION. A *pencil of conics* is a simply infinite system of conics given as loci by an equation of the form $S+\lambda S' = 0$, where S, S' are quadratic forms in x_0, x_1, x_2 and λ is a variable parameter.

A *range of conics* is a simply infinite system of conics given, as envelopes, by an equation of the form

$$\Sigma+\lambda\Sigma' = 0,$$

where Σ, Σ' are quadratic forms in u_0, u_1, u_2 and λ is a variable parameter.

We shall now develop the theory of pencils of conics, leaving it for the most part to the reader to supply the dual properties of ranges. We shall, however, state explicitly some of the more fundamental results for ranges.

Consider, then, the pencil defined by

$$S+\lambda S' = 0. \tag{1}$$

Since the equations $S = 0$ and $S' = 0$ are both quadratic, the two base conics s, s' have either four common points (which need not all be distinct) or an infinity of common points. This latter case occurs if and only if either s and s' coincide, or they are degenerate conics with a line in common. Both these cases are trivial, and we shall exclude them from consideration.

Suppose, then, that the four common points of s and s' are distinct. No three of them can be collinear (since, if they were, the line joining them would form part of both conics), and it follows from Theorem 2 of Chapter V that there is a unique conic through these four points and a fifth general point of the plane. But there is a unique conic $S+\lambda S' = 0$ through the fifth point, and it also passes through the four common points. It follows that *the pencil determined by s and s' is simply the system of all conics through the four common points of s and s'*. Dually, there is a unique conic of a given range that touches any general line of the plane; and the range consists of all conics that touch the four common tangents of the two defining conics.

EXERCISE. Show that if the pencil $S+\lambda S' = 0$ is defined by means of two new base conics, chosen arbitrarily from among the conics of the pencil, the new parameter $\bar{\lambda}$ of a general conic of the pencil is derived from the old parameter λ by a bilinear transformation $\bar{\lambda} = (\alpha\lambda+\beta)/(\gamma\lambda+\delta)$, $\alpha\delta-\beta\gamma \neq 0$.

We shall assume from now on, except when there is an explicit statement to the contrary, that the four common points A, B, C, D of the pencil (1) are distinct; that is to say, we shall confine our attention to the general pencil.

The pencil clearly contains just three degenerate members, the line-pairs (AB, CD), (AC, BD), (AD, BC). Their parameters are the roots of the cubic equation obtained by equating to zero the discriminant of the quadratic form $S+\lambda S'$.

It follows at once from Theorem 4 of Chapter V that the diagonal triangle of the quadrangle $ABCD$ is a common self-polar triangle for all the conics of the pencil. The vertices of this triangle are, of course, the vertices of the three line-pairs of the pencil. If the triangle is taken as triangle of reference, the equation of the general conic $S+\lambda S' = 0$ assumes the form

$$(a_0+\lambda a_0')x_0^2+(a_1+\lambda a_1')x_1^2+(a_2+\lambda a_2')x_2^2 = 0.$$

The equation may be simplified still further by taking advantage of the freedom of choice of the unit point.

(i) Let us change the unit point by applying the non-singular transformation of coordinates (always possible in the complex projective plane)

$$\sqrt{a_0'}\,x_0 = x_0', \qquad \sqrt{a_1'}\,x_1 = x_1', \qquad \sqrt{a_2'}\,x_2 = x_2'.$$

Then the equation may be written

$$(b_0+\lambda)x_0^2+(b_1+\lambda)x_1^2+(b_2+\lambda)x_2^2 = 0.$$

(ii) Let us take A as unit point. Then, as may be verified directly, the coordinates of the four common points are all of the form $(1, \pm 1, \pm 1)$. The equation of a general conic of the pencil is now

$$a_0 x_0^2+a_1 x_1^2+a_2 x_2^2 = 0,$$

where a_0, a_1, a_2 have to satisfy the single symmetrical linear condition $a_0+a_1+a_2 = 0$.

Dualizing the above results, we see that a general range of conics has associated with it a quadrilateral, the sides of which are lines of every conic of the range. The pairs of opposite vertices constitute the three point-pairs of the range; and the diagonal trilateral of the quadrilateral is a common self-polar trilateral for all the conics.

EXERCISE. Prove that, in general, two conics have a unique common self-polar triangle, which is both the diagonal triangle of their quadrangle of common points and the diagonal trilateral of their quadrilateral of common tangents.

§ 2. Possible Types of Pencil and Range

Corresponding to the possible modes of coincidence of the four common points, we have a number of special types of pencil, and dual to each of these is a special type of range. We shall discuss these in turn, obtaining the various special cases from the general case by successive specialization, and at the same time supplying a dual treatment of the corresponding special types of range.

(i) *The general pencil* (i′) *The general range*

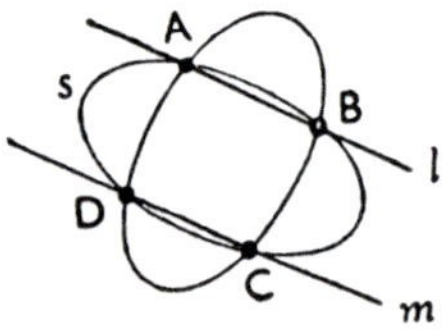

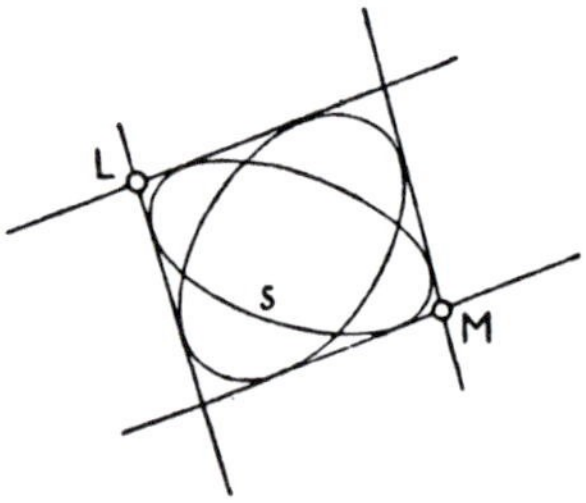

Taking as base conics one proper conic and one degenerate conic, we may represent the systems by the equations

$$S+\lambda lm = 0, \qquad \Sigma+\mu LM = 0$$

respectively.

Each system, as we have already stated, possesses a unique common self-polar triangle, the diagonal triangle of the fundamental quadrangle or quadrilateral.

An important euclidean specialization of (i) is the coaxal system of circles, or set of all circles through two finite fixed points A, B (real or conjugate complex). Let s be a circle, with equation $S = 0$ referred to rectangular axes, and let l be the line AB. If m is the line at infinity, represented by $z = 0$, the equation of the coaxal system with l as radical axis and A, B as common points is

$$S+\lambda lz = 0,$$

or, in non-homogeneous coordinates,

$$S+\lambda l = 0.$$

The three degenerate members of the system are the line-pairs

(AB, IJ), (AI, BJ), (AJ, BI); i.e. the line-circle AB (the radical axis) and two point-circles (the limiting points).

A euclidean specialization of (i′) is the system of all conics confocal with a given central conic. If s is a central conic, the foci of s are the points of intersection of the tangents from I with the tangents from J. The conics confocal with s are simply the conics which touch these four tangents, and they make up the range

$$\Sigma+\lambda IJ = 0.$$

(ii) *The simple-contact pencil* (ii′) *The simple-contact range*

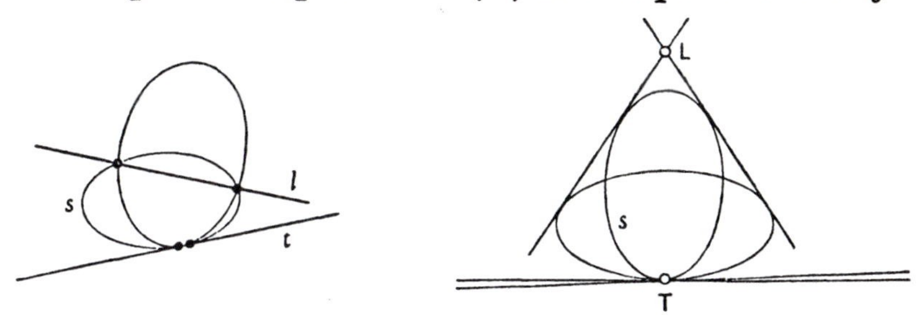

We obtain (ii) as a limiting case of (i) by letting m become a tangent t to s. The equations of (ii) and (ii′) are accordingly

$$S+\lambda lt = 0,$$

where t is a tangent to s;

$$\Sigma+\mu LT = 0,$$

where T is a point of s.

No common self-polar triangle exists for either system.

Parabolas with parallel axes and passing through two common points form a pencil of type (ii); while parabolas with parallel axes and touching two common tangents form a range of type (ii′).

(iii) *The three-point contact pencil* (iii′) *The three-line contact range*

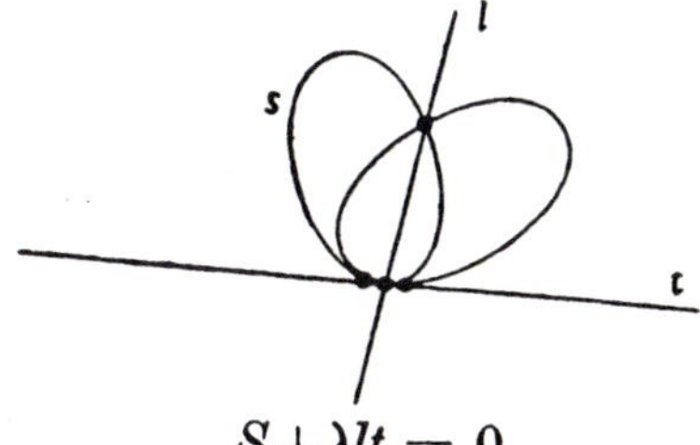

$$S+\lambda lt = 0,$$

where t is a tangent to s and l is a line through its point of contact;

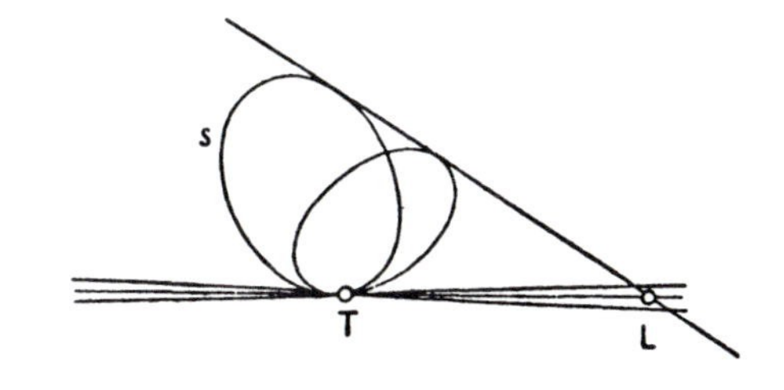

$$\Sigma+\mu LT = 0,$$

where T is a point of s and L is a point on the tangent at T.

These systems are special cases of (ii) and (ii′).

It may be verified analytically that two conics have three-point contact at T if and only if they have three-line contact, with the tangent t at T counting three times as a common tangent. Thus three-point or three-line contact is a self-dual condition; and (iii) and (iii′) are distinguished from each other only by the remaining condition of having either a fourth common point or a fourth common tangent.

Neither system has a common self-polar triangle.

(iv) *The pencil-range with four-point and four-line contact*

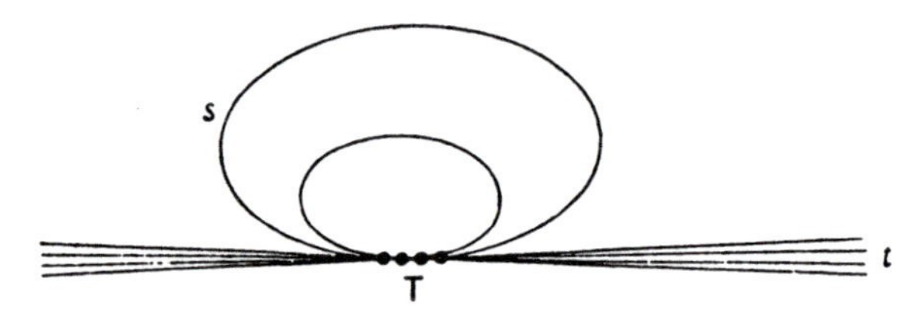

This is a self-dual system, and it may equally well be represented by either of the equations

$$S+\lambda t^2 = 0 \quad \text{and} \quad \Sigma+\mu T^2 = 0,$$

where T is a point of s and t is the tangent at this point. Once again there is no common self-polar triangle.

(v) *The pencil-range with double contact*

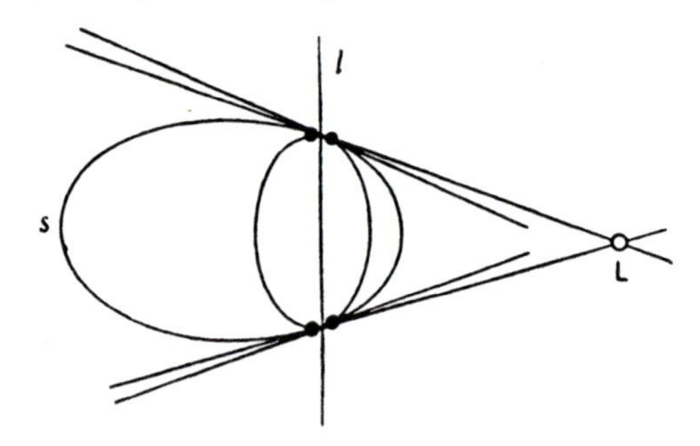

This is another self-dual system (a special case of (ii) or (ii′)) and its equation may be written in either of the forms

$$S+\lambda l^2 = 0 \quad \text{and} \quad \Sigma+\mu L^2 = 0,$$

where l does not touch s and L does not lie on s.

Unlike all the other types of pencil and range, the double-contact system has a simple infinity of common self-polar triangles. To obtain such a triangle, we must take as vertices the point L and any two points of l which are harmonically separated by the two points in which l is met by every conic of the system.

The system of all hyperbolas which have two given lines as asymptotes is a pencil-range of type (v); its cartesian equation, referred to the common asymptotes as axes, is

$$xy+\lambda z^2 = 0.$$

Again, a system of concentric circles is a system of the same kind; its equation, referred to arbitrary rectangular axes, may be written as

$$S+\lambda z^2 = 0.$$

Concentric circles constitute, in fact, a special kind of coaxal system.

EXERCISE. Verify algebraically the statements made in § 2 about self-polar triangles for the special types of pencil and range.

§ 3. GENERAL PROPERTIES OF PENCILS AND RANGES

Having seen what types of pencil and range are possible, we shall now go back to the discussion of the properties of pencils and ranges, establishing a number of fundamental results. These will be proved for the general pencil, and the reader will be left both to supply the duals and also to look into the modifications that are required in special cases.

THEOREM 1 (*Desargues's Theorem*). *The pairs of points in which a fixed line is met by the conics of a pencil are pairs of an involution.*

Proof. If $\mathbf{y}$ and $\mathbf{z}$ represent two fixed points of the given line, a general point of the line is represented by $\mathbf{y}+\theta\mathbf{z}$; and if this point lies on the conic $S+\lambda S' = 0$,

$$S(\mathbf{y}+\theta\mathbf{z})+\lambda S'(\mathbf{y}+\theta\mathbf{z}) = 0,$$

i.e. $$S_{yy}+2\theta S_{yz}+\theta^2 S_{zz}+\lambda(S'_{yy}+2\theta S'_{yz}+\theta^2 S'_{zz}) = 0.$$

The parameters of the points determined by the conic $S+\lambda S' = 0$ are therefore given by this quadratic equation in θ; and hence, by Theorem 15 of Chapter III, the pairs of points corresponding to different values of λ are pairs of an involution.

COROLLARY. *In particular, any line is met by the three pairs of opposite sides of a quadrangle in three pairs of points in involution.*

THEOREM 2. *There are two conics of a pencil which touch a given line.*

Proof. By Desargues's Theorem, the conics of the pencil cut an involution on the line; and this involution has two united points, which are necessarily distinct. Thus two conics of the pencil touch the line, their points of contact being the united points of the Desargues involution.

We see here once again that although the pencil is a linear system of conic loci it is a quadratic system of conic envelopes. The line-equation corresponding to the point-equation

$$S+\lambda S' = 0$$

is, in fact,
$$\Sigma+2\lambda\Phi+\lambda^2\Sigma' = 0,$$

where Σ and Σ' correspond in the usual way to S and S', and Φ is the polarized form of Σ, regarded as a quadratic form in the a_{ik}. In full

$$\Sigma \equiv (a_{11}a_{22}-a_{12}^2)u_0^2+...+2(a_{20}a_{01}-a_{00}a_{12})u_1u_2+...$$

and

$$2\Phi \equiv (a_{11}a'_{22}+a'_{11}a_{22}-2a_{12}a'_{12})u_0^2+...+$$
$$+2(a_{20}a'_{01}+a'_{20}a_{01}-a_{00}a'_{12}-a'_{00}a_{12})u_1u_2+.... $$

EXERCISE. Explain how it comes about that the two types of pencil-range can be self-dual.

THEOREM 3. *A given pencil of conics makes correspond to a general point P of the plane a unique point P^* that is conjugate to P for every conic of the pencil.*

Proof. Let the polars of P with respect to the base conics s, s' of the pencil be p, p', meeting in P^*. Then P and P^* are conjugate points for both s and s'.

Now the conics of the pencil cut an involution on PP^*; and since P, P^* are harmonic with respect to the two pairs cut by s and s', they are the united points of the involution. They are therefore harmonic with respect to all pairs of the involution, i.e. conjugate for all conics of the pencil.

Remarks

(i) A general line of the plane contains just one pair (P, P^*), namely the pair of united points of the Desargues involution cut on it.

(ii) From the fact that the equation $p = 0$ of the polar of a given point P with respect to s is linear in the a_{ik}, it follows at once that the polars of P with respect to the conics of a pencil $S+\lambda S' = 0$ form a pencil of lines $p+\lambda p' = 0$—a pencil with vertex P^*.

(iii) We can express Theorem 3 rather differently by saying that the envelope of polars of a general point P with respect to the conics of a pencil is a point P^*, i.e. the envelope is of the first class. Since, from the dual point of view, the pencil is a quadratic system of conic envelopes, we should expect the locus of poles of a general line p to be a locus of the second order. That this is so is asserted in the next theorem.

THEOREM 4. *The locus of poles of a general line l with respect to the conics of a pencil is a conic k; and this conic is also the locus of the point P^* which corresponds, in accordance with Theorem 3, to a variable point P of l. The conic k passes through the vertices of the line-pairs of the pencil.*

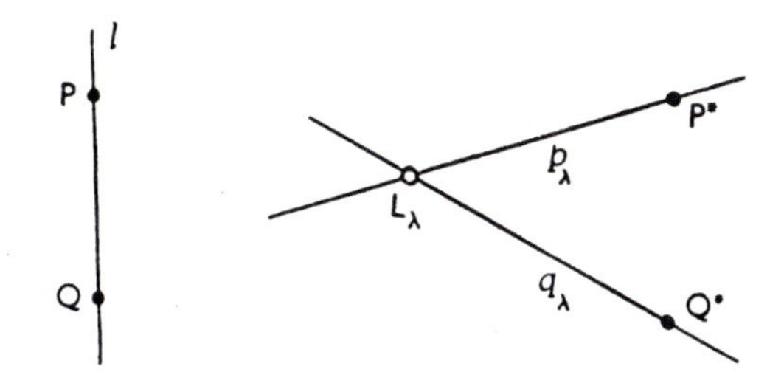

Proof. Take two points P, Q of l. Then the pole L_λ of l with respect to $S+\lambda S' = 0$ is the point of intersection of the polars p_λ, q_λ of P, Q with respect to this conic. But, as λ varies, p_λ and q_λ describe homographic pencils $p+\lambda p' = 0$ and $q+\lambda q' = 0$ with vertices P^* and Q^* respectively; and the locus of L_λ is therefore a conic k through P^* and Q^*.

Since P is an arbitrary point of l, k contains the points P^* corresponding to all points of l.

Finally, the pole of l with respect to a line-pair is the vertex of the line-pair, and k therefore passes through the vertices of the line-pairs of the pencil.

The following argument enables us to prove analytically that the locus of L_λ is a conic, and at the same time to exhibit the equation of the conic as a covariant of s, s', and l. (See Chapter VIII for the notion of covariant.)

Let l and L_λ have coordinates (v_0, v_1, v_2) and (x_0, x_1, x_2) respectively. Then

$$\frac{\partial(S+\lambda S')/\partial x_0}{v_0} = \frac{\partial(S+\lambda S')/\partial x_1}{v_1} = \frac{\partial(S+\lambda S')/\partial x_2}{v_2} = -\rho, \text{ say;}$$

i.e. $$\frac{\partial S}{\partial x_i}+\lambda\frac{\partial S'}{\partial x_i}+\rho v_i = 0 \quad (i = 0,1,2).$$

Eliminating λ and ρ, we have

$$\begin{vmatrix} \dfrac{\partial S}{\partial x_0} & \dfrac{\partial S'}{\partial x_0} & v_0 \\ \dfrac{\partial S}{\partial x_1} & \dfrac{\partial S'}{\partial x_1} & v_1 \\ \dfrac{\partial S}{\partial x_2} & \dfrac{\partial S'}{\partial x_2} & v_2 \end{vmatrix} = 0,$$

i.e. $$\frac{\partial(S,S',l)}{\partial(x_0,x_1,x_2)} = 0,$$

where $$l \equiv \sum_i v_i x_i.$$

The equation just obtained is quadratic in x_0, x_1, x_2, and the locus of L_λ is therefore a conic k.

EXERCISE. When the equation of the pencil is written in the form

$$a_0 x_0^2+a_1 x_1^2+a_2 x_2^2 = 0,$$

with $a_0+a_1+a_2 = 0$, show that if P has coordinates (y_0, y_1, y_2) then P^* has coordinates $(1/y_0, 1/y_1, 1/y_2)$. Obtain the equation of the conic k derived from the line (v_0, v_1, v_2), and verify directly that k passes through the vertices of the line-pairs of the pencil.

§ 4. The Centre-locus of a Range or Pencil

The centres of the conics of a given range or pencil constitute a simple infinity of points lying on a curve, the *centre-locus* of the system. Since the centre of a conic is the pole of the line at infinity, we can find out the nature of this curve by applying the theorems that have just been proved.

Consider first the range. By the dual of Theorem 3, the locus of poles of a general line with respect to the conics of a range is a line, and therefore the centre-locus of a general range is a line. This line contains the centre of the unique parabola of the range; i.e. it is parallel to the axis of the parabola. It also contains the centres of the three point-pairs of the range, and this gives us at once the following well-known theorem.

THEOREM 5. *The mid-points of the three diagonals of a quadrilateral are collinear.*

Turning now to the pencil, we can conclude from Theorem 4 that the centre-locus of a general pencil is a conic. This conic contains the centres of the two parabolas of the pencil; i.e. its asymptotes are parallel to the axes of the parabolas. It also passes through the vertices of the three line-pairs of the pencil; i.e. it circumscribes the common self-polar triangle.

In the particular case in which the pencil of conics is a coaxal system of circles, the line at infinity is a side of the basic quadrangle. In this case the locus of poles breaks up into the line at infinity itself and another line, and we thus arrive at the familiar result that the centres of the circles of a coaxal system are collinear. The line of collinearity must, of course, contain the centres of the two point-circles of the system.

THEOREM 6. *The director circles of the conics of a range form a coaxal system.*

Proof. The theorem follows at once from the fact that, when rectangular axes are used, the equation

$$C(x^2+y^2)-2Fyz-2Gzx+(A+B)z^2 = 0$$

of the director circle of a general conic s is linear in the coefficients of the line-equation of s. The following alternative proof, by means of Desargues's Theorem, is also instructive.

Consider a range determined by two conics s, s', and let the director circles of these conics be k, k', meeting in the finite points P, Q. Then the pairs of tangents from P to the conics of the range belong to an involution; and since two pairs (given by s and s') are orthogonal, the involution is the orthogonal involution. The two tangents from P to any conic of the range are therefore perpendicular, and all the director circles thus pass through P, and similarly also through Q. The circles accordingly form a coaxal system.

COROLLARY. *The three circles on the diagonals of a quadrilateral as diameters are coaxal.*

Theorem 5 is, of course, an immediate consequence of this corollary.

EXERCISE. What are the radical axis and the limiting points of the coaxal system of Theorem 6?

§ 5. Confocal Conics

In this section we shall be concerned with conics in the euclidean plane, and the term 'conic' will be restricted to mean real proper conic.

A conic s has four foci, two of which are real and two complex. If s is a central conic, all its foci are finite points; but if it is a parabola one of the real foci is finite and the other three foci fall on the line at infinity, at I, J and the centre of s. A second conic s' is said to be confocal with s if s and s' have the same four foci, and it is clearly sufficient if the conics have one pair of opposite foci in common. If s is a central conic, every conic confocal with it is central, and if s is a parabola every confocal conic is a parabola.

The system of all conics confocal with a given conic is called a confocal system. There are thus two distinct kinds of confocal system, one consisting entirely of central conics and the other of parabolas. Since, for real proper conics, the relation of being confocal is an equivalence relation, a confocal system is determined uniquely by any one of its members. The conics confocal with s are simply those conics which touch the four tangents from I and J to s, and the confocal system determined by s is therefore the range $\Sigma+\lambda\Omega = 0$, where $\Omega = 0$ stands as usual for the line-equation of the absolute point-pair (I, J).

If s is a central conic, with foci F, F', G, G', we have a figure like that on p. 123, and it is clear from this that C is the centre and CX, CY are the axes of every conic of the confocal system. If the common axes of the system are taken as rectangular axes of reference, the equation of the system assumes a very simple form. Referred to these axes, the point-equation of s is of the form

$$S \equiv ax^2+by^2-z^2 = 0,$$

and the line-equation is therefore

$$\Sigma \equiv \frac{u^2}{a}+\frac{v^2}{b}-w^2 = 0.$$

Since $\Omega \equiv u^2+v^2$, the line-equation of a general conic of the confocal system is

$$\left(\frac{1}{a}+\lambda\right)u^2+\left(\frac{1}{b}+\lambda\right)v^2-w^2 = 0,$$

and the associated point-equation is then

$$\frac{x^2}{(1/a)+\lambda}+\frac{y^2}{(1/b)+\lambda}-z^2 = 0.$$

In particular, the conics confocal with the ellipse $\frac{X^2}{a^2}+\frac{Y^2}{b^2}=1$ are given by the equation $\frac{X^2}{a^2+\lambda}+\frac{Y^2}{b^2+\lambda}=1$.

EXERCISE. Show that, by suitable choice of axes, the equation of a general system of confocal parabolas may be put in the form $Y^2 = 4\lambda(X+\lambda)$.

THEOREM 7. *Through any point P of the plane there pass two conics of a given confocal system, and they cut orthogonally. The tangents at P to these conics bisect the angles between the tangents from P to any other conic of the confocal system.*

Proof. The pairs of tangents from P to the conics of the confocal system are pairs of rays in involution. The involution has two united rays, which are the tangents to the two conics of the system which pass through P. The united rays are separated harmonically by every pair of the involution, in particular by the pair (PI, PJ), and they are therefore perpendicular. The two conics through P therefore cut orthogonally.

Since the united rays are both perpendicular and also harmonic with respect to the pair of tangents to an arbitrarily chosen conic of the system, they are the bisectors of the angles between these tangents.

COROLLARY. *If F, F' are the real foci of the system, the lines PF and PF' are equally inclined to the two tangents from P to any one of the conics.*

For the angles TPT' and FPF' (see fig.) have the same bisectors.

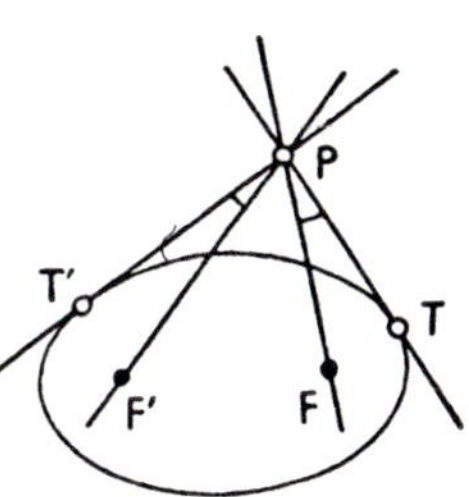

Consider the confocal system

$$\frac{X^2}{a^2+\lambda}+\frac{Y^2}{b^2+\lambda}=1 \quad (a^2>b^2),$$

and suppose that P has coordinates (X_0, Y_0). Then the parameters of the two conics of the system which pass through P are given by the equation

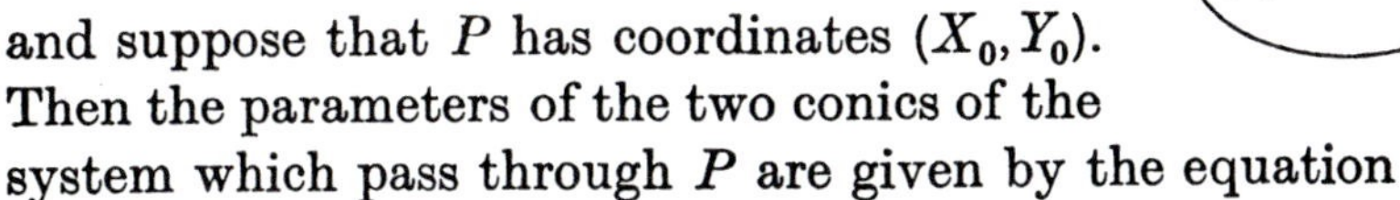

$$f(\lambda) \equiv (a^2+\lambda)(b^2+\lambda)-X_0^2(b^2+\lambda)-Y_0^2(a^2+\lambda) = 0,$$

i.e. $$\lambda^2+\lambda(a^2+b^2-X_0^2-Y_0^2)+a^2b^2-b^2X_0^2-a^2Y_0^2 = 0.$$

Since $f(\infty) > 0$, $f(-b^2) < 0$, and $f(-a^2) > 0$, the roots of $f(\lambda) = 0$ satisfy respectively the conditions $-a^2 < \lambda < -b^2$ and $-b^2 < \lambda$.

It follows from this that *of the two conics through P_0 one is a hyperbola and the other is an ellipse.*

The appearance of a system of confocal central conics is as shown below.

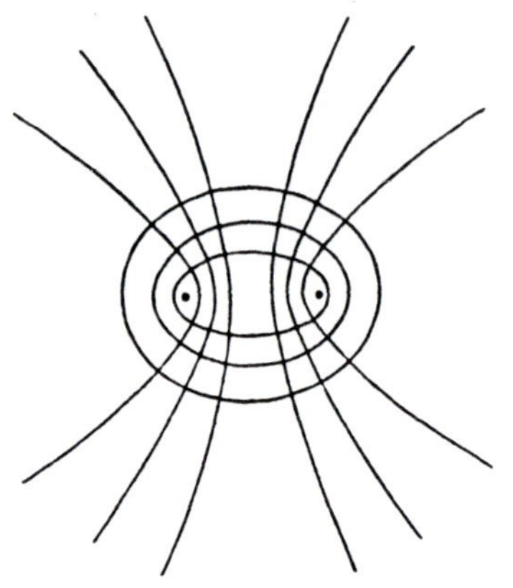

EXERCISE. Show that the appearance of a system of confocal parabolas is as follows.

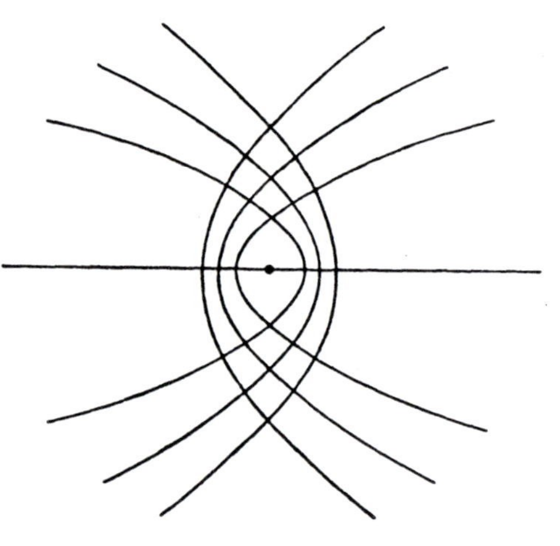

The locus of poles of a line

Since every confocal system of conics is a range, the locus of poles of a line l is a second line l', and the relationship between l and l' is easily characterized geometrically. In order to fix l' we need two points on it, and so we have to find the poles of l with respect to two particular conics of the system. The simplest conics to take are Ω and that conic which touches l. Since l' passes through the pole of l with respect to Ω, it is perpendicular to l; and since it passes through the pole of l with respect to that conic which touches l, it contains the point of contact of this conic with l. Thus *l' is the normal, at the point of contact, to that conic of the confocal system which touches l.*

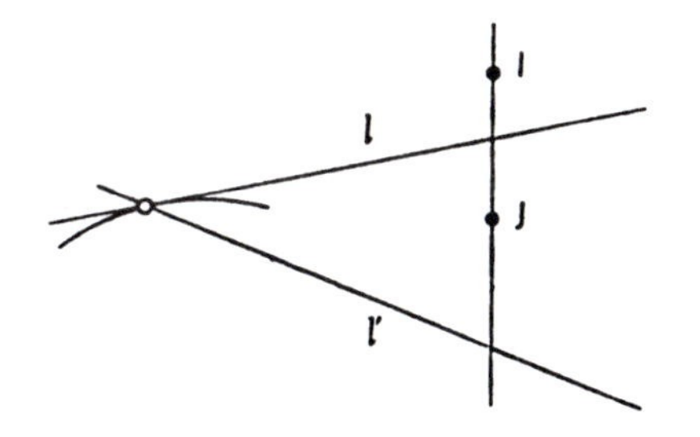

It will be noted how in virtue of the orthogonality property of the confocal system (Theorem 7) the relation between l and l' is symmetrical.

The envelope of polars of a point

The envelope of polars of a given point P is a conic k; and in order to fix this conic we need to find five of its tangents, i.e. the polars of P with respect to five particular conics of the confocal system. The simplest conics to take are the three degenerate ones and the two which pass through P, and the corresponding polars are IJ, FF', GG' (assuming the conics are central conics), and the two tangents at P to the confocal conics through P. *The envelope k is therefore a parabola which touches the axes of the confocal system and these two tangents.* Since the axes of the system and the two tangents at P are both orthogonal pairs of lines, the centre C and the point P both lie on the directrix of k; i.e. *the parabola has CP as directrix.*

If $\Sigma = 0$ is the line-equation of any one of the confocal central conics, the line equation of k is

$$\frac{\partial(\Sigma, \Omega, P)}{\partial(u, v, w)} = 0.$$

Determination of the axes of a conic

An interesting application of the theory of confocal systems is to the following problem: given the equation $\Sigma = 0$ of a conic, referred to an arbitrary rectangular cartesian frame of reference, to find the axes of the conic.

Let (u', v', w') be an axis. Then its pole with respect to the conic

$\Sigma+\lambda\Omega = 0$ is the point

$$u\frac{\partial}{\partial u'}(\Sigma'+\lambda\Omega')+v\frac{\partial}{\partial v'}(\Sigma'+\lambda\Omega')+w\frac{\partial}{\partial w'}(\Sigma'+\lambda\Omega') = 0,$$

where $\Sigma' \equiv \Sigma(u',v',w')$ and $\Omega' \equiv \Omega(u',v',w') \equiv u'^2+v'^2$. But when λ has the value corresponding to the point-pair consisting of the two foci on the axis (u',v',w'), the pole of this axis is indeterminate, and hence

$$\frac{\partial\Sigma'}{\partial u'}+\lambda\frac{\partial\Omega'}{\partial u'} = \frac{\partial\Sigma'}{\partial v'}+\lambda\frac{\partial\Omega'}{\partial v'} = \frac{\partial\Sigma'}{\partial w'}+\lambda\frac{\partial\Omega'}{\partial w'} = 0.$$

Since the value of λ is neither zero nor infinity, we can write the relations as

$$\frac{\partial\Sigma'}{\partial u'}\Big/\frac{\partial\Omega'}{\partial u'} = \frac{\partial\Sigma'}{\partial v'}\Big/\frac{\partial\Omega'}{\partial v'} = \frac{\partial\Sigma'}{\partial w'}\Big/\frac{\partial\Omega'}{\partial w'},$$

and these equations are sufficient to determine $u':v':w'$.

EXERCISE. How is it that we only obtain two solutions, and not four, as the degree of the equations would lead us to expect?

§ 6. NETS OF CONICS

Let s, s', s'' be three conics in S_2 which do not belong to the same pencil. If the point-equations of the three conics, in some allowable representation $\mathscr{R}$, are $S = 0$, $S' = 0$, $S'' = 0$, then the quadratic forms S, S', S'' are linearly independent; and the equation

$$S+\lambda S'+\mu S'' = 0,$$

in which λ, μ are variable parameters, defines an ∞^2 linear system or *net* of conics. The same net is defined by any three of its members which do not all belong to a pencil.

If we pick out any two conics of the net, say s and s', they define a pencil

$$S+\lambda S' = 0,$$

and this pencil is an ∞^1 linear subsystem of the net.

The general conic $S+\lambda S'+\mu S'' = 0$ is degenerate if and only if its discriminant vanishes; and since this restriction imposes a single algebraic condition on the pair of parameters λ, μ, the degenerate conics of the net form a simply-infinite system. This system, however, is not linear. It is selected from the net by an equation of degree 3 in λ and μ.

If we use the mapping of conic loci in S_2 on points of S_5, introduced on p. 106, the conics of a pencil are represented by the points of a line in S_5, and the conics of a net by the points of a plane. The points which represent the degenerate conics of the net are the points of a cubic curve in this plane.

A given net of conics will in general contain no repeated lines, and each degenerate member will therefore have a well-defined vertex. We thus obtain a simply infinite system of points in the plane, the vertices of the line-pairs of the net, and these points form a curve. The equation of the curve is easily found. Suppose, in fact, that $S+\lambda S'+\mu S'' = 0$ represents a line-pair, whose vertex is P. Then the coordinates of P make all the partial derivatives of $S+\lambda S'+\mu S''$ vanish, and they therefore satisfy the equation

$$\begin{vmatrix} \dfrac{\partial S}{\partial x_0} & \dfrac{\partial S'}{\partial x_0} & \dfrac{\partial S''}{\partial x_0} \\ \dfrac{\partial S}{\partial x_1} & \dfrac{\partial S'}{\partial x_1} & \dfrac{\partial S''}{\partial x_1} \\ \dfrac{\partial S}{\partial x_2} & \dfrac{\partial S'}{\partial x_2} & \dfrac{\partial S''}{\partial x_2} \end{vmatrix} = 0.$$

Conversely, if (x_0, x_1, x_2) satisfy this equation, values of λ and μ can be found such that the three partial derivatives of

$$S+\lambda S'+\mu S''$$

are all zero. Thus the point (x_0, x_1, x_2) is the vertex of some line-pair of the net if and only if

$$\frac{\partial(S, S', S'')}{\partial(x_0, x_1, x_2)} = 0.$$

This is therefore the equation of the locus of vertices of line-pairs, or *Jacobian* of the net. This Jacobian j is a curve of order 3, i.e. a plane cubic curve.

THEOREM 8. *The polars of a point P with respect to all the conics* $S+\lambda S'+\mu S'' = 0$ *are concurrent if and only if P lies on the Jacobian j. If this condition is satisfied, the point of concurrence P′ of the polars is also a point of j, and the relation between P and P′ is symmetrical.*

Proof. The polars of P with respect to all the conics are concurrent if and only if the polars with respect to s, s', s'' are concurrent, and these three lines have coordinates

$$\left(\frac{\partial S}{\partial x_0}, \frac{\partial S}{\partial x_1}, \frac{\partial S}{\partial x_2}\right), \quad \left(\frac{\partial S'}{\partial x_0}, \frac{\partial S'}{\partial x_1}, \frac{\partial S'}{\partial x_2}\right) \quad \text{and} \quad \left(\frac{\partial S''}{\partial x_0}, \frac{\partial S''}{\partial x_1}, \frac{\partial S''}{\partial x_2}\right).$$

We thus have the condition

$$\frac{\partial(S, S', S'')}{\partial(x_0, x_1, x_2)} = 0,$$

which means simply that P must lie on j.

If, now, the polars of P are concurrent at P' then P and P' are conjugate for all conics of the net, and the polars of P' are therefore concurrent at P. Thus the relation between P and P' is symmetrical.

Exercise. It can be shown that the equation of a general net of conics can be reduced to the canonical form

$$\lambda_0(x_0^2 - kx_1 x_2) + \lambda_1(x_1^2 - kx_2 x_0) + \lambda_2(x_2^2 - kx_0 x_1) = 0.$$

Find the Jacobian of this net. Prove also that the lines of line-pairs of the net envelop a curve of class three, and find the equation of this envelope.

EXERCISES ON CHAPTER VII

1. Show that all conics which pass through the vertices and orthocentre of a given triangle are rectangular hyperbolas, and that the locus of their centres is the nine-point circle of the triangle. How do these results need to be modified when the triangle is right-angled?

2. If ABC is a triangle and H is its orthocentre, prove that any rectangular hyperbola through A, B, C passes also through H.

By reciprocating the figure with respect to a circle with centre H, or otherwise, prove that if the sides of a triangle all touch a parabola, then the orthocentre lies on the directrix of the parabola.

3. If a line meets a central conic in P, Q and its asymptotes in P', Q', prove that $PP' = QQ'$.

4. The coordinates being rectangular cartesian, find the values of λ which correspond to the parabola, the rectangular hyperbolas, and the point-pairs of the range of conics whose equation is

$$u^2+v^2+w^2+\lambda(u^2+2v^2+2uv-2wu) = 0.$$

Show that the centres of all the conics of the system lie on the x-axis. Distinguish on this line (for real values of λ) the intervals in which lie the centres of (a) hyperbolas, (b) non-virtual ellipses, and (c) virtual ellipses of the system.

5. A variable conic has a given point F as one real focus and touches two fixed lines which meet in T. Show that its centre lies on a fixed line l through the mid-point M of FT, its other real focus describes a line parallel to l, and its conjugate axis envelops a parabola.

Show that the two portions of l which contain the centres of hyperbolas and non-virtual ellipses are separated by M.

6. Four points A, B, C, D, of which no three are collinear, are such that A, B lie on a given conic s and C, D do not lie on s. If a variable conic through

A, B, C, D cuts s again in P and Q, prove that P and Q are mates in a fixed involution on s, and find the vertex of this involution.

Show that if a circle cuts a parabola in points A, B, P, Q then AB and PQ are equally inclined to the axis of the parabola, and also that if a circle cuts a hyperbola in points A', B', P', Q' then $A'B'$ makes the same angle with one asymptote of the hyperbola as $P'Q'$ makes with the other.

7. If $U_i = 0\ (i = 1, 2, 3, 4)$ are the line-equations of four points A, B, C, D, of which no three are collinear, prove that the line-equation of a variable conic s which touches AB, BC, CD, DA is of the form $U_1 U_3 + kU_2 U_4 = 0$.

If s is met by a fixed line through A in P and Q, prove that the tangents to s at P and Q envelop a conic t which touches BC, CD, DB, and the line which joins the points $AB.CD$ and $AD.BC$.

Show also that the tangents at P and Q are mates in a fixed involution of tangents to t.

8. Two conics k, k' touch at two points, and the pole of the common chord of contact is P. If F is any focus of k, prove that FP is one of the bisectors of the angles between the tangents from F to k'.

9. A variable conic passes through two fixed points A, B and has double contact with a fixed conic k. Prove that the chord of contact passes through one or other of two fixed points, namely the two points on AB which are harmonically separated by A, B and also conjugate for k. Deduce that if a circle has double contact with a central conic, then the chord of contact is parallel to one or other of the axes of the conic.

10. Show that any pair of lines through the four points in which a conic is met by a circle are equally inclined to the axes of the conic.

11. Show that two parabolas have their axes parallel and in the same sense and also their latera recta equal if and only if they have three-point contact at infinity. Show further that two such parabolas have the same axis if and only if they have four-point contact at infinity.

12. The locus of centres of a given pencil of conics is a conic c, and the asymptotes of a variable conic s of the pencil meet c at the centre of s and in two further points X, Y. Show that XY passes through a fixed point.

13. Three line-pairs (a, b), (c, d), (e, f) are such that a and b are tangents, at their intersections with a line p, to a conic through the intersections of (c, d) with (e, f). Prove that there is also a conic, through the intersections of (e, f) with (a, b), touching c and d at their intersections with p.

State the dual of the above theorem. Prove that any two points A, B on a parabola k with focus F are opposite foci of a hyperbola which passes through F and has a diameter of k for one asymptote; and that the pole of AB for k is the centre of a circle which touches FA, FB, and the diameters of k which pass through A and B.

14. EFG is the diagonal triangle of a quadrangle $ABCD$, and P is a general point in the plane. Prove that the harmonic conjugates of EP, FP, GP with respect to the pairs of opposite sides of the quadrangle through E, F, G respectively concur at a point Q. Show also that, when P describes a line which does not pass through E, F, or G, then Q describes a proper conic through E, F, and G.

15. If t_1 and t_2 are two of the common tangents of a pair of conics s, s', prove that the point of intersection of the two chords of contact is also the point of intersection of two of the common chords of s and s', and that the two pairs of lines form a harmonic pencil.

Deduce that if two conics have a common real focus, then the two directrices corresponding to this focus and two of the common chords of the conics all meet in a point. Show also that two of the common tangents of the conics meet on the line joining the remaining real foci.

16. Show that the condition for the general conic

$$S \equiv (a,b,c,f,g,h\between x,y,z)^2 = 0$$

to have four-point contact with some conic for which the triangle of reference is self-polar is $FGH+fgh\Delta = 0$, where Δ denotes the discriminant of S.

17. Show that the line-equation of the parabola which has four-point contact with the rectangular hyperbola $XY = c^2$ at the point $(ct, c/t)$ is

$$4c^2uv-w^2+(uct+vc/t+w)^2 = 0,$$

and that the focus of the parabola is the point whose homogeneous coordinates are $(ct^2(t^4+5),\ c(5t^4+1),\ 2t(t^4+1))$.

18. A and B are points of a conic k, and s and t are conics which have four-point contact with k at A and B respectively. If P, Q are two of the common points of s, t, prove that either PQ is conjugate to AB for k or PQ meets the tangents at A and B in points conjugate for k. Show further that in the first case P and Q lie on a conic which touches k at A and B.

19. Show how to find the axes and foci of a conic whose rectangular cartesian line-equation is given.

Find the locus of foci of conics inscribed in a given rectangle; and prove that the polars of a fixed point P for all such conics envelop a parabola whose focus is the inverse of P in the circumcircle of the rectangle.

20. Show that the envelope of polar lines of a fixed point P with respect to the conics of a confocal system is a parabola, and that if Q is the focus of this parabola the relation between P and Q is symmetrical.

If P describes a circle, show that Q describes either a circle or a line, and distinguish between the two cases.

21. Tangents are drawn from a fixed point V to the conics of a confocal system, and the locus of their points of contact is a curve k. Prove that if V lies on an axis of the confocal system then k is a circle through the two foci on the other axis, and also that if V lies on the line at infinity then k is a rectangular hyperbola concentric with the confocal conics.

22. If U and V are two opposite vertices of the quadrilateral formed by the common tangents of a given conic s and a given circle whose centre is P, prove that U and V lie on a conic confocal with s and that the tangents to this conic at U and V are PU and PV.

23. If two conics s, s' cut orthogonally at all four points of intersection, show that the locus of a point whose polars with respect to s and s' are perpendicular is a conic of the pencil determined by s and s'.

If s is a central conic, prove that s' is either confocal with s or belongs to one of three fixed pencils of conics.

CHAPTER VIII

HIGHER CORRESPONDENCES, APOLARITY, AND THE THEORY OF INVARIANTS

So far we have treated the geometry of the conic in a comparatively elementary manner, making use only of the simplest notions of projective geometry and often arguing directly from the definition of the curve. In this chapter we shall show how the theory can be extended by applying more advanced methods, and how the introduction of a more general concept can sometimes unify a large number of apparently disconnected theorems. We shall indicate three separate directions in which such advances are possible, touching successively upon higher correspondences, the relation of apolarity between conics, and the classical theory of invariants.

§ 1. Higher Correspondences

A very useful concept in projective geometry, as has already been made abundantly clear, is that of homographic correspondence between one-dimensional forms; and it owes its usefulness to the fact that we can often generate a homography by geometrical construction, and then carry over the formal properties of homographies into geometrical properties of the figure concerned. Now the homography is a very special kind of correspondence indeed, and we may wonder whether there are more general correspondences, with reasonably simple formal properties, which can also be set up by geometrical construction. Such correspondences do in fact exist—namely algebraic correspondences—and in this section we shall try to give some idea of their place in projective geometry.

The simplest correspondence of the family is the homography itself, and we may conveniently look upon algebraic correspondences as furnishing a natural generalization of the homographic correspondence. To see this we need to reconsider the definition of homography and formulate it in different, though equivalent, terms.

A correspondence between two one-dimensional forms is homographic if the parameters θ and θ' of any two corresponding elements, referred to two arbitrarily chosen allowable parametric representations, are connected by a fixed equation of the form

$a\theta\theta'+b\theta+c\theta'+d = 0$. Instead, however, of requiring θ and θ' to satisfy a *bilinear* equation, we need only require them to satisfy an *algebraic* equation $f(\theta,\theta') = 0$, in which $f(\theta,\theta')$ is a polynomial of unspecified form, as long as we also insist that the correspondence is $(1,1)$. The algebraic theorem involved in this assertion (Theorem 1 below) implies that any $(1,1)$ algebraic correspondence between θ and θ' can be represented by an equation of the form $a\theta\theta'+b\theta+c\theta'+d = 0$. This purely algebraic result allows us to redefine a homography as a $(1,1)$ correspondence that can be represented by an algebraic relation between projective parameters; and when we do this, we see at once how to generalize the concept of homography by introducing (m,n) algebraic correspondences between one-dimensional forms.

DEFINITIONS. We say that two variable numbers θ, θ' of the ground field K (extended by the ideal number ∞) are *in* (m,n) *algebraic correspondence* if they are associated in such a way that:

(i) to every value of θ correspond at most n distinct values of θ', and to some value of θ correspond exactly n distinct values of θ';

(ii) to every value of θ' correspond at most m distinct values of θ, and to some value of θ' correspond exactly m distinct values of θ;

(iii) all the pairs of corresponding values of θ and θ' satisfy a fixed equation $f(\theta,\theta') = 0$, where $f(\theta,\theta')$ is a polynomial in θ and θ' with coefficients in K (and every pair of values which satisfies this equation is a corresponding pair).

Two one-dimensional forms are said to be in (m,n) algebraic correspondence if there is a correspondence between them which is such that, when allowable parametric representations are introduced arbitrarily, the parameters of corresponding elements are in (m,n) algebraic correspondence.

The theory of algebraic correspondences is based on the following algebraic theorem.

THEOREM 1. *If variable numbers θ, θ' of a field*† *K are in (m,n) algebraic correspondence, then there exists a polynomial $g(\theta,\theta')$, of degree m in θ and n in θ', such that α, α' are corresponding values of θ, θ' if and only if $g(\alpha,\alpha') = 0$.*

† It is assumed that the characteristic of K is zero; i.e. that $r.1 \neq 0$ for any positive integer r.

Although it is not at all difficult to give a convincing plausibility argument for the truth of this theorem (making use of the fact that an equation of the nth degree in one unknown has at most n roots), to prove it rigorously demands rather more knowledge of algebra than we wish to presuppose in this book. We therefore refer the reader for the proof to G. T. Kneebone: 'On Algebraic Correspondences' (*Journal of the London Mathematical Society*, **18** (1943), 133–7).

It follows at once from Theorem 1 that our new definition of homography, as a $(1, 1)$ algebraic correspondence, is equivalent to the original definition. The new definition, of course, provides us with an alternative means of showing that a correspondence set up by a specified geometrical procedure is homographic. We first show that it is $(1, 1)$—and this merely involves showing that each step of the construction, forward or backward, leads to a unique result—and then we prove that it is algebraic. To this end, we suppose that the two forms are represented by allowable parameters and then show that by carrying out a series of eliminations we could, in theory at least, arrive ultimately at a polynomial relation between the parameters of corresponding elements. Thus, for example, to prove that the correspondence between the points P, P' in which a variable conic through four fixed points meets a fixed line is an involution, we need only point out that the correspondence, which is obviously $(1, 1)$ and symmetrical, is algebraic in virtue of the fact that the relation between the parameters of the two points is derived from the algebraic condition for six points to lie on a conic. Or again, if a conic s is quadrilaterally circumscribed† to a conic s', each point P of s being one vertex of a quadrilateral $PQRS$ inscribed in s and circumscribed to s', it follows in a similar way that the correspondence between P and R on s is an involution—and hence that the diagonals PR, QS always meet in the same fixed point as P varies on s.

Critical points and united points

In developing the theory of algebraic correspondences we need to introduce a number of special terms, and this we now do. Let us consider an (m, n) algebraic correspondence between two given one-dimensional forms—which we may for definiteness take to be ranges (P) and (P')—and let us suppose it given by an equation

† See Theorem 4 below.

$f(\theta, \theta') = 0$, of degree m in θ and n in θ'. We also take the field of all complex numbers as ground field. Then to any assigned point P_0, with parameter θ_0, correspond the n points whose parameters θ' are the roots of the equation $f(\theta_0, \theta') = 0$; and similarly an assigned point P'_0, with parameter θ'_0, is a point arising from each of the m points whose parameters θ are the roots of the equation $f(\theta, \theta'_0) = 0$. In general the n points arising from a given point P_0 will be distinct, but for special choice of P_0 two or more of them may coincide. Such a special point P_0 is called a *critical point* of the range (P); and in a similar way we define critical points of (P').

The parameters of the critical points of (P) are given by the θ'-discriminant of the equation $f(\theta, \theta') = 0$. For if θ_0 is a critical value of θ, and θ'_0 occurs multiply in the set of corresponding values of θ', then

$$f(\theta_0, \theta'_0) = \frac{\partial}{\partial \theta'_0} f(\theta_0, \theta'_0) = 0, \text{ and conversely.}$$

The critical values of θ are therefore the roots of the equation obtained by eliminating θ' between $f(\theta, \theta') = 0$ and $f_{\theta'}(\theta, \theta') = 0$.

If the ranges (P) and (P') are cobasal, having the same line as axis, it is possible for P to be a self-corresponding or united point. We call P a *united point* if at least one of the n corresponding points P' coincides with P. Thus the parameters of the united points are the roots of the equation $f(\theta, \theta) = 0$; and it follows that, when multiplicities are allowed for and infinite roots are taken into account, *every (m, n) algebraic correspondence has exactly $m+n$ united points.*

An (m, n) correspondence between cobasal ranges may be such that whenever (P, P') is a pair (P', P) is also a pair. This clearly entails $m = n$. If $f(\theta, \theta') = 0$ defines such a correspondence, the conditions $f(\theta, \theta') = 0$ and $f(\theta', \theta) = 0$ are equivalent, so that either

$$f(\theta', \theta) \equiv f(\theta, \theta')$$

or

$$f(\theta' \theta) \equiv -f(\theta, \theta'),$$

i.e. the polynomial $f(\theta, \theta')$ is either symmetric or skew-symmetric.

If $f(\theta, \theta')$ is skew-symmetric, then $f(\theta, \theta) \equiv 0$, so that $f(\theta, \theta')$ has $\theta' - \theta$ as a factor, the residual factor being symmetric. Every point P is a united point of the correspondence.

If $f(\theta, \theta')$ is symmetric, we say that the equation $f(\theta, \theta') = 0$

defines a *symmetrical* (m, m) *correspondence*. It follows then, by a well-known algebraic theorem that $f(\theta, \theta')$ may be written as a polynomial in the elementary symmetric functions $\theta+\theta'$ and $\theta\theta'$ of θ and θ'. In the particular case given by $m = 1$ the correspondence is an involution, and its equation is of the form

$$a\theta\theta'+b(\theta+\theta')+d = 0.$$

EXAMPLE. Let s, s' be fixed proper conics meeting in four distinct points A, B, C, D, and let the polar p with respect to s' of a variable point P of s meet s in P_1' and P_2'. Then the two points P_1', P_2' are related to P by a symmetrical (2, 2) algebraic correspondence on s, namely the correspondence between pairs of points (P, P') of s which are conjugate with respect to s'.

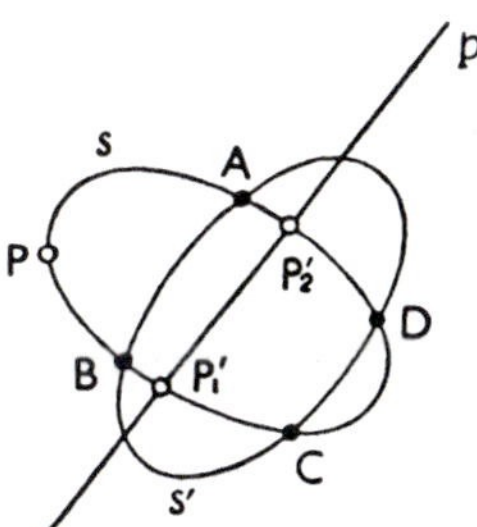

When P is at A, p is the tangent to s' at A; and this line meets s in A and one other point. Thus the four united points of the (2, 2) correspondence are A, B, C, D.

For P to be a critical point, the polar of P with respect to s' must touch s. Now the polars with respect to s' of the points of s envelop a conic $\bar{s}$, the reciprocal of s with respect tc s', and s, $\bar{s}$ have in general four common tangents. The points of s which have these common tangents as their polars for s' are the four critical positions of P.

Symmetrical (2, 2) correspondences on a conic

The above example illustrates the way in which higher correspondences, as well as homographies, can sometimes be set up by a simple geometrical construction. The symmetrical (2, 2) correspondence is associated especially closely with the geometry of the conic. Its equation is

$$a_1(\theta+\theta')^2+a_2(\theta\theta')^2+a_3(\theta+\theta')\theta\theta'+a_4(\theta+\theta')+a_5\,\theta\theta'+a_6 = 0,$$

and the number of coefficients in this equation is the same as the number of coefficients in the equation of the general conic. This makes it possible (in more than one way) to set up a one–one connexion between the ∞^5 symmetrical (2, 2) correspondences and the ∞^5 conics of the plane. Two fundamental connexions of this kind are established by the following theorem.

THEOREM 2. *If the points P and P' correspond in a symmetrical* (2, 2) *algebraic correspondence on a given proper conic s, then*

(*a*) *P, P' are conjugate for a fixed conic s_1, and*

(*b*) *PP' envelops another fixed conic s_2.*

Conversely, pairs of points of s which are conjugate for a fixed conic, or whose joins touch a fixed conic, correspond in a symmetrical $(2, 2)$ *algebraic correspondence on s.*

Proof. If a canonical representation $(\theta^2, \theta, 1)$ of s is taken, the coordinates of the line joining the points P, P', whose parameters are θ, θ', are given by

$$u_0 : u_1 : u_2 = 1 : -(\theta+\theta') : \theta\theta'.$$

Then P, P' are conjugate for a general conic

$$(a, b, c, f, g, h \between x_0, x_1, x_2)^2 = 0$$

if and only if

$$a\theta^2\theta'^2 + b\theta\theta' + c + f(\theta+\theta') + g(\theta^2+\theta'^2) + h\theta\theta'(\theta+\theta') = 0;$$

and PP' touches a general conic $(A, B, C, F, G, H \between u_0, u_1, u_2)^2 = 0$ if and only if

$$A + B(\theta+\theta')^2 + C\theta^2\theta'^2 - 2F\theta\theta'(\theta+\theta') + 2G\theta\theta' - 2H(\theta+\theta') = 0.$$

Since, by giving suitable values to the coefficients, we can make each of these conditions coincide with any given condition of the form

$$a_1(\theta+\theta')^2 + a_2(\theta\theta')^2 + a_3(\theta+\theta')\theta\theta' + a_4(\theta+\theta') + a_5\theta\theta' + a_6 = 0,$$

the theorem is completely proved.

COROLLARY. *The symmetrical* $(2, 2)$ *algebraic correspondence cut on s by tangents to s_2 breaks up into a homography and its inverse if and only if s_2 has double contact with s. If s_2 coincides with s, the correspondence is the identical correspondence taken twice.*

The non-trivial part of this corollary follows at once from Theorems 18 and 19 of Chapter VI.

EXERCISES

(i) Show how it is possible to regard the generation of an involution on s by chords through a fixed point as a degenerate case of the above theorem.

(ii) Examine the way in which the united points and critical points arise when a symmetrical $(2, 2)$ correspondence on s is cut by tangents to a general conic s_2.

THEOREM 3. *The envelope of a variable line which is cut harmonically by two given conics s and s' is another conic k. The eight tangents to s and s' at their common points all touch k.*

Proof. Let l be a variable line, cutting s in P and Q. Then l is cut by s and s' in harmonic pairs of points if and only if P and Q are conjugate for s'. If this condition is satisfied, it follows by

Theorem 2 that P, Q correspond in a certain symmetrical (2, 2) algebraic correspondence on s, and consequently that PQ is a tangent to a fixed conic k.

If l touches s or s' at one of the common points of these conics, three of the four points of intersection coincide, and the condition for a harmonic range is then satisfied (cf. p. 49).

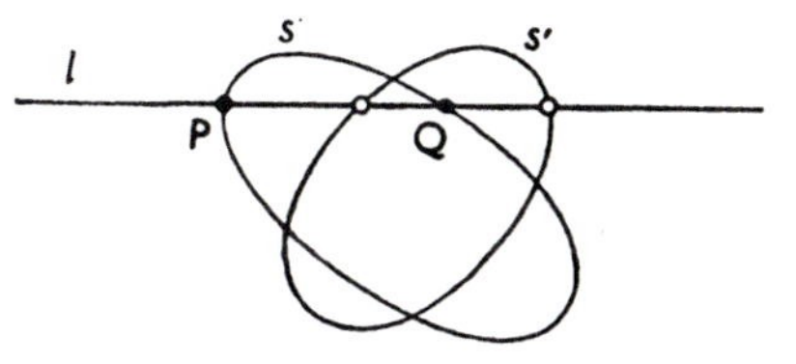

The conic k, just defined, is usually known as the *harmonic envelope* or Φ-*conic* of s and s'. Dually, the locus of a point which moves in such a way that the pairs of tangents drawn from it to s and s' separate each other harmonically is also a conic, the *harmonic locus* or F-*conic* of s and s'.

EXERCISE. By taking s' to be the absolute point-pair Ω, obtain the director circle of s as a harmonic locus.

THEOREM 4. (*Poncelet's Porism*). *If s and s' are proper conics and if, for some fixed integer n exceeding 2, one proper n-gon exists which is both inscribed in s and circumscribed to s', then an infinity of such n-gons exists; and there is one of the n-gons with any general point of s as one of its vertices.*

Proof. Let $A_0 A_1 \ldots A_{n-1}$ be the given inscribed-circumscribed n-gon, and suppose we try to construct another such n-gon, beginning with an arbitrary point P_0 of s as the first vertex. From P_0 we can draw two tangents to s', meeting s again in P_1 and P_1', say; but when one of these points has been chosen as the second vertex, the rest of the construction can be carried out in only one way, unless at some stage we double back. Thus, after $n+1$ steps, we arrive finally at one or other of two points P_n, P_n', and what we have to prove is that at least one of these points always coincides with P_0. Suppose that this is not true. The correspondence between P_0 and P_n must then be symmetrical (2, 2), and algebraic, so that $P_0 P_n$ envelops a fixed conic k with the property that the two tangents from P_0 to k are always the two joins $P_0 P_n$ and $P_0 P_n'$.

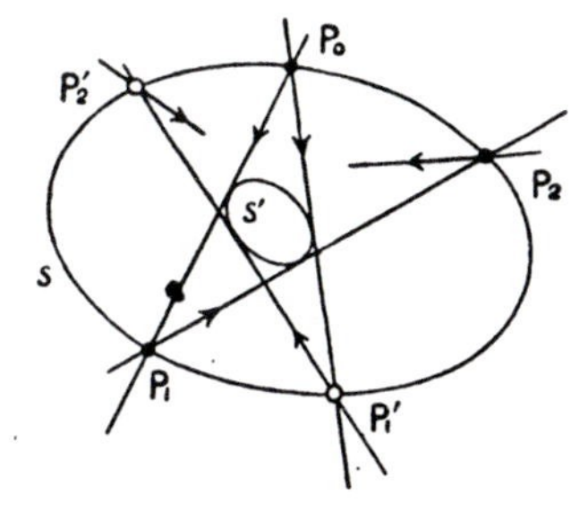

If, now, P_0 falls at a vertex A_i of the given n-gon $A_0 A_1 \ldots A_{n-1}$, both P_n and P_n' also fall at A_i. The two tangents from A_i to k then coincide with the tangent to s at A_i, and k therefore touches s at A_i. Since this happens at at least three distinct points A_0, A_1, A_2, k and s are the same conic. P_n and P'_n thus coincide with P_0 for every position of P_0, in contradiction to our hypothesis that there exist positions of P_0 for which neither P_n nor P'_n coincides with P_0. This proves the theorem.

EXERCISE. If Q_1, Q_2 are two fixed points in the plane of a proper conic s, and $Q_1 Q_2$ meets s in A_1 and A_2, prove that proper $2n$-gons can be inscribed in s with their sides passing alternately through Q_1 and Q_2 if and only if $\{Q_1, Q_2; A_1, A_2\}$ is a primitive nth root of unity. What does this result become in the case $n = 2$?

The method of false position

Theorem 4 brings out the poristic character of the problem of inscribing n-gons in a given conic in such a way that their sides all touch a second given conic—a phenomenon that we have already discussed in the special case of inscribed-circumscribed triangles (Chapter VI, Theorem 21). The possibility of such porisms brings to our notice a danger that is latent in the use of correspondence arguments in order to prove theorems about geometrical constructions; and we shall conclude this section by examining the safeguards that are required when arguments of this kind are employed.

With every problem in projective geometry that asks for the construction of a point or figure satisfying certain stated conditions is associated an existence theorem, which states that the problem has a specified number of solutions—e.g. the theorem that there are in general two conics which pass through four given points and touch a given line—and a powerful method of proving such existence theorems by means of correspondences is the so-called 'method of false position'. Suppose, for example, we are given a conic s and three general points Q_1, Q_2, Q_3 in its plane, and we wish to investigate the possibility of constructing a triangle, inscribed in s, whose sides pass respectively through Q_1, Q_2, and Q_3. If we were to take an arbitrary starting-point P on s and try to construct an inscribed triangle by drawing chords successively through Q_1, Q_2, and Q_3 we should expect the attempt to fail, the reason for its failure being that P was selected at random. If P had been chosen suitably the third chord might have passed through P, so completing the required triangle; and our problem, therefore, is to find the right starting-point. We can solve this problem by observing that, whether the attempted construction solves the original problem or not, it sets up a homographic correspondence on s between P and P', the free point of intersection of the third chord with s, and that the favourable positions for P are simply the united points of this homography. In this way we prove the existence theorem which states that the problem under consideration has in general two solutions (which need not, of course, be real) and, in addition, we are able to find the solutions in any given case; for we need only determine three pairs of the correspondence, by taking arbitrary 'false' positions P_1, P_2, P_3, and then construct the united points in the usual way by drawing the cross axis.

Now suppose we were to apply the method of false position to the problem of finding n-gons that are both inscribed in s and circumscribed to s'. With our earlier notation, P_0 and P_n are connected by a symmetrical (2, 2) algebraic correspondence; and since such a correspondence has four united points, we naturally infer that there are four solutions to the problem. How, then, is this conclusion to be reconciled with the established fact that the problem is poristic? The explanation is that in general, when the conics are not specially related, the four united points arise from degenerate n-gons which double back on themselves. We leave the reader to work out the details, guided by the following figures:†

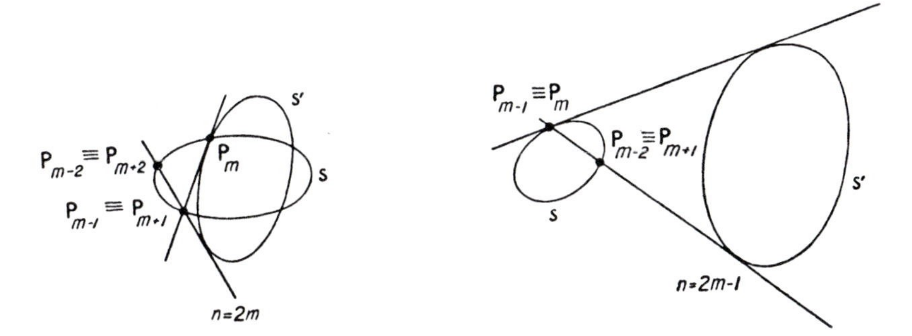

Before arguing, then, by the method of false position that a certain construction problem always has so many solutions (of which an even number may be unreal) we need to satisfy ourselves that the solutions given by the united points of the correspondence are in fact proper ones.

† See Zeuthen, *Abzählende Methoden der Geometrie* (Leipzig, 1914), 77, and van der Waerden, *Einführung in die algebraische Geometrie* (Berlin, 1939), 139.

§ 2. Apolarity

In this section we propose to introduce a new concept, that of apolarity, which provides a powerful means of unifying much of the projective geometry of the conic. It enables us, for example, to bring together under a common head such apparently diverse relations as those which hold between a conic and a pair of conjugate points, a conic and a pair of conjugate lines, and a conic and a second conic to which it is harmonically inscribed or circumscribed. We may conveniently introduce this general notion by going back to the idea of self-polar triangle, by reference to which the relations of harmonic inscription and circumscription were defined in Chapter VI, and trying to see how this idea can be generalized.

A triangle PQR is said to be self-polar for a proper conic s when it stands in a certain self-dual relation to s, both its vertices P, Q, R and its sides p, q, r being conjugate in pairs with respect to s. If the point-equations of the sides of a self-polar triangle of s are respectively $p = 0$, $q = 0$, $r = 0$, and the line-equations of its vertices are $P = 0$, $Q = 0$, $R = 0$, the point-equation and line-equation of s may be written in the forms

$$S \equiv \lambda p^2+\mu q^2+\nu r^2 = 0$$

and

$$\Sigma \equiv \lambda' P^2+\mu' Q^2+\nu' R^2 = 0.$$

This is plain if PQR is taken as triangle of reference. The expressibility of S as a linear combination of p^2, q^2, r^2 is a necessary and sufficient condition for the triangle PQR to be self-polar, and it therefore furnishes an algebraic counterpart of the geometrical relation of self-polarity. The algebra, however, admits of obvious generalization, and this leads us to define the new concepts of polar k-side and polar k-point as follows.

DEFINITION. A set of k lines $(p_1, p_2,..., p_k)$, where k may be 3, 4, or 5, is said to form a *polar k-side* for a given proper conic s if the point-equation of s is expressible in the form $S \equiv \sum_{i=1}^{k} \lambda_i p_i^2 = 0$. Dually, a set of k points $(P_1, P_2,..., P_k)$ is said to form a polar k-point for s if the line-equation of s is expressible in the form

$$\Sigma \equiv \sum_{i=1}^{k} \lambda_i P_i^2 = 0.$$

We need only consider values of k up to 5; for since the conic has five degrees of freedom, the equation of any arbitrary conic can be expressed in the form $\sum_{i=1}^{6}\lambda_i p_i^2 = 0$, $(p_1, p_2,..., p_6)$ being a given set of six general lines.

It will be observed further that it is only in the case $k = 3$ that the polar k-side is a self-dual entity. When k exceeds 3, a set of k lines does not determine an associated set of k points in any simple way.

So far the polar k-side, for k equal to 4 or 5, is connected with the self-polar triangle only by algebraic analogy, and we have still to look for properties that make it geometrically interesting. We may remark, first of all, that there are certain trivial polar k-sides which it is often convenient to leave out of account. Suppose $(p_1, p_2,..., p_{k-1})$ is a polar $(k-1)$-side, and p_k is an arbitrary line. Then

$$S \equiv \lambda_1 p_1^2 + ... + \lambda_{k-1} p_{k-1}^2 + 0 . p_k^2,$$

and $(p_1, p_2,..., p_k)$ counts as a polar k-side. Such polar k-sides will be called *special*. Thus, for example, the sides of a self-polar triangle together with an arbitrary line make up a special polar 4-side. The question that interests us is how non-special polar 4-sides may be characterized geometrically, and we shall now show that the answer is suggested by Hesse's Theorem (Chapter VI, Theorem 24). In virtue of this theorem, if two pairs of opposite vertices of a quadrilateral are conjugate for s, then the third pair is also conjugate for s; and we have here a special relationship that is possible between a quadrilateral and a conic. When the relationship holds, we may call the quadrilateral a *Hesse quadrilateral* for the conic—and Hesse quadrilaterals can now be identified with polar 4-sides.

THEOREM 5. *Four lines, no three of which are concurrent, form a polar 4-side for a given proper conic s if and only if they are the sides of a Hesse quadrilateral of s.*

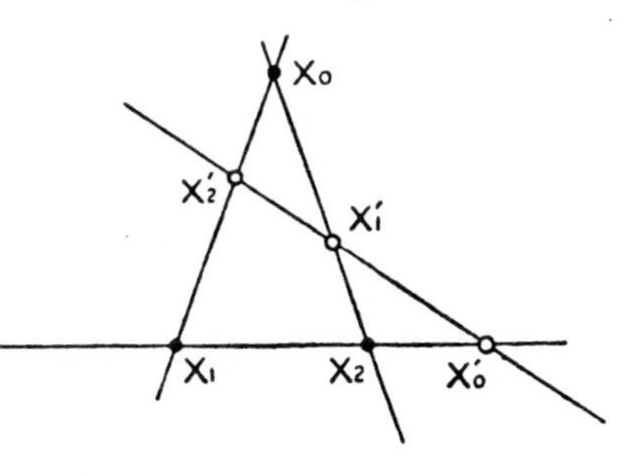

Proof. Let the four given lines be taken as the sides of the triangle of reference and the unit line, and let the unit line meet the other lines in X_0', X_1', X_2' respectively.

(i) If the conic is given by

$$(a, b, c, f, g, h \between x_0, x_1, x_2)^2 = 0$$

and the lines form a Hesse quadrilateral, then, since X_i, X'_i are conjugate for $i = 0, 1, 2$, we have $f = g = h$.

The equation of s may accordingly be written

$$\begin{aligned} S &\equiv ax_0^2+bx_1^2+cx_2^2+2f(x_1x_2+x_2x_0+x_0x_1) \\ &\equiv (a-f)x_0^2+(b-f)x_1^2+(c-f)x_2^2+f(x_0+x_1+x_2)^2 \\ &= 0, \end{aligned}$$

and the four lines form a polar 4-side.

(ii) If, conversely, the lines form a polar 4-side, the equation of s may be written

$$\lambda_0 x_0^2+\lambda_1 x_1^2+\lambda_2 x_2^2+\lambda_3(x_0+x_1+x_2)^2 = 0;$$

and it is immediately verifiable that X_i, X'_i are conjugate points $(i = 0, 1, 2)$.

COROLLARY. *If p_1, p_2, p_3 are three given lines which are not concurrent and which do not form a triangle that is self-polar for a given proper conic s, then there exists a fourth line p_4 such that (p_1, p_2, p_3, p_4) is a polar 4-side for s. In general, p_4 is the axis of perspective of the triangle $p_1p_2p_3$ and its polar triangle.*

EXERCISE. Show that the following is a complete enumeration of all possible types of polar 4-side for a proper conic:

(i) the Hesse quadrilateral;
(ii) the special polar 4-side, consisting of the sides of a self-polar triangle together with an arbitrary line;
(iii) a line and three arbitrary lines through its pole.

We have already seen in Chapter VI that the problems of inscribing in a given conic s and circumscribing about s triangles that are self-polar for a second conic s' are poristic. This fact may now be generalized, polar k-points or k-sides being taken in place of self-polar triangles. The general problem of inscribing in s k-points that are polar for s' is covered by a single comprehensive porism, which involves a certain relation, known as apolarity, that can hold between two conics.

If the two conics s and s' are given, we can in any case choose ∞^1 pairs of points of s which can each be augmented by a third point (not necessarily on s) to a 3-point that is polar for s', and we can choose ∞^3 triads of points of s, each of which can be augmented by a fourth point so as to make it into a polar 4-point for s'. The question is whether the augmenting point does or does not fall on s.

THEOREM 6. *Let s, s' be a conic locus and a conic envelope, represented respectively by the equations*

$$S \equiv (a, b, c, f, g, h \between x, y, z)^2 = 0,$$

and $$\Sigma' \equiv (A', B', C', F', G', H' \between u, v, w)^2 = 0;$$

and let $$I_{s,s'} = aA'+bB'+cC'+2fF'+2gG'+2hH'.$$

Then if $k-1$ points of any k-point polar for s' ($k = 3$, 4, or 5) lie on s, and the $k-1$ points do not themselves form a polar $(k-1)$-point, the kth point either always falls on s or never falls on s, according as $I_{s,s'} = 0$ or $I_{s,s'} \neq 0$.

Furthermore, if $k-1$ lines of any k-side polar for s ($k = 3$, 4, or 5) touch s', and the $k-1$ lines do not by themselves form a polar $(k-1)$-side, the kth line either always touches s' or never touches s', according as $I_{s,s'} = 0$ or $I_{s.s'} \neq 0$.

Proof. Let $P_i \equiv (x_i, y_i, z_i)$ $(i = 1, 2,..., k)$ be the vertices of a k-point polar for s'; and let the points P_1, P_2,..., P_{k-1}, which do not constitute a polar $(k-1)$-point, lie on s. Then there exist $\lambda_1, \lambda_2,..., \lambda_k$ such that

$$\Sigma' \equiv \sum_{i=1}^{k} \lambda_i (ux_i+vy_i+wz_i)^2;$$

and, since $(P_1, P_2,..., P_{k-1})$ is not a polar $(k-1)$-point, $\lambda_k \neq 0$. We therefore have

$$A' = \sum_i \lambda_i x_i^2, \qquad B' = \sum_i \lambda_i y_i^2, \qquad C' = \sum_i \lambda_i z_i^2,$$
$$F' = \sum_i \lambda_i y_i z_i, \qquad G' = \sum_i \lambda_i z_i x_i, \qquad H' = \sum_i \lambda_i x_i y_i;$$

and hence

$$I_{s,s'} = \sum_{i=1}^{k} \lambda_i (a, b, c, f, g, h \between x_i, y_i, z_i)^2 = \sum_{i=1}^{k} \lambda_i S_{ii}.$$

Now $S_{11} = S_{22} = ... = S_{k-1,k-1} = 0$, by hypothesis, and the condition $I_{s,s'} = 0$ therefore reduces simply to $S_{kk} = 0$. This proves the first part of the theorem.

The second part follows at once when we notice that the condition

$$aA'+bB'+cC'+2fF'+2gG'+2hH' = 0$$

remains unaltered when the whole situation is dualized and, at the same time, s and s' exchange roles.

We may remark, incidentally, that since the equation $I_{s,s'} = 0$ represents a projective relation between the two conics, if it holds in one coordinate representation $\mathscr{R}$ it must hold in all.

DEFINITIONS. A conic locus s and a conic envelope s' are said to be *apolar* if $I_{s,s'} = 0$.

A conic k is said to be *outpolar* to a conic k' if k, regarded as a locus, is apolar to k', regarded as an envelope; and in this case we say also that k' is *inpolar* to k.

It is customary and often convenient to associate outpolarity and inpolarity separately with the two aspects of apolarity indicated in the first and second parts of Theorem 6. If we do this, the connexion between the two aspects—a fact of obvious geometrical importance—is that if k is outpolar to k', then k' is inpolar to k. The theorem (already proved by reciprocation in Chapter VI) that if s is harmonically circumscribed to s' then s' is harmonically inscribed in s, is a special instance of this connexion.

We may now summarize the geometrical interpretation of the relation of apolarity between proper conics as follows: If s is outpolar to s', it contains ∞^1 3-points, ∞^3 4-points, and ∞^5 5-points, all polar for s'; and, since s' is then also inpolar to s, it contains ∞^1 3-sides, ∞^3 4-sides, and ∞^5 5-sides, all polar for s.

Properties of the relation of apolarity

(i) The first property of the relation $I_{s,s'} = 0$ that we should note is its linearity, both in the coefficients of the point-equation of s and also in the coefficients of the line-equation of s'. It follows at once from this that if s_1 and s_2 are each outpolar to s_1' and s_2' then every conic of the pencil determined by s_1 and s_2 is outpolar to every conic of the range determined by s_1' and s_2'.

Even more important is the fact that $I_{s,s'} = 0$ is a *general* linear condition on the coefficients of the point-equation of s. When a conic s is made to pass through an assigned point of the plane, a condition of the form

$$ax_1^2+by_1^2+cz_1^2+2fy_1z_1+2gz_1x_1+2hx_1y_1 = 0$$

is imposed upon the coefficients in its point-equation. But while this is a linear condition, it is not the most general linear condition that can be imposed, for the quantities x_1^2, y_1^2, z_1^2, y_1z_1, z_1x_1, x_1y_1, are interconnected. Outpolarity to an assigned conic s' is, however, a general linear condition, since an arbitrary linear condition

$$A'a+B'b+C'c+2F'f+2G'g+2H'h = 0$$

on a, b, c, f, g, h can be interpreted as outpolarity to a conic envelope

$$(A', B', C', F', G', H' \between u, v, w)^2 = 0.$$

We can therefore assert straight away, for example, that there is a unique conic locus that is outpolar to five given linearly independent conic envelopes. If the conic loci of the plane are represented, as described on p. 106, by points of S_5, the condition on a conic s of being outpolar to a fixed conic envelope becomes the condition on the corresponding point of S_5 of lying in a fixed prime (i.e. S_4).

(ii) A conic locus s is outpolar to a point-pair (P, Q) if and only if P and Q are conjugate points for s. For we may choose the coordinate representation in such a way that the line-equation of the point-pair is $vw = 0$; and the condition $I_{s,s'} = 0$ then reduces to $f = 0$, which is the condition for the points $(0, 1, 0)$ and $(0, 0, 1)$ to be conjugate for s.

(iii) A conic locus s is outpolar to a repeated point (P, P) if and only if it passes through P.

(iv) A line-pair (p, q) is outpolar to a conic envelope s' if and only if p and q are conjugate lines for s'; and a repeated line (p, p) is outpolar to s' if and only if p touches s'.

(v) If two conic loci s_1 and s_2, both outpolar for the same conic envelope s', meet in four distinct points, then these points form a polar 4-point for s'. For we can take three of the points and augment them to a polar 4-point for s', and, by Theorem 6, this additional point must then lie on both s_1 and s_2.

We see, then, that the relation of apolarity covers a number of the projective relations between two conics (one or both of which may be degenerate) which we have already learned to handle by more elementary methods, and in this way it proves to be a valuable unifying concept. In addition to this, however, it also admits of some striking euclidean interpretations; and to these we now turn.

Apolarity in the euclidean plane

THEOREM 7. *A circle c is outpolar to a conic envelope s' if and only if it cuts the director circle of s' orthogonally.*

Proof. There exist circles outpolar to s', namely the circumcircles of triangles self-polar for s'. If c_1 and c_2 are any two such circles, every circle of the coaxal system (i.e. the pencil of circles) which they determine is also outpolar to s'. In particular, the two point-circles L, L' of the system are outpolar to s', and this means that the pairs of lines (LI, LJ) and $(L'I, L'J)$ are both conjugate for s'. Thus the tangents from L to s' are perpendicular, and so also are the tangents from L'. The director circle of s' therefore passes through L and L', and so belongs to the conjugate coaxal system. This means that the director circle is orthogonal to c_1, and since c_1 was any circle that is outpolar to s', the director circle of s' is orthogonal to every outpolar circle.

To prove the converse part of the theorem we proceed as follows. Let c be a circle which cuts the director circle of s' orthogonally, and let an arbitrary diameter of c cut the director circle in L and L'. Then, by reversing the previous argument, we can show that the circles of the coaxal system with L, L' as its limiting points —of which c is one—are all outpolar to s'.

COROLLARY 1. *If a triangle is self-polar for a conic s', its circumcircle cuts the director circle of s' orthogonally.*

COROLLARY 2. *If a triangle is circumscribed to a conic s', its polar circle cuts the director circle of s' orthogonally.*

COROLLARY 3. *If a circle is outpolar to a parabola, its centre lies on the directrix of the parabola.*

COROLLARY 4. *If a circle is outpolar to a rectangular hyperbola, it passes through the centre of the hyperbola, and conversely.*

COROLLARY 5. *A rectangular hyperbola is outpolar to a circle if and only if it passes through the centre of the circle.*

Corollary 5 is projectively equivalent to Corollary 4, the roles of the pairs of points in which the two conics cut the line at infinity merely being interchanged.

A condition for six points to lie on a conic

There is one further theorem which it is convenient to insert here although, strictly speaking, it has nothing to do with apolarity. It is concerned, however, with linear combinations of the squares of the equations of points or lines, and it gives useful criteria for six points to lie on a conic and for six lines to touch a conic.

THEOREM 8 (*Serret's Theorem*). *If the equations*

$$P_i \equiv ux_i+vy_i+wz_i = 0 \quad (i = 1, 2,..., 6)$$

represent six points P_i, a necessary and sufficient condition for the six points to lie on a conic is that the six polynomials P_i^2 in u, v, w should be linearly dependent. Dually, six lines $p_i = 0$ are all tangents to a conic if and only if the polynomials p_i^2 in x, y, z are linearly dependent.

Proof. The polynomials P_i^2 are linearly dependent if and only if there exist numbers $\lambda_1, \lambda_2, \ldots, \lambda_6$, not all zero, such that

$$\sum_{i=1}^{6} \lambda_i P_i^2 \equiv 0$$

identically in u, v, w, i.e. if and only if the following equations for $\lambda_1, \ldots, \lambda_6$ are consistent:

$$\sum_i \lambda_i x_i^2 = 0, \ldots; \qquad \sum_i \lambda_i y_i z_i = 0, \ldots .$$

A necessary and sufficient condition for this is

$$\begin{vmatrix} x_1^2 & x_2^2 & . & . & . & x_6^2 \\ . & . & . & . & . & . \\ . & . & . & . & . & . \\ y_1 z_1 & y_2 z_2 & . & . & . & y_6 z_6 \\ . & . & . & . & . & . \\ . & . & . & . & . & . \end{vmatrix} = 0;$$

and this is just the familiar condition for the six points (x_i, y_i, z_i) to lie on a conic.

COROLLARY 1. *If two triangles are self-polar for a conic, their six vertices lie on a conic.*

For if $$S \equiv \lambda_1 P_1^2 + \lambda_2 P_2^2 + \lambda_3 P_3^2,$$

and $$S \equiv \lambda_4 P_4^2 + \lambda_5 P_5^2 + \lambda_6 P_6^2,$$

then $$\lambda_1 P_1^2 + \lambda_2 P_2^2 + \lambda_3 P_3^2 - \lambda_4 P_4^2 - \lambda_5 P_5^2 - \lambda_6 P_6^2 \equiv 0.$$

We thus have an alternative proof of Theorem 22 of Chapter VI.

COROLLARY 2. *If two triangles are both inscribed in a conic k, there exists a conic for which they are both self-polar.*

COROLLARY 3. *If a polar k_1-point and a polar k_2-point for a conic have $k_1 + k_2 - 6$ points in common, then their six vertices lie on a conic.*

§ 3. The Theory of Invariants

In the previous section we expressed in algebraic form $I_{s,s'} = 0$ a projective relation that may possibly hold between two conics, and we inferred (p. 187) that if the algebraic relation holds for the equations of s and s' in any one coordinate representation it must hold for the equations in every allowable representation. This suggests the possibility of a general algebraic investigation of invariant relations of this kind, of which the aim would be, on the one hand, the systematic construction and classification of such relations and, on the other, their geometrical interpretation.

We can only give here a very brief introduction to the subject. The algebra has been worked out fully in the classical theory of invariants, where a technique is developed for handling invariant algebraic relations symbolically; and for the full development we refer the reader who is interested to the following standard works— Grace and Young, *Algebra of Invariants* (Cambridge, 1903); Elliott, *Algebra of Quantics* (Oxford, 1895); Weitzenböck, *Invariantentheorie* (Groningen, 1923).

The discriminant of a quadratic form

As a first example of an invariant relation we may take the condition that a quadratic form should factorize. Consider the form

$$S \equiv (a,b,c,f,g,h \between x,y,z)^2 \equiv \mathbf{x}^T\mathbf{A}\mathbf{x}.$$

The condition for the form to factorize, i.e. for the conic $S = 0$ to be degenerate, is of course $|\mathbf{A}| = 0$. If, now, we change the coordinates according to the scheme

$$\mathbf{x} = \mathbf{P}\bar{\mathbf{x}} \quad (|\mathbf{P}| \neq 0),$$

S is transformed into the quadratic form

$$\bar{S} \equiv \bar{\mathbf{x}}^T\mathbf{P}^T\mathbf{A}\mathbf{P}\bar{\mathbf{x}} = \bar{\mathbf{x}}^T\overline{\mathbf{A}}\bar{\mathbf{x}},$$

where $\overline{\mathbf{A}} = \mathbf{P}^T\mathbf{A}\mathbf{P}$. Clearly, then, $|\overline{\mathbf{A}}| = |\mathbf{P}|^2.|\mathbf{A}|$; and the two discriminants $|\mathbf{A}|$ and $|\overline{\mathbf{A}}|$ are either both zero or both non-zero, as we should expect.

The typical features of this example, from the present point of view, are the following:

(*a*) a geometrical condition is expressed by the vanishing of a certain polynomial in the coefficients of a form;

(*b*) when the variables in the form are subjected to a linear transformation, the value of the polynomial with the new coefficients substituted for the old is equal to the original value multiplied by a power of the determinant of the transformation.

If we now replace the single form by two or more forms in the same set of variables, we have a useful indication of what may be looked for in other cases.

Invariants of a pair of quadratic forms

Let $S \equiv \mathbf{x}^T\mathbf{A}\mathbf{x}$ and $S' \equiv \mathbf{x}^T\mathbf{A}'\mathbf{x}$ be two quadratic forms in x, y, z, which transform respectively into the forms $\bar{S} \equiv \bar{\mathbf{x}}^T\bar{\mathbf{A}}\bar{\mathbf{x}}$ and $\bar{S}' \equiv \bar{\mathbf{x}}^T\bar{\mathbf{A}}'\bar{\mathbf{x}}$. The general linear combination of S and S' may be written as

$$S_{\lambda,\mu} \equiv \lambda S+\mu S' \equiv \mathbf{x}^T(\lambda\mathbf{A}+\mu\mathbf{A}')\mathbf{x},$$

and this form transforms into

$$\bar{S}_{\lambda,\mu} \equiv \bar{\mathbf{x}}^T\mathbf{P}^T(\lambda\mathbf{A}+\mu\mathbf{A}')\mathbf{P}\bar{\mathbf{x}} \equiv \bar{\mathbf{x}}^T(\lambda\bar{\mathbf{A}}+\mu\bar{\mathbf{A}}')\bar{\mathbf{x}}.$$

It follows that

$$|\lambda\bar{\mathbf{A}}+\mu\bar{\mathbf{A}}'| = |\mathbf{P}|^2 . |\lambda\mathbf{A}+\mu\mathbf{A}'|, \tag{1}$$

and this holds for arbitrary values of λ and μ, i.e. identically in λ and μ.

Now

$$|\lambda\mathbf{A}+\mu\mathbf{A}'| \equiv \begin{vmatrix} \lambda a+\mu a' & \lambda h+\mu h' & \lambda g+\mu g' \\ \lambda h+\mu h' & \lambda b+\mu b' & \lambda f+\mu f' \\ \lambda g+\mu g' & \lambda f+\mu f' & \lambda c+\mu c' \end{vmatrix}$$
$$\equiv \Delta\lambda^3+\Theta\lambda^2\mu+\Theta'\lambda\mu^2+\Delta'\mu^3,$$

where $\Delta = |\mathbf{A}|$, $\Delta' = |\mathbf{A}'|$, and Θ, Θ' are certain intermediate expressions. We have in fact

$$3\Delta = aA+bB+cC+2fF+2gG+2hH,$$
$$\Theta = a'A+b'B+c'C+2f'F+2g'G+2h'H,$$
$$\Theta' = aA'+bB'+cC'+2fF'+2gG'+2hH',$$
$$3\Delta' = a'A'+b'B'+c'C'+2f'F'+2g'G'+2h'H'.$$

Since (1) is an identity in λ and μ, we have at once $\bar{\Delta} = |\mathbf{P}|^2\Delta$, $\bar{\Theta} = |\mathbf{P}|^2\Theta$, $\bar{\Theta}' = |\mathbf{P}|^2\Theta'$, $\bar{\Delta}' = |\mathbf{P}|^2\Delta'$. Thus not only Δ and Δ', but also Θ and Θ', which are mixed polynomials in the coefficients of S and S', transform in the special manner described in (*b*) on p. 192. They also have, as it happens, a geometrical interpretation as demanded by (*a*); for Θ and Θ' are the expressions $I_{s',s}$ and $I_{s,s'}$, and their vanishing means that the conic $S = 0$ is inpolar or outpolar to the conic $S' = 0$ as the case may be.

The four expressions Δ, Θ, Θ', Δ' are the four fundamental *invariants* (or *relative invariants*) of S and S'. Each of them is homogeneous both in the coefficients of S and in the coefficients of S'; and since in each case the factor $|\mathbf{P}|$ occurs to the second power on the right-hand side of the equation giving the law of transformation, all four invariants are said to be of *weight* 2.

More generally, we say that a polynomial $\Phi(a,b,\dots,h;a',b',\dots,h')$ in the coefficients of S and S' is an invariant of weight w of the two quadratic forms if

(i) Φ is homogeneous in $a,b,\dots,h$ and in $a',b',\dots,h'$, and

(ii) $\Phi(\bar{a},\bar{b},\dots,\bar{h};\bar{a}',\bar{b}',\dots,\bar{h}') = |\mathbf{P}|^{w}.\Phi(a,b,\dots,h;a',b',\dots,h')$.

A polynomial in the four fundamental invariants, with constant coefficients, say

$$\Phi \equiv \sum_{p,q,r,s} \lambda_{p,q,r,s}\Delta^{p}\Theta^{q}\Theta'^{r}\Delta'^{s},$$

will satisfy (ii) if and only if it is homogeneous, of degree n say, in Δ, Θ, Θ', Δ'. When this is the case $p+q+r+s = n$ for every term of the above expression, and clearly $\bar{\Phi} = |\mathbf{P}|^{2n}.\Phi$. The function Φ will not satisfy (i), however, unless $3p+2q+r$ (and consequently also $q+2r+3s$) is constant for all the terms. When this condition is not satisfied, we can write Φ as a sum $\sum \Phi_i$ of proper invariants Φ_i, all of weight $2n$, by collecting together in one term Φ_i all the terms of Φ for which $3p+2q+r$ has the same value. If Φ satisfies (ii) but not (i), we call it a *pseudo-invariant* of S and S'; and any pseudo-invariant may thus be expressed as a sum of proper invariants.

We may now extend the concept of invariant by admitting not only polynomials Φ in the coefficients of S and S' but also rational functions Φ/Ψ. If Φ and Ψ are invariant polynomials, of weights w_1 and w_2 respectively, we say that the rational function Φ/Ψ is an invariant of weight w_1-w_2 of S and S'. The weight of such an invariant (which must be integral) can be positive, negative, or zero. If it is zero, Φ/Ψ is called an *absolute invariant.*

Invariants of a pair of conics

The algebraic results just obtained may now be interpreted geometrically in terms of the projective geometry of a pair of conics. Let us consider two given conics s and s', represented in some chosen representation $\mathscr{R}$ by point-equations $S = 0$ and $S' = 0$. It should be noted first of all that the conics are associated in this way, not with uniquely defined quadratic forms S and S', but with classes of forms (kS) and $(k'S')$, where k, k' are arbitrary constants. The projective properties of the pair of conics may be expected, therefore, to be reflected in invariant properties of these two classes of forms.

The projective properties of the pair of conics are of two different kinds. In the first place we have *projective relations* of s to s', of which the relation of apolarity is typical; and in the second place, there are numerical invariants or *geometrical moduli* of the two conics, such as, for example, the cross ratio (or some symmetric function of the six cross ratios) of the range of four points cut on s by s'.

The vanishing of any invariant of S and S' clearly corresponds to some projective relation between s and s', for since an invariant of S and S' is homogeneous in the coefficients of the two forms separately, it is also an invariant of kS and $k'S'$ for arbitrary k and k'. No such meaning, nor indeed any meaning at all in terms of the two conics, attaches to the vanishing of a pseudo-invariant of S and S'.

The numerical value of any absolute invariant of S and S' corresponds in a similar manner to a geometrical modulus of s and s', provided that numerator and denominator are of the same degree in the coefficients of S and also in the coefficients of S'.

Applying these results to the known invariants of S and S', we have:

THEOREM 9. *Any algebraic equation* $F(\Delta,\Theta,\Theta',\Delta') = 0$ *represents a projective relation between the conics* $S = 0$ *and* $S' = 0$, *provided that* F *is homogeneous separately in* $a,b,\ldots,h$ *and in* $a',b',\ldots,h'$, *or, what comes to the same thing, provided that* F *is* (*a*) *homogeneous in* Δ, Θ, Θ', Δ', *and* (*b*) *homogeneous also when weights* 3, 2, 1, 0 *are assigned to* Δ, Θ, Θ', Δ'.

Any rational function $F(\Delta,\Theta,\Theta',\Delta')/G(\Delta,\Theta,\Theta',\Delta')$ *represents a projective modulus of the two conics provided that* (*a*) *numerator and denominator are of the same total degree in* Δ, Θ, Θ', Δ', *and* (*b*) *they are both homogeneous and of the same degree in* a, $b,\ldots$, h *and also in* $a',b',\ldots,h'$.

The two simplest absolute invariants of s and s' are

$$\alpha = \frac{\Delta\Theta'}{\Theta^2}, \qquad \beta = \frac{\Delta'\Theta}{\Theta'^2},$$

and it can be shown that any other absolute invariant which is derived from the four fundamental invariants is a rational function of α and β, and also that any irreducible invariant relation (other than $\Theta = 0$ or $\Theta' = 0$) derived in the same way can be expressed in the form $\phi(\alpha,\beta) = 0$, where ϕ is a polynomial. All such relations may therefore be regarded as algebraic relations between the two basic moduli of s and s'. The proof of these statements is left to the reader. It is worth noting, incidentally, that there appear to be no immediate geometrical interpretations, based on cross ratios, of α and β themselves, although many of the obvious moduli of s and s' can be expressed easily enough in terms of these two.

Examples of invariant relations

We have now, in Theorem 9, the means at our disposal for constructing infinitely many types of invariant relation between two conics s and s'. Unfortunately we have no systematic way of interpreting them geometrically, and we might be inclined to say that the algebra has suddenly out-distanced the geometry. What we can do, however, is to select those geometrical relationships between s and s' which interest us and try to express them, by a systematic procedure, in invariant algebraic form. The procedure is as follows. Having settled upon a geometrical relation, we take two conics s and s' which we assume to be connected by this relation only. We then choose a system of reference in which the conics have simple equations, compute for these equations the values of $\Delta, \Theta, \Theta', \Delta'$, and search for a relation between these quantities which satisfies the homogeneity conditions of Theorem 9. If the assumed geometrical relation between s and s' is in fact one which imposes only one condition on the coefficients, and if an invariant relation is found which is irreducible, then this latter is the required algebraic equivalent of the original condition, valid in every coordinate system. The two examples which follow will make the procedure clear.

EXAMPLE 1. *Let s be triangularly circumscribed to s'.* In this case, by suitable choice of the frame of reference, the equations of s and s' can be taken to be

$$S \equiv 2fyz+2gzx+2hxy = 0,$$
$$S' \equiv x^2+y^2+z^2-2yz-2zx-2xy = 0,$$

respectively. Then the values of $A, B,..., H$ are $-f^2, -g^2, -h^2, gh, hf, fg$ and those of $A', B',..., H'$ are 0, 0, 0, 2, 2, 2; and hence

$$\Delta = 2fgh, \qquad \Theta = -(f+g+h)^2, \qquad \Theta' = 4(f+g+h), \qquad \Delta' = -4.$$

The only relation between these quantities which is homogeneous both for weights 1, 1, 1, 1 and for weights 3, 2, 1, 0 of $\Delta, \Theta, \Theta', \Delta'$ is

$$\Theta'^2-4\Delta'\Theta = 0,$$

i.e. $\beta = \frac{1}{4}$; and this is therefore the required invariant relation, valid for every frame of reference.

EXAMPLE 2. *Let s and s' be such that two of their common tangents intersect on a common chord.* When this relation holds, the conics are projectively equivalent to a pair of equal circles; and their equations may be written non-homogeneously as $X^2+Y^2 = 1$ and $(X-d)^2+Y^2 = 1$. Then

$$(a,b,...,h) = (1,1,-1,0,0,0),$$
$$(a',b',...,h') = (1,1,d^2-1,0,-d,0),$$
$$(A,B,...,H) = (-1,-1,1,0,0,0),$$
$$(A',B',...,H') = (d^2-1,-1,1,0,d,0);$$

and hence

$$\Delta = -1, \qquad \Theta = d^2-3, \qquad \Theta' = d^2-3, \qquad \Delta' = -1.$$

To pick out a relation between these quantities which satisfies the necessary homogeneity conditions, it is perhaps simplest to replace them by $\Delta = -k^3$, $\Theta = (d^2-3)k^2$, $\Theta' = (d^2-3)k$, $\Delta' = -1$, and then to seek a homogeneous relation between these new values which is independent of k. In this way we obtain the relation

$$\Delta\Theta'^3-\Delta'\Theta^3 = 0,$$

or, in terms of the absolute invariants, $\alpha = \beta$.

The general theory of invariants

It may be useful at this stage to refer to some more general concepts and results which throw light on what has been said so far, and to indicate very briefly the scope of the general theory of invariants.

We need to define an invariant of a set of forms (i.e. homogeneous polynomials) in $n+1$ indeterminates $x_0,..., x_n$, where the forms are not necessarily assumed to be of the same degree. Now a form of specified type is essentially an array of coefficients, ordered according to the terms to which they belong; and this is what Weitzenböck calls a tensor. We may regard the coefficients in the form as themselves indeterminates, treating the original $x_0,..., x_n$ as auxiliary quantities which serve merely as a device for transforming the given form into another form of the same degree; for the original transformation $\mathbf{x} = \mathbf{P}\bar{\mathbf{x}}$ induces a linear transformation of the coefficients of the form, and the coefficients in this transformation depend only upon the elements of the matrix $\mathbf{P}$. To take a simple example, let S be the quadratic form $\mathbf{x}^T\mathbf{A}\mathbf{x}$, i.e.

$$\sum_{i=0}^{n}\sum_{k=0}^{n} a_{ik}x_i x_k \quad (a_{ki} = a_{ik}).$$

This form defines the symmetric tensor (a_{ik}), and the induced linear transformation of the components is given by $\overline{\mathbf{A}} = \mathbf{P}^T\mathbf{A}\mathbf{P}$, i.e.

$$\mathbf{A} = \mathbf{P}^{-1T}\overline{\mathbf{A}}\mathbf{P}^{-1}.$$

We may say then, in general terms, that every form S (with indeterminate coefficients) defines a tensor or array (a) of coefficients, and that the group $GL(n+1)$ of all non-singular linear transformations $\mathbf{x} = \mathbf{P}\bar{\mathbf{x}}$ gives rise to a group G_S of linear transformations of the array (a). An invariant of a set of forms $S^{(1)},..., S^{(r)}$ may now be defined as a polynomial $\Phi(a^{(1)},..., a^{(r)})$ in all the coefficients of all the forms which has the following properties:

(i) it is homogeneous in the coefficients of each form separately;

(ii) when $S^{(1)},..., S^{(r)}$ are transformed into $\bar{S}^{(1)},..., \bar{S}^{(r)}$ by any given transformation $\mathbf{x} = \mathbf{P}\bar{\mathbf{x}}$ of $x_0,..., x_n$ then

$$\Phi(\bar{a}^{(1)},..., \bar{a}^{(r)}) = K\Phi(a^{(1)},..., a^{(r)}),$$

where K is a quantity which depends only on the matrix $\mathbf{P}$.

It may be proved algebraically† that if Φ is such an invariant then $K = |\mathbf{P}|^w$, where w is a non-negative integer. Φ is accordingly referred to as an invariant of weight w of the r given forms. Similar definitions apply also to invariant rational functions Φ/Ψ.

The fundamental theorem on invariants is the Basis Theorem,‡ due to Gordan. This asserts that if $S^{(1)},..., S^{(r)}$ are arbitrary forms of any assigned degrees in $x_0,..., x_n$, then there exists a finite set of invariants $I_1,..., I_t$, such that every invariant of the forms is expressible as a polynomial in $I_1,..., I_t$.

For a single quadratic form S in x, y, z, the invariant Δ is itself the basis. In the case of a pair of such forms, S and S', it can be shown§ that $\Delta, \Theta, \Theta', \Delta'$ form a basis; and this means that the invariants constructed from these, as described on p. 194, are all the invariants of S and S'.

We refer finally to the formal algebraic operation of *polarization*, which is a simple but extremely powerful instrument for deriving new invariants from those that are already known. The operation may be applied to any homogeneous polynomial $\phi(a_1,..., a_k)$, of degree n, say, in a set of indeterminates $a_1,..., a_k$, and it is defined as follows. We take a second set of k indeterminates, $a'_1,..., a'_k$, and then convert ϕ into the polynomial

$$\phi^*(a_1,..., a_k; a'_1,..., a'_k) \equiv \frac{1}{n}\sum_{i=1}^{k} a'_i \frac{\partial\phi}{\partial a_i}$$

† Grace and Young, p. 21; Weitzenböck, p. 11.

‡ The original proof of Gordan, later simplified by Hilbert, is discussed by Weitzenböck, pp. 143–8.

§ Weitzenböck, p. 61.

by means of the polarizing operator

$$\Omega \equiv \frac{1}{n}\sum_{i=1}^{k} a_i' \frac{\partial}{\partial a_i}.$$

The usefulness of this operation in the theory of invariants arises from the fact—which the reader should have little difficulty in verifying—that if ϕ is an invariant of r forms of which one, say S, has $a_1,\ldots, a_k$ as coefficients, and if S' is a new form derived from S by replacing the coefficients $a_1,\ldots, a_k$ respectively by $a_1',\ldots, a_k'$, then ϕ^* is necessarily an invariant of the $r+1$ forms obtained by adding S' to the original set of r (or of the original set, if this already includes S').

An example of this process lies ready to hand, for polarization of the invariant Δ of the quadratic form $S(x,y,z)$ gives us the invariant $\frac{1}{3}\Theta$ of the two forms S and S'. The invariant Θ is itself quadratic in $a, b,\ldots, h$ and linear in $a', b',\ldots, h'$, and if we now polarize Θ (with respect to the pair of forms S, S') we change it into an invariant that is linear in $a, b,\ldots, h$ and quadratic in $a', b',\ldots, h'$, namely Θ'. Finally, polarization of Θ' gives the invariant $3\Delta'$. From the single invariant Δ of S, therefore, we can obtain by successive polarization all the four fundamental invariants of S and S'. The same procedure also gives us a new invariant Φ of three quadratic forms S, S', S'', obtained by polarizing Θ with respect to the pair of forms S, S''. Φ is linear in the coefficients of each of the forms.

Covariants and contravariants

We have now seen how it is possible to define and manipulate algebraically the invariants of a set of forms, and we have also given some examples of the application of this general theory to the projective geometry of conics. In the applications so far envisaged, the conics have been regarded primarily as loci; but it would be a simple matter to give a dual treatment of properties of conic envelopes, simply by applying the same algebraic theory to quadratic forms in u, v, w instead of x, y, z. Unfortunately, however, the self-duality of the projective plane, with properties of points and properties of lines symmetrically related to each other, cannot readily be exhibited by means of the theory of invariants, In applying this theory we have to begin by taking either point-coordinates or line-coordinates as the primary variables, and this choice gives a bias to the whole of the subsequent development. If we regard x, y, z as primary, all the transformations $a \to \bar{a}$ of coefficients are defined in terms of the one basic transformation

$$\mathbf{x} = \mathbf{P}\bar{\mathbf{x}}.$$

This transformation of point-coordinates induces an associated transformation of line-coordinates, represented algebraically by the so-called contragredient transformation

$$\mathbf{u} = \mathbf{P}^{-1T}\bar{\mathbf{u}},$$

but the elements p_{ik} of $\mathbf{P}$ enter differently into the two equations, and in consequence the algebra fails to reflect the symmetry of the geometry. We do not wish, however, to go into this question here, but rather to look for a moment at a generalization of invariant theory that is possible in a different direction.

In projective geometry we are often interested in studying not only properties of given figures but also new figures that are projectively related to given ones, or possibly derivable from them by projective construction. The harmonic locus and harmonic envelope of two conics, for instance, are cases in point. A derived figure of this kind will have an equation which involves point-coordinates or line-coordinates as well as the coefficients in the equations of the original figures, and this equation will of necessity be related in some invariant manner to the original equations. The question is how to exhibit algebraically the invariant relation —and this brings us to the notion of concomitants.

Roughly speaking, a *concomitant* of a set of forms

$$S^{(1)}(x_0,\ldots, x_n),\ldots, S^{(r)}(x_0,\ldots, x_n)$$

is an 'invariant' which involves not only the coefficients in the $S^{(i)}$ but also point-coordinates and dual coordinates as well. It has the property that when the coefficients in the $S^{(i)}$ are replaced by the corresponding coefficients in the $\bar{S}^{(i)}$, and $x_0,\ldots, x_n$ and $u_0,\ldots, u_n$ by the quantities $\bar{x}_0,\ldots, \bar{x}_n$ and $\bar{u}_0,\ldots, \bar{u}_n$ defined by the equations $\mathbf{x} = \mathbf{P}\bar{\mathbf{x}}$, $\mathbf{u} = \mathbf{P}^{-1T}\bar{\mathbf{u}}$, its value is unaltered except possibly for multiplication by some power of $|\mathbf{P}|$. If the concomitant involves $x_0,\ldots, x_n$ but not $u_0,\ldots, u_n$ it is called a *covariant* of $S^{(1)},\ldots, S^{(r)}$; and if it involves $u_0,\ldots, u_n$ but not $x_0,\ldots, x_n$ it is called a *contravariant*. If it involves neither set of variables it is, of course, an invariant in the sense previously defined.

In order to make these ideas more concrete let us now consider a single form

$$S(x,y,z) \equiv (a,b,c,f,g,h \between x,y,z)^2.$$

The transformed coefficients $\bar{a}, \bar{b},..., \bar{h}$ are defined by the condition

$$(\bar{a}, \bar{b}, \bar{c}, \bar{f}, \bar{g}, \bar{h} \between \bar{x}, \bar{y}, \bar{z})^2 \equiv (a, b, c, f, g, h \between x, y, z)^2,$$

and this means that S is itself a covariant of S of weight zero.

We write the line-equation associated with the point-equation $S = 0$ as

$$\Sigma(u, v, w) \equiv (A, B, C, F, G, H \between u, v, w)^2 = 0,$$

or, in matrix notation,

$$\Sigma \equiv |\mathbf{A}| . \mathbf{u}^T \mathbf{A}^{-1} \mathbf{u} = 0.$$

Then, if $\bar{\Sigma}$ denotes the form obtained by putting $\bar{u}, \bar{v}, \bar{w}$ and $\bar{a}, \bar{b},..., \bar{h}$ in place of u, v, w and $a, b,..., h$ in Σ,

$$\begin{aligned} \bar{\Sigma} &\equiv |\bar{\mathbf{A}}| . \bar{\mathbf{u}}^T \bar{\mathbf{A}}^{-1} \bar{\mathbf{u}} \\ &\equiv |\mathbf{P}|^2 . |\mathbf{A}| . (\mathbf{P}^T \mathbf{u})^T (\mathbf{P}^T \mathbf{A} \mathbf{P})^{-1} (\mathbf{P}^T \mathbf{u}) \\ &\equiv |\mathbf{P}|^2 . |\mathbf{A}| . \mathbf{u}^T \mathbf{A}^{-1} \mathbf{u} \\ &\equiv |\mathbf{P}|^2 . \Sigma; \end{aligned}$$

and Σ is therefore a contravariant of S of weight 2.

If, now, we begin with two conics s, s', represented by point-equations $S = 0$, $S' = 0$, they serve to define a pencil of conics; and the general conic s_λ of this pencil has point-equation

$$S + \lambda S' = 0$$

and line-equation $\Sigma + 2\lambda\Phi + \lambda^2\Sigma' = 0$. Σ may be regarded as a quadratic form in $a, b,..., h$ whose coefficients involve u, v, w, and Φ is obtained from this quadratic form by polarization with respect to $a', b',..., h'$. If fixed values are given to u, v, w, the roots λ_1, λ_2 of the quadratic equation $\Sigma + 2\lambda\Phi + \lambda^2\Sigma' = 0$ are the parameters of the two conics of the pencil which touch the line (u, v, w); and since when the linear substitutions $\mathbf{x} = \mathbf{P}\bar{\mathbf{x}}$ and $\mathbf{u} = \mathbf{P}^{-1T}\bar{\mathbf{u}}$ are made simultaneously the parameters of the conics which touch the line remain unaltered, the equations $\Sigma + 2\lambda\Phi + \lambda^2\Sigma' = 0$ and $\bar{\Sigma} + 2\lambda\bar{\Phi} + \lambda^2\bar{\Sigma}' = 0$ have the same roots. We have already shown, however, that $\bar{\Sigma} = |\mathbf{P}|^2\Sigma$; and therefore $\bar{\Phi} = |\mathbf{P}|^2\Phi$—i.e. Φ is a contravariant of weight 2 of S and S'. Thus the line-equation $\Phi = 0$ represents a conic envelope that is projectively related to s and s'. This envelope is actually the harmonic envelope of s and s', as we may show by the following argument.

Let (u, v, w) be a line, selected once for all, and let the matrix $|\mathbf{P}|$ be chosen in such a way that the transformed coordinates

$(\bar{u}, \bar{v}, \bar{w})$ of this line are $(1, 0, 0)$. Then, in the new coordinate representation, the pairs of points of intersection of the line with s and s' are given by the equations

$$\bar{b}\bar{y}^2+2\bar{f}\bar{y}\bar{z}+\bar{c}\bar{z}^2 = 0 = \bar{x},$$

and

$$\bar{b}'\bar{y}^2+2\bar{f}'\bar{y}\bar{z}+\bar{c}'\bar{z}^2 = 0 = \bar{x},$$

respectively. The pairs are therefore harmonic if and only if

$$\bar{b}\bar{c}'+\bar{b}'\bar{c}-2\bar{f}\bar{f}' = 0,$$

i.e. if and only if $\bar{\Phi}(1, 0, 0) = 0$; and, in view of the contravariance of Φ, this condition is equivalent to $\Phi(u, v, w) = 0$.

By duality, the point-equation of the harmonic locus of s and s' is $F = 0$, where $S+2\mu F+\mu^2 S' = 0$ is the point-equation associated with the line-equation $\Sigma+\mu\Sigma' = 0$.

What has been said on the preceding pages may serve as an introduction to the classical theory of invariants, with particular reference to its application to the projective geometry of conics. The theory admits of immediate extension to quadrics, and the invariants which play a central part there are analogous to those which we have met above. We shall not give details of this extension, but refer the reader who is interested to Chapter VII of Todd, *Projective and Analytical Geometry*, where a full account is to be found.

EXERCISES ON CHAPTER VIII

1. Show that the equation of any (2, 2) algebraic correspondence between parameters θ and θ' can be written in matrix form $\boldsymbol{\theta}'^T\mathbf{A}\boldsymbol{\theta} = 0$, where $\boldsymbol{\theta}$ and $\boldsymbol{\theta}'$ are the column-vectors whose components are $\theta^2, \theta, 1$ and $\theta'^2, \theta', 1$ respectively, and $\mathbf{A}$ is a fixed 3×3 matrix. Show also that the correspondence is symmetrical if and only if $\mathbf{A}$ is a symmetric matrix.

If $\mathbf{A}$ is symmetric, and $\boldsymbol{\theta}$ and $\boldsymbol{\theta}'$ are interpreted as the coordinate vectors of points P, P' of the conic $x_1^2 = x_2 x_0$, show that P and P' are conjugate with respect to the conic $\mathbf{x}^T\mathbf{A}\mathbf{x} = 0$.

2. A triangle is circumscribed to a fixed conic s, and two of its vertices lie on another fixed conic s'. Show that the locus of the third vertex is a conic, and find the equation of this conic when the equations of s and s' are

$$x^2+y^2+z^2 = 0 \quad \text{and} \quad ax^2+by^2+cz^2 = 0$$

respectively.

3. If the normals at points P, P' of a central conic meet in a point of the conic, different, in general, from P and P', determine the nature of the correspondence between P and P', and prove that PP' envelops a conic.

4. A variable conic through three fixed points A, B, C meets a fixed conic in P, Q, R, S. If PQ passes through a fixed point, prove that in general RS envelops a conic inscribed in the triangle ABC.

5. The tangents to a conic t from a variable point P of a conic s meet s again at Q and R. Prove that QR envelops a conic k, that k touches the second tangent to t from a point of contact of s with a common tangent of s and t, and that k passes through the intersections of s and t.

State the duals of these results.

6. A variable chord PP' of a conic s envelops a second conic s' and is such that P and P' are conjugate for a third conic s''. Show that s is quadrilaterally circumscribed to s' if and only if s'' is a line-pair, and that in this case s'' is a pair of tangents to s' whose intersection has the same polar for s as for s'.

Give a euclidean interpretation of this result when s' and s are a conic and its director circle.

7. Show that the harmonic envelope of two orthogonal circles consists of the two pencils of lines whose vertices are the centres of the circles. Show also that the harmonic locus is a pair of lines perpendicular to the line of centres.

8. If A, B, C, D are concyclic, show that the focus of the parabola which touches AB, BC, CD, DA is the foot of the perpendicular from the intersection of BD and AC on to the line joining the other diagonal points of the quadrangle $ABCD$.

9. If P, Q correspond in a symmetrical $(2, 2)$ algebraic correspondence on a conic k, prove that PQ touches a fixed conic s and P, Q are conjugate with respect to another fixed conic t.

If T is the pole of PQ for k, prove that the point of contact of PQ with s is the harmonic conjugate, with respect to P and Q, of the intersection of PQ with the polar of T for t.

10. If the variables θ and ϕ are connected by a $(2, 2)$ algebraic correspondence, show that the four critical values of θ and the four critical values of ϕ are homographically related when they are paired together suitably.

11. Two conics k, k' are such that k is outpolar to k'. Prove that the points of contact of tangents to k' from any point of k are conjugate with respect to k.

12. Show that a parabola is outpolar to a circle if and only if the diameter of the circle which is perpendicular to the axis of the parabola is divided harmonically by the parabola.

13. Show that the harmonic locus of two conics s and s' is a rectangular hyperbola if and only if the director circle of s is outpolar to s'.

14. Being given a quadrilateral $abcd$, prove that the orthocentres of the four triangles abc, abd, acd, bcd and the circumcentre of the diagonal triangle of the quadrilateral lie on a line perpendicular to the line joining the mid-points of the three diagonals.

Show that two rectangular hyperbolas can be inscribed in the quadrilateral, and that they intersect in the incentre and the three excentres of the diagonal triangle.

15. If two conics k, k' have double contact and are each outpolar to the same conic t, prove that their chord of contact touches t.

16. Two conics s, s' touch a conic k at the same point A, and both are outpolar to k. Show that s and s' have three-point contact at A.

If a conic is required to have three-point contact with k at A and also to be outpolar to k, show that it must break up into the common tangent at A and another line through A.

17. Show that if a conic k is both outpolar and inpolar to a conic k' then it is also triangularly circumscribed to k'.

18. A circle c, whose centre is C, is outpolar to a central conic k. If T is the point of contact of c with a common tangent to c and k, supposed distinct from the point of contact of this tangent with k, prove that CT lies along a principal axis of any conic that is outpolar to k and has three-point contact with c at T.

19. If two of the common tangents of two conics divide a third common tangent harmonically, prove that they also divide the fourth common tangent harmonically. Show that, for two conics which are related in this way, $\Delta\Delta' = \Theta\Theta'$. Show, further, that two conics which are related in this way are also related in the dual way.

20. If a conic s is quadrilaterally circumscribed to a conic s', show that their mutual invariants satisfy the relation $8\Delta\Delta'^2 - 4\Theta\Theta'\Delta' + \Theta'^3 = 0$.

21. If two vertices of a triangle which is self-polar for the conic $S = 0$ lie on the conic $S' = 0$, show that the locus of the third vertex is the conic

$$\Theta S - \Delta S' = 0.$$

22. Show that, with the usual notation, the point-equation of the harmonic envelope of the conics whose equations are $S = 0$ and $S' = 0$ is

$$\Theta S' + \Theta' S - 2F = 0.$$

23. Show that the polar reciprocal of s with respect to s' has the covariant equation $\Theta S' - 2F = 0$.

CHAPTER IX

TRANSFORMATIONS OF THE PLANE

§ 1. PLANE COLLINEATIONS

WE have seen already, in Chapter II, § 7, that a set of algebraic equations

$$x'_i = \sum_{k=0}^{2} a_{ik} x_k \quad (i = 0, 1, 2),$$

which defines a non-singular linear transformation, admits of two distinct geometrical interpretations:

(i) as a transformation of coordinates from one projective system to another;

(ii) as a projective transformation of points, referred to a fixed frame of reference.

In the development of the formal theory we have concentrated so far upon the first interpretation, and have treated equations (1) as specifying the transformation from one allowable representation $\mathscr{R}$ to a second such representation $\mathscr{R}'$. In the present chapter we shall turn to the second interpretation, and see how projective transformations of points—or collineations, as we shall now call them—also have an important place in the formal system.

Let $\mathscr{R}$, $\mathscr{R}'$ be allowable representations in two distinct projective planes S_2, S'_2, and let

$$x'_i = \sum_{k=0}^{2} a_{ik} x_k \quad (i = 0, 1, 2)$$

or

$$\mathbf{x}' = \mathbf{A}\mathbf{x},$$

be a fixed transformation, making correspond to the point $\mathbf{x}$ of S_2 the point $\mathbf{x}'$ of S'_2. Such a transformation will be called a *collineation*.

We may also consider self-collineations of the single plane S_2, obtained by letting S'_2 coincide with S_2. In this case it is usual to take the same representation as both $\mathscr{R}$ and $\mathscr{R}'$.

If $|\mathbf{A}| \neq 0$, the transformation $\mathbf{x}' = \mathbf{A}\mathbf{x}$ has an inverse

$$\mathbf{x} = \mathbf{A}^{-1}\mathbf{x}',$$

and every point of S'_2 then arises from a unique point of S_2. The collineation thus sets up a (1, 1) correspondence between the two planes. Although singular collineations are not without geometrical interest and significance, it is desirable at this stage to avoid making the theory more complicated than is absolutely necessary, and for this reason we shall confine ourselves exclusively to non-singular collineations. *Whenever we use the word 'collineation' without qualification it is to be understood as meaning 'non-singular collineation'.*

When a collineation ϖ is represented by an equation $\mathbf{x}' = \mathbf{A}\mathbf{x}$, the matrix $\mathbf{A}$ defines the collineation uniquely, but the collineation only defines the matrix to within a scalar factor; i.e. $\mathbf{A}_1$ and $\mathbf{A}_2$ define the same collineation if and only if $\mathbf{A}_1 = \lambda\mathbf{A}_2$ for some value of λ.

Collineations between S_2 and S_2', as we have defined them, are the simplest generalization of homographic correspondences between lines S_1 and S_1'; but they are not, as it happens, the only simple generalization that can usefully be considered. The transformation

$$x_0' : x_1' : x_2' = x_1 x_2 : x_2 x_0 : x_0 x_1,$$

for example, is not a collineation; but it is algebraic, and it is (1, 1) everywhere except for points on the sides of the two reference triangles (cf. the characterization of homographies on p. 176). Transformations of this kind—to be considered in § 3—are said to be (1, 1) *in general*, i.e. except on certain specified curves. The principal geometric property that distinguishes collineations inside this wider class of transformations is that they transform lines into lines.

To show this we consider a general collineation ϖ, given by

$$\mathbf{x} \to \mathbf{x}' = \mathbf{A}\mathbf{x} \quad (|\mathbf{A}| \neq 0).$$

If the point $\mathbf{x}$ describes a fixed line $\mathbf{u}$, then

$$\mathbf{u}^T\mathbf{x} = 0,$$

i.e.
$$\mathbf{u}^T\mathbf{A}^{-1}\mathbf{x}' = 0,$$

i.e.
$$\mathbf{u}'^T\mathbf{x}' = 0,$$

where $\mathbf{u}'^T = \mathbf{u}^T\mathbf{A}^{-1}$ and hence $\mathbf{u}' = \mathbf{A}^{-1T}\mathbf{u}$. Thus the collineation transforms the line $\mathbf{u}$, regarded as the locus of a variable point, into the line $\mathbf{u}'$, and the transformation $\mathbf{x}' = \mathbf{A}\mathbf{x}$ of points into points induces the contragredient transformation $\mathbf{u}' = \mathbf{A}^{-1T}\mathbf{u}$ of lines into lines. By duality, a transformation

$$\mathbf{u}' = \mathbf{B}\mathbf{u} \quad (|\mathbf{B}| \neq 0)$$

of lines into lines induces a transformation

$$\mathbf{x}' = \mathbf{B}^{-1T}\mathbf{x}$$

of points into points; and so we have the important result:

THEOREM 1. *Every collineation ϖ is a self-dual transformation, which transforms points into points and lines into lines. It is represented, in terms of arbitrary allowable representations $\mathscr{R}$ and $\mathscr{R}'$, by equations of the form*

$$\mathbf{x}' = \mathbf{A}\mathbf{x}, \qquad \mathbf{u}' = \mathbf{A}^{-1T}\mathbf{u} \quad (|\mathbf{A}| \neq 0),$$

each of which determines the other.

It follows, therefore, that a necessary condition for a (1, 1) correspondence between two planes to be a collineation is that it makes lines correspond to lines. To give an adequate discussion of the sufficiency of this condition would take us too far out of our way, but it is worth referring to a form of argument that has played an important part in the historical development of projective geometry. This argument applies only to real projective geometry,† and it does not belong to the systematic development of our present theory because it is not purely algebraic. Let σ be a (1, 1) correspondence between two real planes S_2 and S_2' which is representable analytically by continuous functions and which has the further property of making lines correspond to lines. Then, in virtue of the quadrangle construction for harmonic conjugates, the range corresponding to any harmonic range is also harmonic. If, now, we select three distinct points A, B, C, of a line l in S_2, and take them as reference points and unit point in a parametric representation of l, and if we take all points of l that can be obtained by repeated application of the quadrangle construction to the three initial points or to points already constructed, the set of all these points (called a *Möbius net* or net of rationality) is simply the set of all points whose parameters are rational numbers. Every point of the line may then be obtained as the limit point of a sequence of points of the Möbius net. Now suppose σ transforms l into l', and A, B, C into A', B', C'. Then, by what we have said, the parameter of any point P on l is equal to the parameter of the

† Indeed the condition is not sufficient when the ground field is unrestricted. In the complex plane, for example, we have the anti-collineation $\mathbf{x} \rightarrow \bar{\mathbf{x}}$, which transforms every coordinate vector into its complex conjugate. The transformation is (1, 1) and it transforms lines into lines, but it is not a collineation. It does not, in fact, belong to the projective geometry of the complex plane.

corresponding point P' on l', referred to A', B' as reference points and C' as unit point, and the cross ratio of any ordered set of four points of l is therefore equal to the cross ratio of the corresponding points of l'.

To complete the proof, we need only take an arbitrary representation $\mathscr{R}$ of S_2, defined by X_0, X_1, X_2, E, and the particular representation $\mathscr{R}'$ of S_2' that is defined by the corresponding points X_0', X_1', X_2', E'. Then, in virtue of the invariance of cross ratio, if (P, P') is a general pair of corresponding points,

$$X_0\{E, P; X_1, X_2\} = X_0'\{E', P'; X_1', X_2'\}, \quad \text{etc.}$$

Thus the coordinates of P and P' are connected by the relation

$$x_0 : x_1 : x_2 = x_0' : x_1' : x_2',$$

and σ is accordingly a collineation.

We now return to the general collineation ϖ, defined by the equation $\mathbf{x}' = \mathbf{A}\mathbf{x}$. Let new coordinates be introduced in S_2 and S_2' by means of the transformations

$$\mathbf{x} = \mathbf{P}\bar{\mathbf{x}} \quad \text{and} \quad \mathbf{x}' = \mathbf{Q}\bar{\mathbf{x}}'.$$

Then, from the equation $\mathbf{x}' = \mathbf{A}\mathbf{x}$, we have

$$\mathbf{Q}\bar{\mathbf{x}}' = \mathbf{A}\mathbf{P}\bar{\mathbf{x}},$$

i.e.

$$\bar{\mathbf{x}}' = \mathbf{Q}^{-1}\mathbf{A}\mathbf{P}\bar{\mathbf{x}}.$$

Thus if the matrix of ϖ referred to $\mathscr{R}$ and $\mathscr{R}'$ is $\mathbf{A}$, the matrix of the same collineation referred to the new representations $\bar{\mathscr{R}}$ and $\bar{\mathscr{R}}'$ is $\mathbf{Q}^{-1}\mathbf{A}\mathbf{P}$.

If S_2' is distinct from S_2, we can introduce new coordinates in S_2' by the transformation

$$\mathbf{x}' = \mathbf{A}\bar{\mathbf{x}}'$$

while retaining the original coordinates in S_2. In this case the equation of ϖ reduces to

$$\bar{\mathbf{x}}' = \mathbf{A}^{-1}\mathbf{A}\mathbf{I}\bar{\mathbf{x}},$$

i.e.

$$\bar{\mathbf{x}}' = \bar{\mathbf{x}}.$$

In other words, by adapting the coordinate representations to a given collineation between distinct planes we can arrange matters so that two points correspond if and only if they have the same coordinates. This is a consequence of the twofold interpretation of the algebraic transformation $\mathbf{x}' = \mathbf{A}\mathbf{x}$, to which we referred at the beginning of the chapter.

When S_2' coincides with S_2, and the same representation $\mathscr{R}$ is used twice over, the reduction of the equation of ϖ to the simple form $\bar{\mathbf{x}}' = \bar{\mathbf{x}}$ is not possible. If $\mathscr{R}$ is changed to $\bar{\mathscr{R}}$ by the transformation $\mathbf{x} = \mathbf{P}\bar{\mathbf{x}}$, the transformed equation of ϖ is

$$\bar{\mathbf{x}}' = \mathbf{P}^{-1}\mathbf{A}\mathbf{P}\bar{\mathbf{x}}.$$

Thus the matrices which represent a given self-collineation of S_2 in the different allowable representations $\mathscr{R}$ are all similar. By choosing the representation (that is to say, the transforming matrix $\mathbf{P}$) suitably, we can obtain any one of a set of similar matrices. In particular, we can obtain various simple canonical forms, as will be shown later in this section. The reduction of the equation of a self-collineation to canonical form should be compared with the corresponding reduction of the equation of a homographic correspondence between S_1 and itself.

General properties of collineations

Before going into the classification of the possible types of collineation, we shall first enumerate a few theorems that are valid for collineations in general.

THEOREM 2. *There is a unique collineation which transforms four given points, no three of which are collinear, into four given points, no three of which are collinear. Dually, there is a unique collineation which transforms four given lines, no three of which are concurrent, into four given lines, no three of which are concurrent.*

This is, of course, Theorem 1 of the Appendix once again.

THEOREM 3. *If a collineation ϖ transforms two points P, Q, into P', Q', then it transforms the line PQ into the line $P'Q'$. If ϖ transforms two lines p, q into p', q', then it transforms the point pq into the point $p'q'$.*

THEOREM 4. *A collineation ϖ transforms a variable point P of a fixed line p into a variable point P' of a second fixed line p', and the ranges described by P and P' are homographically related.*

Proof. If $\mathbf{x} = \bar{\mathbf{x}}^{(1)} + \theta\mathbf{x}^{(2)}$, then $\mathbf{A}\mathbf{x} = \mathbf{A}\mathbf{x}^{(1)} + \theta\mathbf{A}\mathbf{x}^{(2)}$.

COROLLARY. *If ϖ is a collineation between S_2 and itself, and m is a self-corresponding line for ϖ, every point P of m is transformed by ϖ into another point P' of m, and $(P) \barwedge (P')$; i.e. ϖ induces a homography on every self-corresponding line of the plane.*

EXERCISE. Dualize Theorem 4 and its corollary.

Since a collineation is a linear transformation, it transforms any locus into a locus of the same order and any envelope into an envelope of the same class. In particular, conics are transformed into conics. Quite generally, every projectively generated figure goes over into an equivalent projectively generated figure.

If c, c' are two curves which touch at P, two of the points of intersection of the curves coincide at P. After transformation, therefore, c and c' become two curves which intersect twice in the point ϖP. This means that the collineation preserves tangency.

Self-collineations of the plane

When we restrict the theory of collineations to transformations of one plane into itself two important developments take place.

(i) Two collineations ϖ_1 and ϖ_2 now have a product $\varpi_2 \varpi_1$ which is also a collineation, and the matrix of $\varpi_2 \varpi_1$ is the product $\mathbf{A}_2 \mathbf{A}_1$ of the matrices of ϖ_2 and ϖ_1. In fact the set of all (non-singular) self-collineations of S_2 is a group, isomorphic with the projective group $PGL(2)$.

(ii) If ϖ is a self-collineation of S_2 there may possibly be united points, which are left invariant by ϖ. If ϖ is the identical transformation ϵ, every point of S_2 is a united point; and whatever collineation is taken as ϖ there is always at least one united point. We shall find that the properties of ϖ and the canonical form of its equation are closely connected with the united points of ϖ, as was the case with homographies in Chapter III.

Consider a self-collineation ϖ of S_2, with equation $\mathbf{x}' = \mathbf{A}\mathbf{x}$. If the coordinate vector $\mathbf{x}$ represents a united point of ϖ, then $\mathbf{x}' = \lambda\mathbf{x}$ for some λ, and therefore

$$\mathbf{A}\mathbf{x} = \lambda\mathbf{x}.$$

In order to determine $x_0 : x_1 : x_2$ we have to solve the three equations

$$\sum_{k=0}^{2} (a_{ik} - \lambda\delta_{ik})x_k = 0 \quad (i = 0, 1, 2),$$

and this is only possible if $|a_{rs} - \lambda\delta_{rs}| = 0$, i.e. $|\mathbf{A} - \lambda\mathbf{I}| = 0$. In other words, λ must be a characteristic root of the matrix $\mathbf{A}$, and the coordinate vector $\mathbf{x}$ must be a characteristic vector corresponding to this root.

To find the united points of ϖ, then, we first obtain the characteristic roots λ_0, λ_1, λ_2 by solving the cubic equation $|\mathbf{A} - \lambda\mathbf{I}| = 0$. To each of these roots there corresponds at least one characteristic vector $\mathbf{x}$ (or, more precisely, at least one set of proportional characteristic vectors), and every such vector gives the coordinates of a united point. The number of linearly independent vectors $\mathbf{x}$ which satisfy the equation

$$\mathbf{A}\mathbf{x} = \lambda_i \mathbf{x},$$

where λ_i is a given characteristic root of $\mathbf{A}$, depends upon the rank of the matrix $\mathbf{A}-\lambda_i \mathbf{I}$. If this rank $\rho[\mathbf{A}-\lambda_i \mathbf{I}]$ is denoted by $\rho(\lambda_i)$, the number of linearly independent vectors is $3-\rho(\lambda_i)$.

There is a well-known connexion between the rank $\rho(\lambda_i)$ and the multiplicity of λ_i as a characteristic root of $\mathbf{A}$, given by the algebraic theorem: *If* $\mathbf{A}$ *is an* $n \times n$ *matrix and* $\rho[\mathbf{A}-\lambda_i \mathbf{I}] = n-p$, *then the multiplicity of* λ_i *as a characteristic root of* $\mathbf{A}$ *is at least* p. In our special case, therefore, the multiplicity $\mu(\lambda_i)$ of λ_i is not less than $3-\rho(\lambda_i)$, and so there are six cases which can arise.

(i) λ_i is a simple root. Then $\mu = 1$, and ρ can only be 2.

(ii) λ_i is a double root. Then $\mu = 2$, and either (*a*) $\rho = 2$, or (*b*) $\rho = 1$.

(iii) λ_i is a triple root. There are then three possibilities: (*a*) $\rho = 2$, (*b*) $\rho = 1$, (*c*) $\rho = 0$.

In cases (i), (ii *a*), (iii *a*) there is only one linearly independent vector $\mathbf{x}$, and λ_i gives rise to an isolated united point. In cases (ii *b*) and (iii *b*) there are two independent vectors $\mathbf{x}$, and λ_i gives rise to a line of united points. Finally, in case (iii *c*) there are three independent vectors, and every point of the plane is a united point. In the last case, ϖ is the identical collineation ϵ, and so we have only five non-trivial cases to consider. We shall now take these one by one.

I. *The general collineation*

In the general case, the characteristic roots λ_0, λ_1, λ_2 of $\mathbf{A}$ are all different, and each is therefore a simple root. They give rise to three isolated united points, which are distinct since $\mathbf{A}\mathbf{x} = \lambda_0 \mathbf{x}$ and $\mathbf{A}\mathbf{x} = \lambda_1 \mathbf{x}$ would give $(\lambda_0-\lambda_1)\mathbf{x} = \mathbf{0}$, i.e. $\mathbf{x} = \mathbf{0}$. Furthermore, the three united points are not collinear, for if they all belonged to a line l the collineation would induce on l a homography with three distinct united points (Theorems 3 and 4) and every point of l would then be self-corresponding. The united points may therefore be taken as vertices of the triangle of reference, and when this is done the equations of ϖ assume the form

$$x_0' = \alpha_0 x_0,$$
$$x_1' = \alpha_1 x_1,$$
$$x_2' = \alpha_2 x_2.$$

Since the matrices $\mathbf{A}$ and $\mathbf{P}^{-1}\mathbf{AP}$ have the same characteristic roots, $\alpha_0, \alpha_1, \alpha_2$ are proportional to $\lambda_0, \lambda_1, \lambda_2$; and we may accordingly write the equations of the general self-collineation of the plane in the canonical form

$$x_0' = \lambda_0 x_0,$$
$$x_1' = \lambda_1 x_1,$$
$$x_2' = \lambda_2 x_2.$$

This form may be compared with the canonical equation $\theta' = k\theta$ for a homography with distinct united points.

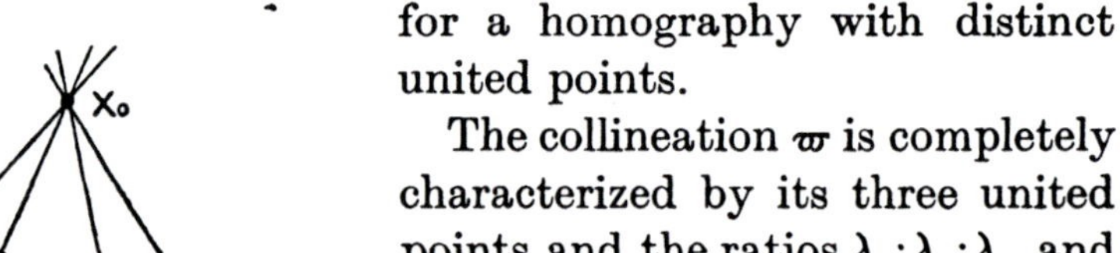

The collineation ϖ is completely characterized by its three united points and the ratios $\lambda_0 : \lambda_1 : \lambda_2$, and these ratios admit of a simple geometrical interpretation. Let P, P' be corresponding points, with coordinate vectors $\mathbf{y}$, $\mathbf{y}'$ respectively. Then the equations of $X_0 P$ and $X_0 P'$ are

$$x_1 - \frac{y_1}{y_2} x_2 = 0 \quad \text{and} \quad x_1 - \frac{y_1'}{y_2'} x_2 = 0,$$

and the parameters of $X_0 P$ and $X_0 P'$ in their pencils are

$$\frac{y_1}{y_2} \quad \text{and} \quad \frac{y_1'}{y_2'} = \frac{\lambda_1 y_1}{\lambda_2 y_2}.$$

Thus $X_0 P$ and $X_0 P'$ are corresponding rays in a homography with $X_0 X_1$, $X_0 X_2$ as united rays and λ_1/λ_2 as modulus. The ratios between λ_0, λ_1, and λ_2 may thus be interpreted geometrically as moduli of certain homographies, and they are therefore projectively invariant—a result which can also be inferred algebraically from the theorem that similar matrices have the same characteristic roots.

We see, then, that if (P, P') is any corresponding pair of ϖ, $X_0 P$ and $X_0 P'$ correspond in the homography just discussed and $X_1 P$ and $X_1 P'$ correspond similarly in a homography with united rays X_1X_2, X_1X_0 and modulus λ_2/λ_0; and it follows at once that the point P' corresponding to any assigned point P can be found by geometrical construction.

EXERCISE. If A and B are two fixed points in the plane, p and p' are variable rays through A which correspond in a homography ϖ_1, and q and q' are variable rays through B which correspond in a homography ϖ_2, the ray AB being self-corresponding in both homographies, show that the points pq and $p'q'$ are connected by a collineation.

II. *The collineation with two united points*

If the characteristic roots of $\mathbf{A}$ are $\lambda_0, \lambda_0, \lambda_2$, and $\rho[\mathbf{A}-\lambda_0\mathbf{I}] = 2$, then ϖ has only two united points. If these points are taken as X_0 and X_2, the matrix of $\mathbf{A}$ may be written in the form

$$\begin{pmatrix} \lambda_0 & \beta & 0 \\ 0 & \lambda_0 & 0 \\ 0 & \gamma & \lambda_2 \end{pmatrix}.$$

If P, P' are general corresponding points, with coordinate vectors $\mathbf{y}$, $\mathbf{y}'$, the equations of X_0P and X_0P' are

$$x_2 = \frac{y_2}{y_1}x_1 \quad \text{and} \quad x_2 = \frac{\gamma y_1+\lambda_2 y_2}{\lambda_0 y_1}x_1,$$

and the equations of the homography ϖ_0 which relates them may be written as

$$\theta' = \frac{\lambda_2}{\lambda_0}\theta+\frac{\gamma}{\lambda_0}, \quad \text{where} \quad \theta = \frac{y_2}{y_1}.$$

Similarly the equation of the homography ϖ_2 which relates X_2P and X_2P' may be written as

$$\theta' = \theta+\frac{\beta}{\lambda_0}, \quad \text{where} \quad \theta = \frac{y_0}{y_1}.$$

Since $\lambda_2/\lambda_0 \neq 1$, the united rays of ϖ_0 are $x_1 = 0$ and another line distinct from $x_1 = 0$. If we take this other line as the side $x_2 = 0$ of the triangle of reference, then $\gamma = 0$. The united rays of ϖ_2 both coincide with the line $x_1 = 0$.

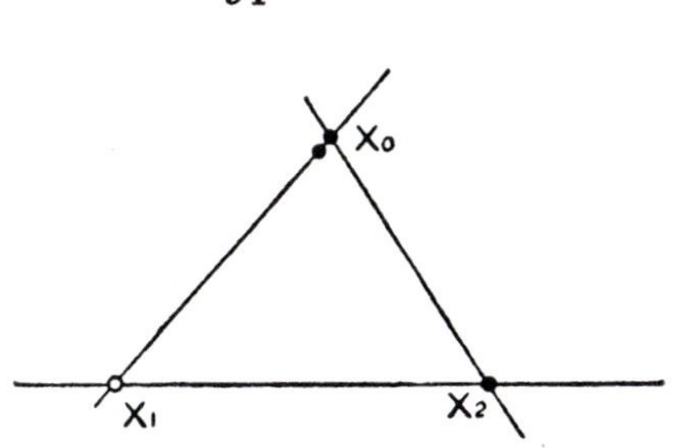

We may look upon this type of collineation as a specialization of type I, in which two of the united points happen accidentally to coincide at X_0, while their join X_0X_1 remains a determinate line.

By the above choice of the triangle of reference, the matrix **A** has been put in the form

$$\begin{pmatrix} \lambda_0 & \beta & 0 \\ 0 & \lambda_0 & 0 \\ 0 & 0 & \lambda_2 \end{pmatrix},$$

and it can be simplified still further by suitable choice of the unit point. We have $(0,1,0) \to (\beta, \lambda_0, 0)$, and if we choose the unit point so that the point ϖX_1 has coordinates $(1, \lambda_0, 0)$ we can make β equal to 1. Then the equations of ϖ assume the canonical form

$$\begin{aligned} x_0' &= \lambda_0 x_0 + x_1, \\ x_1' &= \qquad\quad \lambda_0 x_1, \\ x_2' &= \qquad\qquad\qquad \lambda_2 x_2. \end{aligned}$$

III. *The collineation with one united point*

If the characteristic roots of **A** are $\lambda_0, \lambda_0, \lambda_0$, and $\rho[\mathbf{A} - \lambda_0 \mathbf{I}] = 2$, ϖ has only one united point. This case arises from the previous one by further specialization. It leads to the canonical form

$$\begin{aligned} x_0' &= \lambda_0 x_0 + x_1, \\ x_1' &= \qquad\quad \lambda_0 x_1 + x_2, \\ x_2' &= \qquad\qquad\qquad \lambda_0 x_2. \end{aligned}$$

IV. *The plane homology*

If the characteristic roots of **A** are $\lambda_0, \lambda_0, \lambda_2$, and $\rho[\mathbf{A} - \lambda_0 \mathbf{I}] = 1$, ϖ has an isolated united point and a line of united points. The isolated point cannot lie on the line of united points, as it would then arise from two distinct characteristic roots.

If the isolated united point is taken as X_2 and two points of the line of united points as X_0 and X_1, the equations of ϖ may be written as

$$\begin{aligned} x_0' &= \lambda_0 x_0, \\ x_1' &= \qquad \lambda_0 x_1, \\ x_2' &= \qquad\qquad \lambda_2 x_2. \end{aligned}$$

The same transformation of points is equally well given by

$$\begin{aligned} x_0' &= x_0, \\ x_1' &= \quad x_1, \\ x_2' &= \qquad k x_2, \end{aligned}$$

where $k = \lambda_2/\lambda_0$, and we shall take this as the canonical form.

If P is the point (y_0, y_1, y_2), then the corresponding point P' is (y_0, y_1, ky_2), and $X_2 P$, $X_2 P'$ are the same line. This line $X_2 PP'$ meets $X_0 X_1$ in the united point M, given by $(y_0, y_1, 0)$, and

$$\{X_2, M; P, P'\} = k.$$

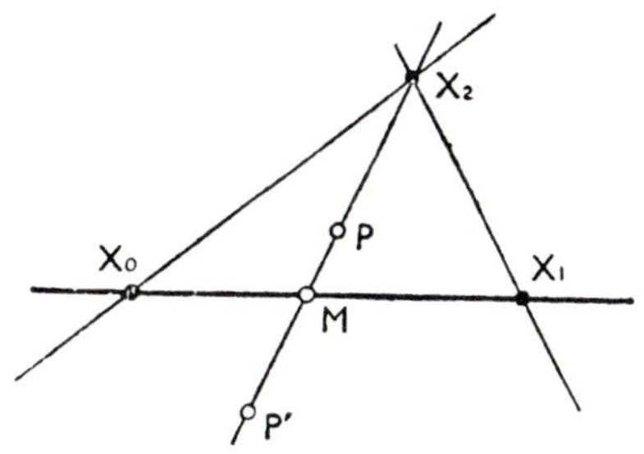

P and P' therefore correspond in the homography on $X_2 M$ which has united points X_2, M and modulus k.

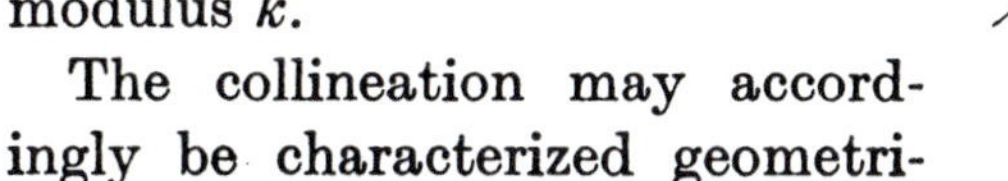

The collineation may accordingly be characterized geometrically as follows: Let P be a general point of the plane, and let $X_2 P$ meet $X_0 X_1$ in M. Then P' is the unique point such that (i) P' lies on $X_2 M$, and (ii) $\{X_2, M; P, P'\} = k$.

In this case ϖ is said to be a *plane homology* with X_2 as *vertex*, $X_0 X_1$ as *axis*, and k as *modulus*. The vertex, axis, and modulus are sufficient to determine it completely. In the special case in which the modulus k is -1, ϖ is called a *harmonic homology*. It is then involutory, i.e. $\varpi^2 = \epsilon$.

V. *The special plane homology*

If the characteristic roots of $\mathbf{A}$ are λ_0, λ_0, λ_0, and

$$\rho[\mathbf{A}-\lambda_0 \mathbf{I}] = 1,$$

ϖ has a line of united points and no other united point. A collineation of this type, called a *special homology*, may be looked upon as arising from the general homology by the accidental incidence of the vertex with the axis. For suppose we take two of the united points as X_0 and X_1, and an arbitrary point which is not collinear with them as X_2. Then the equations of ϖ may be written

$$\begin{aligned} x_0' &= \lambda_0 x_0 \qquad\qquad + a_{02} x_2, \\ x_1' &= \qquad\qquad \lambda_0 x_1 + a_{12} x_2, \\ x_2' &= \qquad\qquad\qquad\qquad \lambda_0 x_2, \end{aligned}$$

or, alternatively,

$$\begin{aligned} x_0' &= x_0 \qquad + a x_2, \\ x_1' &= \qquad x_1 + b x_2, \\ x_2' &= \qquad\qquad x_2. \end{aligned}$$

If P is the point (y_0, y_1, y_2), P' is $(y_0 + a y_2, y_1 + b y_2, y_2)$, and PP'

therefore passes through the point $(ay_2, by_2, 0)$, i.e. the fixed united point $(a, b, 0)$. This point is the vertex of the special homology.

The homography induced on any line through the vertex has coincident united points, and there is therefore no modulus in this case. The collineation may be defined geometrically by means of its axis a, its vertex A, and one general corresponding pair (P_0, P_0'). The point P' corresponding to any given point P is found by joining PP_0 to meet a in M, and then joining MP_0' to meet AP in P'.

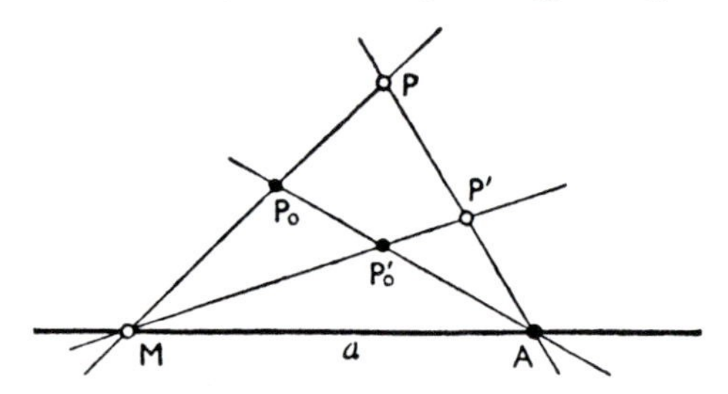

Exercise. Prove this construction.

Real collineations

When the ground field is the field of real numbers every collineation has a matrix whose elements are all real, and the characteristic roots of this matrix are either (*a*) all real, or (*b*) one real and two conjugate complex. In case (*b*) the characteristic roots, and hence also the associated united points, are necessarily distinct; and of these points one is real, while the remaining two, being conjugate complex points, are joined by a real self-corresponding line. The coordinates (x, y, z) of the conjugate complex united points can be changed by a real transformation into $(1, i, 0)$ and $(1, -i, 0)$ respectively, and hence any real collineation of type (*b*) is projectively equivalent to a collineation which leaves invariant the absolute points I, J and one real (finite) point.

We consider now various elementary euclidean transformations of the extended euclidean plane, with the object of characterizing them projectively as collineations of the real projective plane.

A *translation* is a special homology with the line at infinity i as axis. For if the vertex V of such a homology is $(a, b, 0)$, the equations of the homology are of the form

$$\begin{aligned} x' &= x \quad\quad + az, \\ y' &= \quad\quad y + bz, \\ z' &= \quad\quad\quad\quad z, \end{aligned}$$

and these represent a general translation.

A *radial expansion* or dilatation (in an obviously extended sense) from a centre O is a homology with O as vertex and i as axis. If the homology is harmonic, the collineation reduces to reflection in the point O. From this can easily be deduced (by Menelaus's Theorem, for example) the projective theorem that if two (real) homologies τ_1, τ_2 have the same axis a and distinct vertices O_1, O_2, then $\tau_2 \tau_1$ is in general a third homology whose axis is a and whose vertex is some point O_3 of $O_1 O_2$; but it may reduce (if the moduli of τ_1 and τ_2 are reciprocal) to a special homology whose vertex is at the point of intersection of $O_1 O_2$ with a.

A *rotation about a centre* O is a collineation with O, I, J as united points and with characteristic roots in the ratios $1:e^{i\theta}:e^{-i\theta}$, where θ is real. In fact the equations of the most general real collineation with O, I, J as united points can be written in the form

$$x' = c(x\cos\theta + y\sin\theta),$$
$$y' = c(-x\sin\theta + y\cos\theta),$$
$$z' = z,$$

where c is real and positive. (Compare Theorem 17 of Chapter IV.) The characteristic roots are 1, $ce^{i\theta}$, $ce^{-i\theta}$, and the collineation is a rotation if $c = 1$. If $c \neq 1$, the collineation is a combination of a rotation and a radial expansion from O; and every real collineation of the type (*b*) referred to above is projectively equivalent to such a collineation.

Reflection in a line a is a harmonic homology whose axis is a and whose vertex is the point at infinity in the perpendicular direction. More generally, a homology whose axis is a and whose vertex is the point at infinity in the direction of some other line d is a *rabatment* on a in the direction of d.

The above analysis provides a basis for the projective generalization of many theorems of euclidean geometry. We see, for instance, how our almost intuitive perception and use of the symmetry of a euclidean figure, or of similitude between figures, can be translated into statements concerning the existence of certain collineations which transform a projective figure into itself, or one such figure into another.

Exercises

1. Interpreting the plane collineation ϖ as a transformation of lines into lines, work out the details of the dual classification of collineations into five types. Show that the types of collineation obtained are the same as before, and that each type is self-dual.

2. Deduce the self-duality of the five types of collineation algebraically from the fact that if ϖ is given by $\mathbf{x}' = \mathbf{A}\mathbf{x}$, ϖ^{-1} is given by $\mathbf{u} = \mathbf{A}^T\mathbf{u}'$.

3. ϖ is a collineation of the plane with itself, (A, A') and (B, B') are fixed corresponding pairs, and (P, P') is a variable corresponding pair; and ϖ is resolved into the pair of homographies

$$A(P) \barwedge A'(P'), \qquad B(P) \barwedge B'(P'),$$

with $A'B'$ corresponding to AB in each of them. Show how the united points of ϖ may be found as points of intersection of two conics, one through A and A' and the other through B and B'. Examine further how these conics are related when ϖ is one of the special collineations.

4. By reversing the order of the argument used in the previous question, show how any two conics may be used to define a collineation.

Cyclic collineations

If ϖ is a self-collineation of the plane and P is any point, we may construct the sequence of points: P, ϖP, $\varpi^2 P$,... . We may also extend the sequence backwards, so that we have

$$\ldots\, \varpi^{-2}P, \varpi^{-1}P, P, \varpi P, \varpi^2 P, \ldots .$$

There are now two possibilities: (i) all the points of the sequence are distinct, and (ii) some point occurs twice in the sequence. In the latter case, the sequence consists of a finite cycle of points, say $P_0, P_1, \ldots, P_{r-1}$, recurring again and again,

$$\ldots\, P_{r-2}, P_{r-1}, P_0, P_1, \ldots, P_{r-1}, P_0, \ldots;$$

i.e. $\varpi^m P = P_\mu$, where μ is the residue of m modulo r.

The points $\varpi^m P$ may recur either because P is a special point for ϖ or because the collineation ϖ is itself special. The first alternative occurs, for example, when P is a united point of ϖ; and then the cycle generated by P consists of one term only. If, on the other hand, we take as ϖ a harmonic homology, every point of the plane is either a united point or else generates a cycle of two points. If ϖ is such that a general point of the plane generates a cycle of length r, we say that ϖ is cyclic with period r. A cyclic collineation of period 2—a harmonic homology for instance—is said to be involutory.

THEOREM 5. *If a collineation ϖ has two involutory pairs of corresponding points which do not lie on the same line it is a harmonic homology.*

Proof. Let (P, P'), (Q, Q') be involutory pairs of ϖ, lying on lines p, q which meet in A. Then, since P and P' transform into P' and P,

p is a self-corresponding line. Similarly q is self-corresponding, and A is therefore a self-corresponding point. Now ϖ induces on p a homography with an involutory pair (P, P'), i.e. an involution.

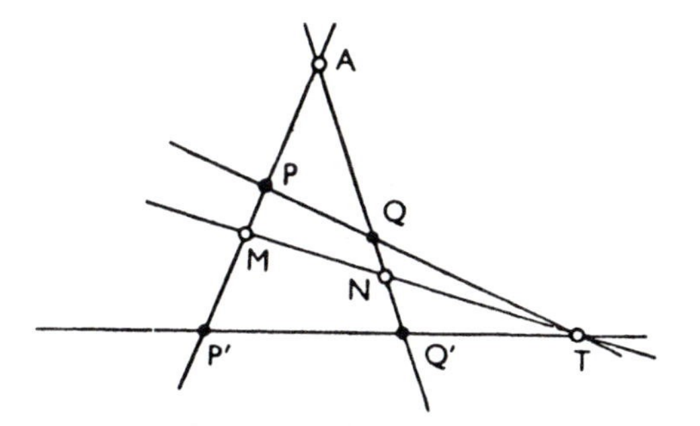

The united points of this involution are A and another point M. If N is defined similarly on QQ', then

$$\{A, M; P, P'\} = \{A, N; Q, Q'\} = -1,$$

and the ranges (A, M, P, P') and (A, N, Q, Q') are in perspective. Thus MN passes through the common point T of PQ and $P'Q'$. But PQ and $P'Q'$ are transformed into each other by ϖ, and T is therefore self-corresponding. The line MN now contains three distinct united points M, N, T, of ϖ, and the homography induced on it by ϖ is consequently the identical homography. Every point of MN is therefore a united point of ϖ, and ϖ is a homology with this line as axis and A as vertex. Since $\{A, M; P, P'\} = -1$, the homology is harmonic.

COROLLARY. *The only kind of involutory plane collineation is the harmonic homology.*

THEOREM 6. *If a collineation has one cyclic triad of non-collinear points, it is a cyclic collineation of period* 3.

Proof. Let ϖ be a collineation with a cyclic triad (P_0, P_1, P_2) of non-collinear points. The collineation must have at least one united point, M say; and M cannot lie on a side of the triangle $P_0 P_1 P_2$, for if it were to lie on $P_0 P_1$, this line $P_0 M \equiv P_1 M$ would be self-corresponding, and P_1 could not transform into P_2. But the collineation ϖ^3 now has the four pairs (P_0, P_0), (P_1, P_1), (P_2, P_2), (M, M) in common with ϵ, and so, by Theorem 2, $\varpi^3 = \epsilon$.

If $P_0 P_1 P_2$ is taken as triangle of reference and M as unit point, the equations of ϖ are

$$\frac{x_0'}{x_2} = \frac{x_1'}{x_0} = \frac{x_2'}{x_1}.$$

The united points of ϖ are M and the two points $(1,\omega,\omega^2)$ and $(1,\omega^2,\omega)$.

A special case of such a collineation is a rotation through an angle $2\pi/3$ in the euclidean plane.

If a collineation ϖ has a cyclic tetrad (P_0, P_1, P_2, P_3), forming the vertices of a proper quadrangle, it necessarily has three isolated united points. In order to show this let us suppose that the figure is labelled in the manner indicated.

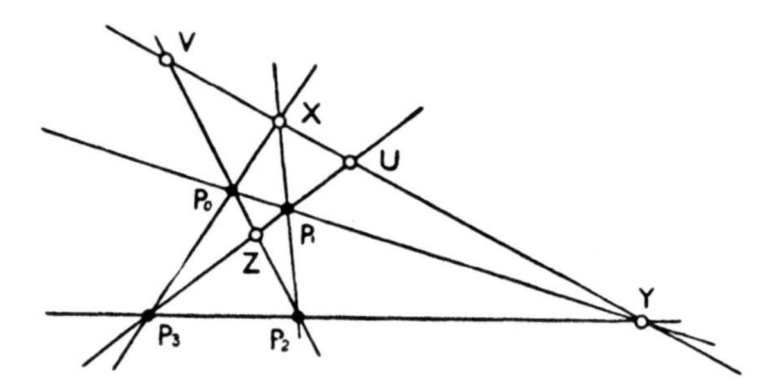

Since $P_0 P_2$ and $P_1 P_3$ are transformed into each other, Z is a united point. Also, X and Y are transformed into each other, and XY is therefore a united line. The points U and V in which this line is met by $P_1 P_3$ and $P_0 P_2$ are transformed into each other. The homography induced on XY is therefore an involution with (X, Y) and (U, V) as two pairs, and the united points of this involution are the remaining united points of ϖ.

If XYZ is taken as triangle of reference and P_0 as unit point, the coordinates of the four points P_i are all of the form $(1, \pm 1, \pm 1)$, and the equations of ϖ are

$$\frac{x_0'}{x_1} = \frac{x_1'}{-x_0} = \frac{x_2'}{x_2}.$$

It may be seen from these equations that ϖ is cyclic of period 4. This could be inferred, of course, by an argument similar to that used in proving Theorem 6.

THEOREM 7. *If $P_0 P_1 P_2 P_3 P_4$ is a given proper pentagon in the euclidean plane, there exists a collineation ϖ with $(P_0, P_1, P_2, P_3, P_4)$ as a cyclic pentad if and only if the given pentagon is projectively equivalent to a regular pentagon.*

Proof. (i) Suppose the given pentagon can be transformed projectively into a regular pentagon by a collineation σ. Then a rotation through $2\pi/5$ about the circumcentre of the regular pentagon is a collineation ϖ' with the vertices of the regular pentagon as a cyclic pentad. The collineation $\varpi = \sigma^{-1}\varpi'\sigma$ is then cyclic of period 5, since $\varpi^5 = \sigma^{-1}\varpi'^5\sigma = \epsilon$, and it transforms $(P_0, P_1, P_2, P_3, P_4)$ into $(P_1, P_2, P_3, P_4, P_0)$.

(ii) Suppose, conversely, that ϖ is a collineation with

$$(P_0, P_1, P_2, P_3, P_4)$$

as a cyclic pentad. If k is the unique conic through the five points, ϖ transforms k into itself and induces a homography on it;† and since the homography is cyclic it has distinct united points M, N. These points are then united points of ϖ, and the pole of MN with respect to k is also a united point of ϖ. If, now, σ is a collineation which transforms M, N into I, J, the pentagon is transformed by σ into a pentagon inscribed in a circle; and since this cyclic pentagon is transformed into itself by the rotation $\sigma\varpi\sigma^{-1}$, which is cyclic of period 5, it is regular.

Collineations which leave a conic invariant

The set of all self-collineations of the plane is a group, and the set of all those collineations which have a given set of invariants is, of course, a subgroup of this group. We have already seen, for instance, that those collineations of the euclidean plane which leave invariant a finite point O and the two absolute points I, J make up the group generated by the radial expansions and rotations about O. We now come to another important type of subgroup of the full group of self-collineations of S_2, the group of collineations which leave a fixed conic invariant.

Let k_0 be a fixed proper conic, and let ϖ be a collineation which transforms k_0 into itself. If V is a fixed point of k_0 and P a variable point of k_0, transformed respectively into V' and P', then

$$(P) \barwedge V(P) \barwedge V'(P') \barwedge (P').$$

The points P and P' are thus related by a homography ϖ_0 on k_0, subordinate to the collineation ϖ.

Not only is ϖ_0 uniquely determined by ϖ, but conversely ϖ is the only collineation which transforms k_0 into itself and induces ϖ_0 on it. For if ϖ_0 is given and P_0, P_1, P_2, P_3 are four points of k_0, no three of these points and no three of the points $\varpi_0 P_0$, $\varpi_0 P_1$, $\varpi_0 P_2$, $\varpi_0 P_3$ can be collinear; and ϖ is uniquely determined by the four corresponding pairs $(P_i, \varpi_0 P_i)$. We shall now show that there is a

† See the section immediately below for the proof of this.

simple algebraic connexion between the equation of ϖ and that of ϖ_0.

THEOREM 8. *If k_0 is the conic $x_0:x_1:x_2 = \theta^2:\theta:1$, and ϖ_0 is a homography on k_0 whose equation is $\theta' = (\alpha\theta+\beta)/(\gamma\theta+\delta)$, the matrix equation*

$$\begin{pmatrix} x_0' & x_1' \\ x_1' & x_2' \end{pmatrix} = \begin{pmatrix} \alpha & \beta \\ \gamma & \delta \end{pmatrix} \begin{pmatrix} x_0 & x_1 \\ x_1 & x_2 \end{pmatrix} \begin{pmatrix} \alpha & \gamma \\ \beta & \delta \end{pmatrix}$$

defines a collineation ϖ which transforms k_0 into itself and induces the homography ϖ_0 on it.

Proof. When we multiply out the product of the three matrices on the right-hand side we find that the given matrix equation is equivalent to the three equations

$$\begin{aligned} x_0' &= \alpha^2 x_0 + 2\alpha\beta x_1 + \beta^2 x_2, \\ x_1' &= \alpha\gamma x_0 + (\alpha\delta+\beta\gamma)x_1 + \beta\delta x_2, \\ x_2' &= \gamma^2 x_0 + 2\gamma\delta x_1 + \delta^2 x_2, \end{aligned}$$

and it therefore defines a collineation ϖ (necessarily non-singular). By taking determinants of both sides of the original matrix equation we see immediately that ϖ leaves k_0 invariant.

Now the general point $(\theta^2, \theta, 1)$ of k_0 is transformed by ϖ_0 into the point $((\alpha\theta+\beta)^2,\ (\alpha\theta+\beta)(\gamma\theta+\delta),\ (\gamma\theta+\delta)^2)$, i.e.

$$(\alpha^2\theta^2+2\alpha\beta\theta+\beta^2,\ \alpha\gamma\theta^2+(\alpha\delta+\beta\gamma)\theta+\beta\delta,\ \gamma^2\theta^2+2\gamma\delta\theta+\delta^2),$$

and this is the point into which it is transformed by ϖ. The theorem is therefore completely proved.

There are two types of collineation which can transform a given conic k_0 into itself, corresponding to the cases in which the united points of the induced homography are distinct and coincident respectively.

(i) Let ϖ_0 have distinct united points. If we take these points as X_0, X_2, and the pole of X_0X_2 as X_1, then, since the tangents at X_0 and X_2 transform into themselves, X_1 is also a united point of ϖ. The equations of ϖ are then of the form

$$\frac{x_0'}{x_0} = \frac{x_1'}{\alpha x_1} = \frac{x_2'}{\alpha^2 x_2},$$

where α is arbitrary.

In the particular case in which $\alpha = -1$, ϖ_0 is an involution with vertex X_1 and ϖ is the harmonic homology with X_1 as vertex and X_0X_2 as axis.

(ii) If ϖ_0 has coincident united points at X_0, the equation of ϖ_0 is $\theta' = \theta+\alpha$ and the equations of ϖ are

$$\begin{aligned} x_0' &= x_0+2\alpha x_1+\alpha^2 x_2, \\ x_1' &= \qquad\quad x_1+\alpha x_2, \\ x_2' &= \qquad\qquad\quad x_2. \end{aligned}$$

The characteristic equation is $(1-\lambda)^3 = 0$; and there is only one characteristic root 1, with $\rho[\mathbf{A}-\mathbf{I}] = 2$ (unless $\alpha = 0$). Thus, unless ϖ_0 is the identical homography, ϖ has only one united point, namely X_0.

The above collineations admit of non-euclidean interpretation when k_0 is taken as absolute conic (cf. p. 97).

Homologies which transform one conic into another

We have seen that there exist collineations, and indeed even harmonic homologies, which transform any given proper conic into itself. In this section we shall examine the possibility of transforming one given conic into another by a suitably chosen homology. It is convenient to begin with the converse problem of seeing how a given conic behaves when it is transformed by a given homology.

Let k be a given conic and ϖ a non-special homology whose vertex A does not lie on k and whose axis a does not touch k. The

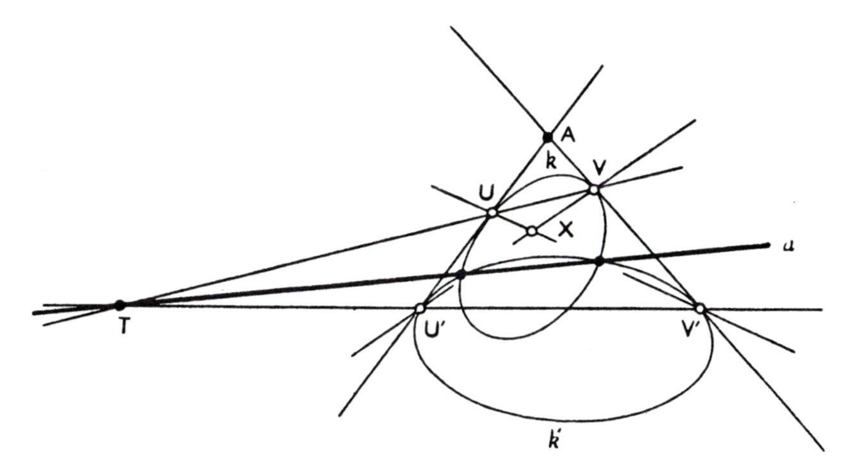

homology transforms k into another proper conic k', which passes through the points in which k is met by a and which touches the tangents from A to k. The chords of contact UV and $U'V'$ of these tangents with the two conics meet in a point T of a.

EXERCISE. How are the above conclusions modified (*a*) when ϖ is a special homology, and (*b*) when k passes through A, or touches a, or does both at once?

We can now turn to our original problem of determining ϖ when k and k' are given, and we have the following existence theorem.

THEOREM 9. *If k and k' are two proper conics of general position there exist twelve homologies each of which transforms k into k'.*

Proof. We first prove that if UU' and VV' (figure as on p. 223) are two common tangents of k and k', meeting in A, then the chords of contact UV and $U'V'$ are concurrent with two of the six common chords of k and k'.

Let UV, $U'V'$ meet in T. Then A and T are conjugate points for both conics. If, further, UV', $U'V$ meet in X, AX is the harmonic conjugate of AT with respect to AU, AV, and is therefore the polar of T for each conic. Now the conics of the pencil determined by k and k' cut an involution on AX, with united points M, N, and these points are then conjugate for every conic of the pencil. TMN is therefore a common self-polar triangle for k and k', and it is consequently the diagonal triangle of the quadrangle formed by the four common points of the conics. Thus two of the common chords of k and k' pass through each vertex of the triangle, and in particular two of them pass through T.

The homology ϖ with A as vertex, one of these two common chords as axis a, and (U, U') as a corresponding pair, clearly transforms k into a conic which touches AU and AV at U' and V' respectively and which passes through two of the common points of k and k', i.e. into the conic k'. Since there are six possible choices of A, and for each of these there are two choices of a, we have in all twelve homologies which transform k into k'.

EXERCISES

(i) Examine the special cases which arise when the common points of k and k' are not all distinct.

(ii) Obtain euclidean specializations of the various results by taking A or a at infinity.

§ 2. PLANE CORRELATIONS

The collineation, as we have seen in § 1, is a linear transformation of points into points and lines into lines, and it has the property of transforming every figure into a projectively equivalent figure, leaving all its projective properties invariant. We now turn to a different kind of transformation which, instead of leaving projective properties unaltered, systematically dualizes them. This transformation is known as the correlation.

DEFINITION. A *correlation* is a transformation of the points of a plane S_2 into the lines of a plane S_2' which is represented, in terms of arbitrary allowable representations $\mathscr{R}$ and $\mathscr{R}'$ of the two planes, by an equation of the form

$$\mathbf{u}' = \mathbf{A}\mathbf{x}.$$

In what follows we shall assume always that $|\mathbf{A}| \neq 0$, so that by 'correlation' we shall always mean 'non-singular correlation'. Let us consider, then, the correlation κ given by

$$\mathbf{x} \to \mathbf{u}' = \mathbf{A}\mathbf{x} \quad (|\mathbf{A}| \neq 0).$$

To any point P of S_2 corresponds a unique line $p' = \kappa P$ of S_2', and conversely, any line p' of S_2' arises from a unique point $P = \kappa^{-1}p'$ of S_2.

If P describes a line p of S_2, whose equation is $\mathbf{u}^T\mathbf{x} = 0$, then p' envelops the point P' of S_2' whose equation is $\mathbf{u}^T\mathbf{A}^{-1}\mathbf{u}' = 0$, i.e. the point $\mathbf{x}' = \mathbf{A}^{-1T}\mathbf{u}$. We say then that κ carries p into P'; and this means that the original point–line correlation carries with it an associated† line–point correlation, given by

$$\mathbf{u} \to \mathbf{x}' = \mathbf{A}^{-1T}\mathbf{u}.$$

Both correlations are uniquely reversible, and each determines the other (cf. Theorem 1). We shall regard them as aspects of the same duality correspondence between S_2 and S_2', and the essential connexion between them may be expressed by saying that if a point and a line are incident in S_2 then the corresponding line and point are incident in S_2'.

We say that a point P of S_2 is *conjugate*, with respect to κ, to a point P' of S_2' if P' lies on the line $p' = \kappa P$ corresponding to P —which implies also that P lies on the line $p = \kappa^{-1}P'$ corresponding to P'. When this is the case, the coordinate vectors $\mathbf{x}$, $\mathbf{x}'$ of P, P' are connected by the bilinear relation

$$W(\mathbf{x}, \mathbf{x}') \equiv \mathbf{x}^T\mathbf{A}^T\mathbf{x}' \equiv \mathbf{x}'^T\mathbf{A}\mathbf{x} = 0;$$

and, if either P or P' is fixed, this relation gives the equation of the corresponding line κP or $\kappa^{-1}P'$. The relation $W(\mathbf{x}, \mathbf{x}') = 0$ thus completely defines κ.

† This associated correlation is called by Hodge and Pedoe the 'dual correlation' belonging to κ. See *Methods of Algebraic Geometry*, p. 365.

In the same way, two lines p, p' of S_2, S_2' are said to be conjugate if each passes through the point corresponding to the other; and, if this is so, their coordinate vectors are connected by the bilinear relation

$$\Omega(\mathbf{u},\mathbf{u}') \equiv \mathbf{u}^T\mathbf{A}^{-1}\mathbf{u}' \equiv \mathbf{u}'^T\mathbf{A}^{-1T}\mathbf{u} = 0,$$

which likewise defines κ completely.

Up to this point we have treated S_2 and S_2' as distinct planes, and for the sake of clarity we shall continue to preserve the formal distinction as far as possible, even when S_2' is superimposed on S_2. When S_2 and S_2' are actually distinct, however, there is very little to be said, for in this case we are able to apply a transformation of coordinates $\bar{\mathbf{x}} = \mathbf{A}\mathbf{x}$ to S_2 while leaving the representation $\mathscr{R}'$ of S_2' unaltered. The equations of κ then reduce to

$$\mathbf{u}' = \bar{\mathbf{x}}, \qquad \mathbf{x}' = \bar{\mathbf{u}},$$

and the conjugacy relations become

$$W(\mathbf{x},\mathbf{x}') \equiv \bar{x}_0 x_0' + \bar{x}_1 x_1' + \bar{x}_2 x_2' = 0$$

and

$$\Omega(\mathbf{u},\mathbf{u}') \equiv \bar{u}_0 u_0' + \bar{u}_1 u_1' + \bar{u}_2 u_2' = 0.$$

We assume, therefore, from now on that S_2 and S_2' are coincident, and that the same coordinate representation $\mathscr{R}$ is used for both of them. In this case three topics of considerable interest present themselves for discussion:

(i) the collineation κ^2, obtained by applying κ twice over;
(ii) the coincidence locus W and the coincidence envelope Ω, i.e. the locus of a variable point P which lies on the corresponding line κP and the envelope of a variable line p which passes through the corresponding point κp;
(iii) the classification of correlations, and the canonical forms to which their equations may be reduced.

We shall not attempt here to deal in detail with all the problems which arise, and we refer the reader to Hodge and Pedoe, *Methods of Algebraic Geometry*, for a full discussion.

Let us consider first of all the way in which a correlation κ defines a collineation κ^2. The correlation transforms points into lines and lines into points according to the equations

$$\mathbf{x} \to \mathbf{u}' = \mathbf{A}\mathbf{x}, \qquad \mathbf{u} \to \mathbf{x}' = \mathbf{A}^{-1T}\mathbf{u}.$$

If, now, we apply κ twice in succession, every point $\mathbf{x}$ is transformed into a point $\mathbf{x}''$ and every line $\mathbf{u}$ into a line $\mathbf{u}''$, where

$$\mathbf{x}'' = \mathbf{A}^{-1T}\mathbf{u}' = \mathbf{A}^{-1T}\mathbf{A}\mathbf{x}$$

and

$$\mathbf{u}'' = \mathbf{A}\mathbf{x}' = \mathbf{A}\mathbf{A}^{-1T}\mathbf{u}.$$

But these are just the equations of a collineation

$$\mathbf{x} \to \mathbf{x}'' = \mathbf{B}\mathbf{x}, \qquad \mathbf{u} \to \mathbf{u}'' = \mathbf{B}^{-1T}\mathbf{u},$$

which is non-singular since $|\mathbf{B}| = |\mathbf{A}^{-1T}\mathbf{A}| = 1$; and we naturally denote this collineation by κ^2.

More generally, any two correlations κ_1, κ_2, with matrices $\mathbf{A}_1$, $\mathbf{A}_2$, define a collineation $\kappa_2\kappa_1$ whose matrix is $\mathbf{A}_2^{-1T}\mathbf{A}_1$. We may say, therefore, that although the self-correlations of the plane clearly do not form a group, the set of all (non-singular) collineations and correlations together is a group of transformations of the plane into itself. The identity element of the group is the collineation ϵ given by

$$\mathbf{x} \to \mathbf{x}' = \mathbf{x}, \qquad \mathbf{u} \to \mathbf{u}' = \mathbf{u};$$

and the set of all collineations is a subgroup of the full group. We see here how desirable it is to treat the plane as a single self-dual system, made up of ∞^2 points and ∞^2 lines at the same time.

We now turn back once again to the general correlation κ, given by $\mathbf{u}' = \mathbf{A}\mathbf{x}$, $\mathbf{x}' = \mathbf{A}^{-1T}\mathbf{u}$, in order to investigate the properties of its coincidence conics W and Ω. A point P belongs to W if it lies on its corresponding line $p' = \kappa P$, i.e. if it is self-conjugate for κ, and the algebraic condition for this is simply

$$W \equiv W(\mathbf{x}, \mathbf{x}) \equiv \mathbf{x}^T\mathbf{A}\mathbf{x} = 0.$$

$W = 0$ is therefore the equation of the coincidence locus of κ. The same conic is also the coincidence locus of κ^{-1}, for this correlation is given by $\mathbf{x}' \to \mathbf{u} = \mathbf{A}^T\mathbf{x}'$, and its matrix is therefore $\mathbf{A}^T$.

In the same way, the equation of the common coincidence envelope of κ and κ^{-1} is seen to be

$$\Omega \equiv \Omega(\mathbf{u}, \mathbf{u}) \equiv \mathbf{u}^T\mathbf{A}^{-1}\mathbf{u} = 0.$$

If, now, P is a point of W, the two lines κP and $\kappa^{-1}P$ both pass through P, and it follows that both these lines are tangents to Ω. Thus if P lies on W the tangents from P to Ω are κP and $\kappa^{-1}P$; and dually, if p touches Ω the points of intersection of p with W are κp and $\kappa^{-1}p$.

In the special case in which the matrix $\mathbf{A}$ is symmetric, W and Ω are the same proper conic k, regarded on the one hand as a locus and on the other as an envelope. The correlation κ is then simply the polarity defined by k, i.e. the correspondence of pole and polar with respect to this conic, and it is self-inverse. In general, however, when $\mathbf{A}$ is not symmetric, the two coincidence conics are distinct proper conics, but they are related in a remarkably simple manner.

THEOREM 10. *The two coincidence conics of a general self-correlation κ of the plane have double contact.*

Proof. We assume that κ is general in the sense that W and Ω are distinct proper conics. If P is a variable point of W, κP is one of the tangents from P to Ω, and $\kappa^2 P$ is then the second point of intersection of κP with W. Thus the collineation κ^2 transforms W into itself, and it therefore induces a homography on it (p. 221). Since P and $\kappa^2 P$ correspond in this homography on W, their join envelops a conic having double contact with W (Chapter VI, Theorem 18); but this join is a variable tangent to Ω, and the theorem is therefore proved.

Suppose we choose the triangle of reference in such a way that X_0 and X_2 are the points of contact of W and Ω, and X_1 is the common pole of the line $X_2 X_0$. Then clearly both κ and κ^{-1} transform the points X_0, X_1, X_2 respectively into the lines $X_0 X_1$, $X_2 X_0$, $X_1 X_2$. The matrix of κ is therefore of the form

$$\begin{pmatrix} 0 & 0 & a \\ 0 & b & 0 \\ c & 0 & 0 \end{pmatrix},$$

and we may, without loss of generality, represent κ by the equations

$$\begin{aligned} u_0' &= ax_2, \\ u_1' &= x_1, \\ u_2' &= cx_0 \end{aligned} \quad \text{and} \quad \begin{aligned} x_0' &= \frac{1}{a}u_2, \\ x_1' &= u_1, \\ x_2' &= \frac{1}{c}u_0. \end{aligned} \qquad \text{(A)}$$

The corresponding coincidence conics then have equations

$$\begin{aligned} W &\equiv x_1^2+(a+c)x_2 x_0 = 0, \\ \Omega &\equiv u_1^2+\left(\frac{1}{a}+\frac{1}{c}\right)u_2 u_0 = 0; \end{aligned} \qquad \text{(B)}$$

and since, by suitable choice of the triangle of reference and the constants a and c, the equations of any two proper conics with double contact can be represented by equations (B), it follows that equations (A) represent a general correlation κ. It is easily verified that the correlation κ given by (A) and its inverse κ^{-1} are the only correlations whose coincidence conics are given by (B). *Equations (A), then, give the canonical form to which the equations of a general self-correlation of the plane can be reduced.*

The collineation which is the square of the correlation κ defined by (A) is now easily obtained. When we apply κ twice over, the point (x_0, x_1, x_2) is first transformed into the line (ax_2, x_1, cx_0), and this line is then transformed into the point $\left(\frac{c}{a}x_0, x_1, \frac{a}{c}x_2\right)$. Thus the equations of the collineation are

$$x_0' = \frac{c}{a}x_0,$$
$$x_1' = x_1,$$
$$x_2' = \frac{a}{c}x_2.$$

It follows immediately that this collineation leaves the coincidence conics W and Ω invariant. We can infer also from the form of the equations of κ^2 that *only a collineation whose characteristic roots are in geometrical progression can be expressed as the square of a correlation.*

We may note, in conclusion, a euclidean specialization of the general correlation which is not without interest. If W is a circle whose centre is O, then the polar line of a point P with respect to W is obtained by drawing through the inverse point P^* a line perpendicular to OP^*. If we generalize this procedure by drawing always through P^* a line p_α making a constant angle α (in the positive sense) with OP^*, we may call p_α the polar at angle α of P for the circle. It is easy to show then that the correspondence $P \to p_\alpha$ is a correlation, projectively equivalent for general α to the general correlation, whose coincidence conics are the circle W and a concentric circle Ω. For $\alpha = \frac{1}{2}\pi$ or $\alpha = -\frac{1}{2}\pi$ the correlation is symmetrical, and is simply the polarity defined by W.

The reader should follow out, in this concrete case, the general properties of correlations that we have outlined above.

Our treatment of correlations in this chapter has been essentially algebraic, but in the older books on projective geometry such transformations were often introduced in a more geometrical manner. The polarity determined by a conic k is still sometimes referred to as reciprocation with respect to k, and this usage recalls an interesting chapter in the history of projective geometry. In the first major treatise on projective geometry—Poncelet's *Traité des propriétés projectives des figures*, published in 1822—there is an important section on the 'transformation by reciprocal polars'. Poncelet did not know of the principle of duality, but he saw that reciprocation provided a method of passing from any projective theorem involving points and lines to an equivalent theorem involving lines and points; and so he prepared the way for the eventual formulation, by Gergonne and Steiner in 1832, of the general principle of duality.

§ 3. Cremona Transformations

The transformations we have considered so far have all been linear, and for this reason they can be handled conveniently by the well-known methods of linear algebra and, in particular, by matrices. A more general class of transformations—of one plane into another or, more generally, of one algebraic manifold into another—consists of those algebraic transformations which are merely (1, 1), except possibly at special points. These are the so-called birational transformations, which are of fundamental importance in the more advanced parts of the geometry of algebraic manifolds. To treat them in a purely algebraic manner we would need more elaborate algebraic technique than is presupposed in this book, namely the theory of polynomials, and in this section, therefore, we shall only attempt to indicate one direction in which the notion of collineation can be generalized.

A *rational transformation* of a plane into another plane, or into itself, is a transformation which can be represented by equations of the form

$$\lambda y_i = \xi_i(x_0, x_1, x_2) \quad (i = 0, 1, 2), \tag{1}$$

where the ξ_i are homogeneous polynomials of the same degree $m \geqslant 1$ in x_0, x_1, x_2. In matrix notation we may write

$$\lambda \mathbf{y} = \boldsymbol{\xi}(\mathbf{x}). \tag{1'}$$

If the equations admit of rational solution in the form

$$\mu x_i = \eta_i(y_0, y_1, y_2) \quad (i = 0, 1, 2), \tag{2}$$

i.e.

$$\mu \mathbf{x} = \boldsymbol{\eta}(\mathbf{y}), \tag{2'}$$

where the η_i are polynomials of the same degree $n \geqslant 1$ in y_0, y_1, y_2, we say that the transformation is *birational*. Any such transformation of one plane into another, or into itself, is called a *Cremona transformation.*

The simplest example of a Cremona transformation—excluding the collineation, which is to be regarded in this connexion as trivial—is the so-called *reciprocal transformation*

$$y_0 : y_1 : y_2 = x_1 x_2 : x_2 x_0 : x_0 x_1,$$

or, as it is usually written,

$$y_0 : y_1 : y_2 = \frac{1}{x_0} : \frac{1}{x_1} : \frac{1}{x_2}.$$

The inverse of this transformation is of exactly the same form. We have already encountered the reciprocal transformation of a plane into itself, in geometrical guise, when considering the correspondence between points P and P^* which are conjugate for all the conics of a general pencil (p. 164); and the correspondence between harmonic pole and polar with respect to a triangle was seen on p. 78 to be a point–line transformation that admits of equations of the same reciprocal form.

Returning to the general Cremona transformation ϖ of a plane ρ into a plane σ, given by equations of the form

$$\lambda y_i = \xi_i(x_0, x_1, x_2) \quad (i = 0, 1, 2), \tag{1}$$

where x_0, x_1, x_2 and y_0, y_1, y_2 are coordinates in ρ, σ respectively, we note that the solved equations

$$\mu x_i = \eta_i(y_0, y_1, y_2) \quad (i = 0, 1, 2) \tag{2}$$

represent the Cremona transformation ϖ^{-1} of σ into ρ.

A general line of ρ, whose equation is

$$u_0 x_0 + u_1 x_1 + u_2 x_2 = 0,$$

corresponds to the curve in σ whose equation is

$$u_0 \eta_0(y_0, y_1, y_2) + u_1 \eta_1(y_0, y_1, y_2) + u_2 \eta_2(y_0, y_1, y_2) = 0,$$

i.e. $$\mathbf{u}^T \boldsymbol{\eta}(\mathbf{y}) = 0.$$

When u_0, u_1, u_2 are allowed to vary, we obtain the complete system of ∞^2 lines in ρ and, corresponding to them, a *linear net* or ∞^2 linear system of curves in σ. Thus *the transformation ϖ carries the lines of ρ into the curves of a linear net (η) in σ; and, in the same way, the inverse transformation ϖ^{-1} carries the lines of σ into the curves of a linear net (ξ) in ρ.*

Since the correspondence set up by ϖ is $(1, 1)$, except at special points, and since two lines in ρ have a unique point in common, the net of curves (η) is *homaloidal*; that is to say, two general curves of the net have one and only one free point of intersection (not common to all curves of the net). For a similar reason, the net (ξ) is also homaloidal.

A Cremona transformation, as we have remarked, may have exceptional points, for which the transformed point is not uniquely defined. Indeed, unless the transformation is a collineation there must be exceptional points. For, if the polynomials ξ_i are not all linear, the order of a general ξ-curve will be greater than 1; and the net (ξ) can only be homaloidal if the curves have fixed points in common, i.e. if the polynomials ξ_0, ξ_1, ξ_2 vanish simultaneously for at least one non-zero vector $\mathbf{x}$. If A is a base point of the net (ξ)—i.e. a point common to all the curves of the net—the ratios of the coordinates of the transformed point ϖA are indeterminate. We may say that the base points of (ξ) are *fundamental* for ϖ, and similarly that those of (η) are fundamental for ϖ^{-1}. The existence of fundamental points is thus characteristic of non-trivial Cremona transformations.

A general line of ρ and a general ξ-curve transform respectively into a general η-curve and a general line in σ, and the number of points of intersection remains invariant. It follows, therefore, that the ξ-curves and the η-curves are of the same order. This common order is called the *order* of the Cremona transformation.

When a Cremona transformation ϖ is given, the associated net (ξ) is uniquely determined as the system of curves obtained by applying the transformation ϖ^{-1} to the lines of the plane σ; but we are not able to assert that, conversely, ϖ is uniquely determined by the net of curves. The net defined by the three curves

$$\xi_i(x_0, x_1, x_2) = 0$$

may, in fact, be defined equally well by any triad of curves

$$\xi_i'(x_0, x_1, x_2) = 0 \quad (i = 0, 1, 2),$$

where the ξ_i' are three linearly independent linear combinations of the ξ_i. In other words, if we replace equation (1') by the new equation

$$\mathbf{y} = \boldsymbol{\xi}'(\mathbf{x})$$
$$= \mathbf{A}\boldsymbol{\xi}(\mathbf{x}) \quad (|\mathbf{A}| \neq 0),$$

we shall obtain a Cremona transformation ϖ' with the same generating net (ξ) as ϖ. We thus arrive at the theorem:

THEOREM 11. *Two Cremona transformations ϖ, ϖ' have the same generating net (ξ) if and only if*

$$\varpi' = \varpi_0 \varpi,$$

where ϖ_0 is a collineation.

From this point onwards we shall confine our attention to Cremona transformations of order 2, which we call *quadratic transformations*. For any such transformation, (ξ) and (η) must both be homaloidal nets of conics, i.e. systems of conics through three fixed points; and we shall consider only cases in which the base points in question are distinct. We shall assume finally, for the sake of simplicity, that the $\mathbf{x}$-plane ρ and the $\mathbf{y}$-plane σ are the same, and that $\mathbf{x}$ and $\mathbf{y}$ are coordinate vectors referred to the same representation $\mathscr{R}$.

Suppose, then, that ξ_0, ξ_1, ξ_2 are three quadratic forms in x_0, x_1, x_2, which all vanish simultaneously at three non-collinear base points X_0, X_1, X_2. We may take these points as reference points, and the net (ξ) then has the equation

$$\lambda_0 x_1 x_2 + \lambda_1 x_2 x_0 + \lambda_2 x_0 x_1 = 0.$$

One transformation based on this net is the reciprocal transformation

$$y_0 : y_1 : y_2 = x_1 x_2 : x_2 x_0 : x_0 x_1;$$

and hence, by Theorem 11, *any quadratic transformation with three distinct fundamental points is of the form $\varpi_0 \varpi$, where ϖ is a reciprocal transformation and ϖ_0 is a collineation.* In view of this fact, we can limit our discussion to the reciprocal transformation itself.

Let us consider, therefore, the reciprocal transformation ϖ just defined. Clearly, the correspondence which it sets up is involutory. It is $(1, 1)$ in general, but we observe that (i) if $\mathbf{x}$ is at a fundamental point it has no transform, and (ii) the points of $X_1 X_2$, except for X_1 and X_2, all transform into the same point X_0, and similarly for the points of $X_2 X_0$ and $X_0 X_1$. The three lines $X_1 X_2$, $X_2 X_0$, $X_0 X_1$, which join the fundamental points in pairs, are called *fundamental lines* of ϖ.

If the point $\mathbf{x}$ approaches the fundamental point X_0 along a specified curve, we may write,† in terms of a parameter t,

$$x_1/x_0 = \lambda t + o(t), \qquad x_2/x_0 = \mu t + o(t),$$

where λ and μ are constants. Then

$$y_0 : y_1 : y_2 = \lambda\mu t^2 + o(t^2) : \mu t + o(t) : \lambda t + o(t),$$

and in the limit, as t approaches zero, the point $\mathbf{y}$ approaches a determinate point $(0, \mu, \lambda)$ of $X_1 X_2$.

Conversely it is simple to show that if a variable point approaches any assigned point of $X_1 X_2$, other than X_1 or X_2, then its transform approaches X_0 in a definite 'direction'. Thus we have the following important property of ϖ.

THEOREM 12. *If X_0, X_1, X_2 are the fundamental points of a reciprocal transformation ϖ, then to points in the first neighbourhood of X_i (i.e. directions of approach to X_i) there correspond homographically the individual points of the opposite fundamental line; and, conversely, to points (other than fundamental points) of a fundamental line there correspond the related directions of approach to the opposite fundamental point.*

We leave the reader to supply the modifications required in regard to directions of approach to fundamental points along fundamental lines.

In considering the transformation of curves by ϖ, we exclude altogether the fundamental lines. That is to say, we investigate only *proper* transforms of curves, i.e. the actual loci of $\mathbf{y}$ corresponding to loci described by $\mathbf{x}$, disregarding any factors in the equation of a transformed curve which represent fundamental lines. In this sense, a line which does not pass through any fundamental point transforms into a conic through X_0, X_1, X_2, and conversely, a conic through these three points transforms into a line. A line through a fundamental point, however, yields another line through the same fundamental point. The equation $\lambda x_1 + \mu x_2 = 0$, for instance, leads to $y_0(\lambda y_2 + \mu y_1) = 0$; and the transformed line meets $X_1 X_2$ in the point corresponding to the direction of the original line at X_0.

Consider now the transform of a general curve C^n, of order n, which does not pass through X_0, X_1, or X_2. Since C^n meets a general ξ-conic in $2n$ points, the transformed curve meets a general line in

† We introduce non-algebraic notions here for simplicity, but the developments of modern algebra in fact make it possible to handle the problem by purely algebraic means.

$2n$ points, and it is therefore of order $2n$. But C^n meets each fundamental line in n points, and so its transform passes n times through each fundamental point, i.e. it has an n-fold point at each of X_0, X_1, X_2. We say that the transform is a curve of type

$$C^{2n}(X_0^n, X_1^n, X_2^n).$$

If it happens that two of the points of intersection of C^n with $X_1 X_2$ coincide, two of the nodal tangents of the transformed curve at X_0 clearly coincide also.

Now suppose that C^n passes simply through X_0, but not at all through X_1 or X_2. Then a general ξ-conic cuts C^n in the fixed point X_0 and in $2n-1$ free points; and the transformed curve is of order $2n-1$. This curve now passes only $n-1$ times through X_1 and X_2, although it still passes n times through X_0, and it is in fact of type $C^{2n-1}(X_0^n, X_1^{n-1}, X_2^{n-1})$.

In the general case, in which C^n has assigned multiplicities at the fundamental points of the quadratic transformation, we can readily establish the nature of the transformed curve by considering what happens to the equation of C^n when the appropriate substitution is made. In this way we obtain the following comprehensive theorem.

THEOREM 13. *If C^n is a curve of type $C^n(X_0^{\alpha_0}, X_1^{\alpha_1}, X_2^{\alpha_2})$, its proper transform by ϖ is a curve of type*

$$C^{2n-\alpha_0-\alpha_1-\alpha_2}(X_0^{n-\alpha_1-\alpha_2}, X_1^{n-\alpha_2-\alpha_0}, X_2^{n-\alpha_0-\alpha_1}).$$

Proof. Let the equation of C^n be $f(x_0, x_1, x_2) = 0$, where $f(x_0, x_1, x_2)$ is a polynomial of degree n which does not have x_0, x_1, or x_2 as a factor. Then the equation of the transformed curve is obtained from the equation

$$f(x_1 x_2, x_2 x_0, x_0 x_1) = 0$$

by dropping all factors which are powers of x_0, x_1, or x_2.

Since C^n has an α_0-fold point at X_0, we may write

$$f(x_0, x_1, x_2) \equiv x_0^{n-\alpha_0} u_{\alpha_0}(x_1, x_2) + x_0^{n-\alpha_0-1} u_{\alpha_0+1}(x_1, x_2) + \ldots + u_n(x_1, x_2),$$

where $u_i(x_1, x_2)$ is homogeneous of degree i in x_1 and x_2 together ($i = \alpha_0, \ldots, n$) and neither $u_{\alpha_0}(x_1, x_2)$ nor $u_n(x_1, x_2)$ vanishes identically. It is immediately evident that the polynomial

$$f(x_1 x_2, x_2 x_0, x_0 x_1)$$

contains the factor x_0 to the power α_0 exactly; and so, by symmetry, we have

$$f(x_1 x_2, x_2 x_0, x_0 x_1) \equiv x_0^{\alpha_0} x_1^{\alpha_1} x_2^{\alpha_2} g(x_0, x_1, x_2),$$

where $g(x_0, x_1, x_2)$ is a polynomial, of degree $2n-\alpha_0-\alpha_1-\alpha_2$, which has no factor x_i $(i = 0, 1, 2)$.

Furthermore, the degree of $g(x_0, x_1, x_2)$ in x_0 is clearly $n-\alpha_0$, and the multiplicity of X_0 as a point of the transformed curve is therefore $(2n-\alpha_0-\alpha_1-\alpha_2)-(n-\alpha_0)$, i.e. $n-\alpha_1-\alpha_2$. By symmetry, then, the curve has the stated multiplicities at X_0, X_1, X_2; and this completes the proof of the theorem.

By applying Theorem 13 to various simple curves we are now able to write down a large number of interesting results, of which those given immediately below are typical. The reader will find it a useful exercise to verify each of the statements from first principles, by direct substitution in the equation of the curve, and also to show that the converse of each statement is valid.

(i) A conic through X_0 and X_1 (but not through X_2) becomes a conic through X_0 and X_1.

(ii) A conic through X_0 only becomes a cubic which passes simply through X_1 and X_2 and has a node† at X_0.

(iii) A general conic becomes a quartic with nodes at X_0, X_1, and X_2.

(iv) A conic inscribed in the triangle $X_0 X_1 X_2$ becomes a quartic with cusps† at X_0, X_1, and X_2.

(v) A cubic through X_0 and X_1 becomes a quartic which passes simply through X_2 and has nodes at X_0 and X_1.

The reader will now have no difficulty in seeing how to derive many properties of nodal cubics and trinodal quartics from known properties of conics by applying a quadratic transformation.

To conclude this section, we may point out that the reader already knows one special type of quadratic transformation, namely inversion with respect to a circle in the euclidean plane. This symmetrical transformation of the plane into itself has an abundance of striking applications, many of which will already be familiar.

To see how inversion fits into the general scheme, we first form the equations, in homogeneous rectangular cartesian coordinates x, y, z, of the reciprocal transformations whose fundamental points are the origin O and the absolute points I, J. Clearly the equations

$$\frac{x'+iy'}{a} = \frac{a}{x+iy}, \qquad \frac{x'-iy'}{a} = \frac{a}{x-iy}, \qquad z' = \frac{1}{z}$$

† Nodes and cusps are double points with distinct and coincident nodal tangents respectively.

define such a transformation for any non-zero value of a; and these equations reduce to

$$\frac{x'}{z'} = \frac{a^2zx}{x^2+y^2}, \qquad \frac{y'}{z'} = -\frac{a^2yz}{x^2+y^2}.$$

If we now superimpose on this transformation the collineation

$$x'' : y'' : z'' = x' : -y' : z',$$

we obtain the equations

$$\frac{x''}{z''} = \frac{a^2zx}{x^2+y^2}, \qquad \frac{y''}{z''} = \frac{a^2yz}{x^2+y^2},$$

which represent inversion with respect to the circle $X^2+Y^2 = a^2$. We have, therefore, the theorem:

THEOREM 14. *Inversion is a quadratic transformation, with the centre of inversion O and the absolute points I, J as fundamental points. It can be generated by a reciprocal transformation based on O, I, J, followed by reflection in a line through O (a collineation which leaves O invariant and interchanges I and J).*

The homaloidal net (ξ) belonging to the quadratic transformation is the net of all circles through O. These circles invert, as we know, into the lines of the plane. Also, from the first of the five properties of quadratic transformations listed on p. 236, we see at once that circles not through O transform into circles.

EXERCISES ON CHAPTER IX

1. The coordinates being cartesian, find the equations of the plane collineation that leaves the three points $(2, 0)$, $(-1, \sqrt{3})$, $(-1, -\sqrt{3})$ invariant and transforms the point $(0, 0)$ into the point at infinity on the X-axis.

2. Find all the united points of the collineations

(i) $x' : y' : z' = 3x-2z : 3y : x-y$;
(ii) $x' : y' : z' = x : z : y$;
(iii) $x' : y' : z' = y+z : z+x : x+y$;
(iv) $x' : y' : z' = y : -x+2y : -2x+2y+z$.

Find also, in each case, the equations of the corresponding line–line transformation.

Show that (ii) is a harmonic homology, (iii) is a homology of modulus -2, and (iv) is a special homology. Find the vertex of (iv).

3. Show that the equations of any homology (general or special) whose vertex is (X, Y, Z) and whose axis is $l \equiv ux+vy+wz = 0$ are of the form

$$x' : y' : z' = px+Xl : py+Yl : pz+Zl,$$

where p is a constant.

Show also that the homology is special if $uX+vY+wZ = 0$, and that if it is not special its modulus is $(uX+vY+wZ+p)/p$.

4. Prove that any collineation which leaves invariant the reference point Z and the reference line XY has equations of the form

$$x' : y' : z' = ax+by : cx+dy : z,$$

where $ad-bc \neq 0$; and prove that this collineation is a homology, with vertex on XY and axis through Z, if $(a-1)(d-1) = bc$. Show also that the collineation is a harmonic homology which interchanges X and Y if

$$a = d = bc-1 = 0.$$

If ϖ is the collineation

$$x' : y' : z' = \lambda_1 x : \lambda_2 y : z,$$

where $\lambda_1 \lambda_2 = 1$, prove that ϖ can be expressed, in infinitely many ways, as the product $\tau_2 \tau_1$ of two harmonic homologies τ_1, τ_2 which interchange X and Y. Interpret this result in the special case in which ϖ, τ_1, τ_2 are all real collineations and X, Y are the absolute points.

5. The equations in non-homogeneous coordinates of two collineations σ and τ are

$$X' = X/Y, \qquad Y' = 1/Y, \quad \text{and} \quad X' = 1/X, \qquad Y' = -Y/X$$

respectively. Show that $\sigma^2 = \tau^2 = \epsilon$, where ϵ is the identical transformation, and that σ and τ generate the group of six collineations whose members are $\epsilon, \sigma, \tau, \sigma\tau, \tau\sigma, \sigma\tau\sigma$.

Discuss the character of these collineations, and show that if n is any odd integer the curve $X^n - Y^n = 1$ is left invariant by every transformation of the group.

6. Show that a $(1, 1)$ correspondence between points P, P' of the plane is a collineation if it carries lines into lines in such a way that P and P' always describe homographic ranges on corresponding lines.

7. A, B, A', B' are four fixed points and l is a fixed line through the intersection of AB and $A'B'$; and a transformation $P \rightarrow P'$ is defined by the condition that PA, $P'A'$ meet on l and PB, $P'B'$ also meet on l. Show that when P describes a line P' also describes a line.

Find the self-corresponding points of the transformation.

8. Show, in each of the following cases, that the correspondence between P and P' is a collineation, and discuss in each case the united points and the geometrical character of the transformation.

(i) the pairs of tangents from P and P' to a fixed conic k meet a fixed line a (not a tangent to k) in the same pair of points;

(ii) the pairs of tangents from P and P' to two given conics k and k' respectively meet a given common tangent of k and k' in the same pair of points;

(iii) P and P' are poles of the same line with respect to two given conics k and k'.

State also, in each of the three cases, the corresponding dual result.

9. A, B being a given pair of points and a, b a given pair of lines, both of general position, a correspondence between points P, P' of the plane is determined by the condition that AP, BP' meet on a and AP', BP meet on b. Show that the correspondence is a collineation, and find its simplest equations, using the triangle formed by a, b, and AB as triangle of reference. Investigate the united points of the collineation.

10. Show that, in the euclidean plane, the product of two radial expansions is either a radial expansion or a translation, distinguishing between the two cases. Under what condition is the product a reflection in a fixed point?

State the general projective theorems which correspond to these results, and prove that two homologies with the same axis cannot commute with each other unless they are both special or both have the same vertex.

11. Show that the circles which are transformed into circles by a given general collineation of the euclidean plane form a coaxal system.

If a coaxal system (c) is transformed into itself by a plane collineation ϖ, which transforms neither of the absolute points into an absolute point, prove that ϖ is either (i) a harmonic homology with one limiting point as vertex and a line through the other as axis, or (ii) one or other of two collineations whose squares are each the reflection in the line of centres of (c).

12. If two plane collineations ϖ, σ satisfy the condition $\varpi\sigma = \sigma\varpi$, prove that σ transforms any united point of ϖ into another, or the same, united point of ϖ.

If ϖ has only three united points A, B, C, prove that for σ to commute with ϖ, it is necessary and sufficient that either σ has A, B, C as united points or each of ϖ, σ is cyclic of period three and each permutes the united points of the other cyclically.

13. A variable line p meets a fixed conic k in U, V, and the lines joining U, V to a fixed point A, not on k, meet k again in points U', V' whose join is p'. Prove that p, p' correspond in a harmonic homology with A as vertex and the polar line of A with respect to k as axis.

If ϖ_A, ϖ_B are the two harmonic homologies derived as above from different points A, B and the same conic k, find the united points of the collineation $\varpi_A \varpi_B$, and prove that this collineation is also a harmonic homology if A, B are conjugate points for k.

14. Two chords AB, CD of a conic k meet in O. Prove that there are four collineations each of which leaves O and k invariant and transforms the line AB into CD. Show that two of these collineations are harmonic homologies whose axes meet in O, and that if ϖ_1, ϖ_2 are the other two then $\varpi_2 \varpi_1^{-1}$ is a harmonic homology with vertex O.

15. Two conics k, k' touch at O, meet in two further points A, B, and have two further common tangents u, v. Show that there exist two homologies, one with O as vertex and AB as axis and the other with uv as vertex and the common tangent at O as axis, which transform k into k'. Obtain simple euclidean cases of these results.

16. Show that if there exists a special homology which transforms a conic k into a conic k' (these two conics having four distinct common points), then there exists also a harmonic homology which transforms k into k'.

Show also that in this case k and k' are projectively equivalent to a pair of equal circles.

17. If k' is the transform of a conic k by a special homology ϖ, describe the relation of k' to k when (i) k is of general position, (ii) k passes through the vertex of ϖ, (iii) k touches the axis of ϖ, (iv) k touches the axis at the vertex.

If two conics have their asymptotes parallel and possess a pair of parallel common tangents, prove that one of them can be obtained from the other by means of a translation.

18. Show that a general plane collineation does not leave any conic invariant, but that if the characteristic roots of the matrix of the collineation are in geometric progression the collineation leaves invariant every conic of a simply infinite family.

19. Show that any $(2, 2)$ algebraic correspondence on a conic is associated with a unique correlation of the plane, the two points corresponding to any point P of the conic being those in which the conic is met by the line into which P is transformed by the correlation. (Cf. Chapter VIII, Exercise 1.)

20. If ϖ_0 is a non-involutory homographic correspondence between points P, P' on a conic k, show that the line PP' corresponds to P in a fixed correlation with k as coincidence locus. If the joins of two pairs P_1, P_1' and P_2, P_2' meet in T, show that $P_1' P_2'$ is the line corresponding to T in the correlation in question.

21. If k is the conic whose equation is $x^2+y^2+z^2 = 0$, write down the equations of the general correlation whose coincidence locus is k. Find the associated coincidence envelope k', and verify that k and k' have double contact.

22. If a rotation through a fixed angle α about a fixed point O carries any point P into a point P', and the polar of P' with respect to a fixed circle of centre O is p, show that the correspondence $P \to p$ is a correlation; and investigate its coincidence locus and envelope.

If α is a right angle, show that the correlation is special, having the line at infinity counted twice as its coincidence locus and the point O counted twice as its coincidence envelope.

23. Four circles c_1, c_2, c_3, c_4 cut at O, and c_i cuts c_j a second time at P_{ij}. Show that the pairs of lines which join O to opposite pairs of points P_{ij} are in involution.

Show that a second involution has the tangents at O to c_1, c_2, c_3 as mates of the lines OP_{23}, OP_{31}, OP_{12} respectively.

24. Show that the three cuspidal tangents of a tricuspidal quartic curve meet in a point, and that the curve has a unique double tangent, which is the harmonic polar of the point of concurrence in question with respect to the triangle whose vertices are the cusps.

25. Show that a circular cubic curve (i.e. a cubic through the absolute points) which also possesses a node has two systems of bitangent circles, and that the circle through the node and the points of contact of any one of these circles touches one or other of two fixed lines at the node.

26. Show that a nodal cubic curve is of class four, and prove that it has three points of inflexion, which are collinear.

CHAPTER X

PROJECTIVE GEOMETRY OF THREE DIMENSIONS

IF we give n the value 3 in the general definition at the beginning of Chapter III, we obtain formal definitions of three-dimensional projective space S_3 and projective geometry of three dimensions. The projective properties of S_3 can then all be deduced from the fundamental definition of a three-dimensional projective domain, for the most part by methods of linear algebra, but in the present chapter we do not propose to go into all the details of this development. Many of the subsidiary definitions and the enunciations and proofs of theorems are strictly analogous to those already given for S_2 in Chapter IV, and the reader can easily supply them for himself. We shall accordingly pass quickly over this part of the theory, merely calling attention to certain features that arise for the first time when the number of dimensions attains the value 3.

S_3 is much richer in projective properties than S_1, or even S_2; and by the time we are ready to study so complex a geometrical system we must be able to think in geometrical rather than algebraic terms. In the informal treatment of projective geometry outlined in Part I we relied upon geometrical intuition in the naïve sense. Then, as we began the formal deductive treatment of the same subject in the early chapters of Part II, we found it necessary to proceed slowly and cautiously by small algebraic steps. Now we can begin to detach ourselves from the details of the algebra, making use once more of geometrical intuition, but this time less naïvely. We have seen how projective geometry can be provided with a solid algebraic foundation, and we know how to build it up in this way, but we do not now need always to be conscious of the presence of the algebra. We are free, in fact, to think geometrically once again, going back to algebraic symbols only when added precision has to be given to geometrical notions.

§ 1. THE CONSTRUCTION AND GENERAL PROPERTIES OF THE SYSTEM

Let us now summarize the essential steps in the construction of a formal system of three-dimensional projective geometry.

The points of S_3 are represented, in any allowable representation

$\mathscr{R}$, by tetrads of homogeneous coordinates (x_0, x_1, x_2, x_3) belonging to the ground field K (which we take once again to be the complex field); and the different allowable representations are connected by the complete group of non-singular linear transformations with coefficients in K. Thus the transformation from one allowable representation to another may be written as

$$x'_i = \sum_{k=0}^{3} a_{ik} x_k \quad (i = 0, 1, 2, 3),$$

or

$$\mathbf{x}' = \mathbf{A}\mathbf{x},$$

where

$$|a_{rs}| \equiv |\mathbf{A}| \neq 0.$$

Linear dependence of points

As in Chapter IV, any relation of linear dependence between vectors in $V_4(K)$ may be carried over to the points that are represented by these vectors in a given representation $\mathscr{R}$. More precisely, a point P is said to be linearly dependent on a set of points

$$P_1, P_2, \ldots, P_m$$

if a coordinate vector representing P in some representation $\mathscr{R}$ is linearly dependent on coordinate vectors representing $P_1, P_2, \ldots, P_m$ in $\mathscr{R}$. If the relation holds between vectors in $\mathscr{R}$ it also holds between the corresponding vectors in every other allowable representation, and linear dependence is therefore an intrinsic property of points.

Now suppose X_0, X_1, X_2, X_3, E are five given points of S_3, no four of which are linearly dependent. Then, by Theorem 1 of the Appendix, there exists a unique allowable representation $\mathscr{R}$ in which the five points have respectively the coordinates $(1, 0, 0, 0)$, $(0, 1, 0, 0)$, $(0, 0, 1, 0)$, $(0, 0, 0, 1)$, $(1, 1, 1, 1)$. We call X_0, X_1, X_2, X_3 the reference points of $\mathscr{R}$, or the vertices of the tetrahedron of reference, and E the unit point. Specifying the reference points and the unit point is, as a rule, the most convenient way of fixing a representation of S_3. We shall now assume that this has been done, so that we have a definite representation $\mathscr{R}$ to work with.

Let P_1, P_2, P_3 be three given points, represented by coordinate vectors $\mathbf{x}^{(1)}$, $\mathbf{x}^{(2)}$, $\mathbf{x}^{(3)}$ respectively. These vectors may be taken as columns of a 4×3 matrix $(x_r^{(s)})$, and the rank ρ of this matrix has a simple interpretation, for the three points are linearly independent if and only if $\rho = 3$. If this condition is satisfied, the three sets of coordinates satisfy one and (effectively) only one linear equation

$$\sum_{i=0}^{3} u_i x_i = 0,$$

and all the points whose coordinates satisfy this equation, and only these points, are linearly dependent on P_1, P_2, P_3. We call the set of all such points the *plane* determined by P_1, P_2, and P_3. It follows at once that a plane is determined by any three of its points, provided only that they are linearly independent. Four points are said to be coplanar if they belong to a common plane; and this is the case if and only if the determinant of their coordinates is zero.

EXERCISE. Write down the equation of the plane determined by the points P_1, P_2, P_3 above.

Now consider a pair of given points P_1, P_2, with coordinate vectors $\mathbf{x}^{(1)}$, $\mathbf{x}^{(2)}$. The points are linearly independent if and only if the 4×2 matrix $(x_r^{(s)})$ is of rank 2; and in this case their coordinates satisfy two (and not more than two) linearly independent linear equations

$$\sum_{i=0}^{3} u_i x_i = 0, \qquad \sum_{i=0}^{3} v_i x_i = 0.$$

The points which are linearly dependent on P_1 and P_2 are the points whose coordinates satisfy these two equations simultaneously, and these points are said to make up the *line* determined by P_1 and P_2. Every point of the line is said to be collinear with P_1 and P_2; and, of course, the roles of P_1 and P_2 may be taken over by any two linearly independent points of the line.

Relations of incidence

A point P is incident with (or belongs to) a plane when its coordinates satisfy the equation of the plane, and it is incident with a line when its coordinates simultaneously satisfy the two equations of the line. We thus have the two fundamental incidence relations

$$\sum_{i=0}^{3} u_i x_i = 0 \quad \text{and} \quad \sum_{i=0}^{3} u_i x_i = 0 = \sum_{i=0}^{3} v_i x_i;$$

and from these relations numerous incidence properties of planes and lines follow at once by the theory of linear equations. Thus a line either has a unique point in common with a plane or it lies wholly in the plane. Two distinct planes have a unique line in common. Three distinct planes have either a single point or a line in common. Two lines either lie in a common plane, when they also have a unique common point, or they neither lie in a common plane nor possess a common point; and in the latter case the lines are said to be *skew*.

EXERCISE. Prove that through a given point which does not lie on either of two given skew lines there is a unique transversal line which meets both of the skew lines.

The representation of planes and lines

A plane may be represented algebraically in two different ways.

(i) If the plane is determined, as above, by three points P_1, P_2, P_3, a general point P in it has a coordinate vector

$$\mathbf{x} = \lambda_1 \mathbf{x}^{(1)} + \lambda_2 \mathbf{x}^{(2)} + \lambda_3 \mathbf{x}^{(3)};$$

and so we have a parametric representation of the plane by a triad of homogeneous parameters $(\lambda_1, \lambda_2, \lambda_3)$.

(ii) Alternatively, the plane may be represented by an equation

$$\sum_{i=0}^{3} u_i x_i = 0.$$

This equation is determined by its coefficients, which may be taken as a tetrad of homogeneous *plane-coordinates* of the plane. If the column-vector with components (u_0, u_1, u_2, u_3) is denoted by $\mathbf{u}$, the equation of the plane may be expressed in terms of the inner product of the vectors $\mathbf{u}$ and $\mathbf{x}$:

$$(\mathbf{u}, \mathbf{x}) \equiv \mathbf{u}^T \mathbf{x} = 0.$$

A line is fixed by any two of its points P_1, P_2; and the equation $\mathbf{x} = \lambda_1 \mathbf{x}^{(1)} + \lambda_2 \mathbf{x}^{(2)}$, which gives the coordinate vector of a general point of the line, leads at once to the homogeneous parametric representation of the line by the pair of parameters (λ_1, λ_2).

Line-coordinates

Not only the points and the planes of S_3 but also the lines may be represented by suitably chosen sets of homogeneous coordinates; and this representation is of some interest because it is different in kind from any that we have met so far. The coordinates of a line, as will shortly appear, form a redundant set and are connected by an identical relation.

Let p be a general line of space. The line may be fixed by any two of its points—say the points P_1 and P_2, whose coordinate vectors in the representation $\mathscr{R}$ are $\mathbf{x}^{(1)}$ and $\mathbf{x}^{(2)}$. If we now put

$$p_{ij} = x_i^{(1)}x_j^{(2)} - x_j^{(1)}x_i^{(2)} \quad (i,j = 0,..., 3)$$

we obtain a set of numbers which can function as coordinates of the line p. There are sixteen of the p_{ij}, but, since they are clearly the elements of a skew-symmetric 4×4 matrix (i.e. p_{ji} is always equal to $-p_{ij}$) their number reduces effectively to six. So we arrive at the set of six numbers

$$p_{23}, p_{31}, p_{12}, p_{01}, p_{02}, p_{03}$$

which we take to be components of the coordinate vector $\mathbf{p}$ of p.

The fundamental property of these line-coordinates is that their ratios are independent of the choice of the two points P_1, P_2 on p; i.e. they are a unique set of coordinates of p. To see this it is only necessary to observe that if

$$\bar{\mathbf{x}}^{(1)} = \lambda \mathbf{x}^{(1)} + \mu \mathbf{x}^{(2)}$$

and

$$\bar{\mathbf{x}}^{(2)} = \lambda' \mathbf{x}^{(1)} + \mu' \mathbf{x}^{(2)}$$

then

$$\bar{p}_{ij} \equiv \bar{x}_i^{(1)}\bar{x}_j^{(2)} - \bar{x}_j^{(1)}\bar{x}_i^{(2)} = (\lambda\mu' - \lambda'\mu)p_{ij}.$$

In this way every line yields a well-defined set of ratios

$$p_{23} : p_{31} : p_{12} : p_{01} : p_{02} : p_{03}.$$

The converse of this statement, on the other hand, that is to say the statement that every such set of ratios arises from some line of space, cannot possibly be true. There are ∞^5 sets of ratios of the p_{ij}, but since a line is uniquely determined by the points in which it meets two fixed planes there are only ∞^4 lines. The coordinates of every line must therefore be connected by one and the same identical relation, which reduces their effective number to four; and this relation is not hard to discover. Expanding the vanishing determinant

$$\begin{vmatrix} x_0^{(1)} & x_1^{(1)} & x_2^{(1)} & x_3^{(1)} \\ x_0^{(2)} & x_1^{(2)} & x_2^{(2)} & x_3^{(2)} \\ x_0^{(1)} & x_1^{(1)} & x_2^{(1)} & x_3^{(1)} \\ x_0^{(2)} & x_1^{(2)} & x_2^{(2)} & x_3^{(2)} \end{vmatrix}$$

in terms of the first two rows, we obtain the identity

$$\Omega_{pp} \equiv p_{01}p_{23} + p_{02}p_{31} + p_{03}p_{12} = 0,$$

and this quadratic relation holds between the coordinates p_{ij} of every line. The six numbers p_{ij}, connected by the relation $\Omega_{pp} = 0$, are usually referred to as the *Grassmann* (or *Plücker*) *coordinates* of the line p.

Instead of defining the coordinates of a line in terms of the point-coordinates of two points on the line we could equally well begin with the plane-coordinates of two planes through the line, and this dual procedure leads to the introduction of the dual coordinates

$$\pi_{ij} = u_i^{(1)}u_j^{(2)} - u_j^{(1)}u_i^{(2)} \quad (i,j = 0,..., 3).$$

The same line p now has two distinct representations, by vectors $\mathbf{p}$ and $\boldsymbol{\pi}$ respectively, but it is easy to show that they come ultimately to the same thing. The points $\mathbf{x}^{(1)}$, $\mathbf{x}^{(2)}$ both lie in each of the planes $\mathbf{u}^{(1)}$, $\mathbf{u}^{(2)}$, and from the equations

$$u_0^{(1)}x_0^{(1)} + u_1^{(1)}x_1^{(1)} + u_2^{(1)}x_2^{(1)} + u_3^{(1)}x_3^{(1)} = 0,$$

$$u_0^{(2)}x_0^{(1)} + u_1^{(2)}x_1^{(1)} + u_2^{(2)}x_2^{(1)} + u_3^{(2)}x_3^{(1)} = 0$$

we have, by eliminating $x_0^{(1)}$,

$$\pi_{01}x_1^{(1)} + \pi_{02}x_2^{(1)} + \pi_{03}x_3^{(1)} = 0.$$

Similarly $$\pi_{01}x_1^{(2)} + \pi_{02}x_2^{(2)} + \pi_{03}x_3^{(2)} = 0,$$

and from these two equations

$$\frac{\pi_{01}}{p_{23}} = \frac{\pi_{02}}{p_{31}} = \frac{\pi_{03}}{p_{12}}.$$

The two sets of coordinates are therefore connected by the relations

$$\pi_{01} : \pi_{02} : \pi_{03} : \pi_{23} : \pi_{31} : \pi_{12} = p_{23} : p_{31} : p_{12} : p_{01} : p_{02} : p_{03}.$$

We shall not make very much use of line-coordinates until we take up the systematic study of line-geometry in Chapter XV, but before leaving the subject at this stage it is worth while obtaining the condition for two lines to intersect.

Let p, q be two lines in space, determined by the pairs of points P_1, P_2 and Q_1, Q_2 respectively. The lines intersect if and only if the four points P_1, P_2, Q_1, Q_2 are coplanar, and a necessary and sufficient condition for this is

$$\begin{vmatrix} x_0^{(1)} & x_1^{(1)} & x_2^{(1)} & x_3^{(1)} \\ x_0^{(2)} & x_1^{(2)} & x_2^{(2)} & x_3^{(2)} \\ y_0^{(1)} & y_1^{(1)} & y_2^{(1)} & y_3^{(1)} \\ y_0^{(2)} & y_1^{(2)} & y_2^{(2)} & y_3^{(2)} \end{vmatrix} = 0,$$

where $\mathbf{x}^{(1)}$, $\mathbf{x}^{(2)}$, $\mathbf{y}^{(1)}$, $\mathbf{y}^{(2)}$ are coordinate vectors of P_1, P_2, Q_1, Q_2. Expanding the determinant in terms of its first two rows enables us to write the condition in terms of line-coordinates in the form

$$2\Omega_{pq} \equiv p_{01}q_{23}+p_{02}q_{31}+p_{03}q_{12}+p_{23}q_{01}+p_{31}q_{02}+p_{12}q_{03} = 0,$$

Ω_{pq} being derived from the quadratic form Ω_{pp} by the process of polarization.

Subordinate projective geometries

We now consider again the parametric representation of the plane $P_1P_2P_3$. The numbers $(\lambda_1, \lambda_2, \lambda_3)$ may be regarded as coordinates of a general point P in the plane; and since choosing three other points Q_1, Q_2, Q_3 to fix the plane amounts to applying a non-singular linear transformation to these coordinates, the projective geometry of S_3 induces a subordinate projective geometry of two dimensions in the plane. This is true of every plane in S_3, and in the same way a one-dimensional projective geometry is induced on every line of S_3. When looked at from this point of view, a plane is often referred to as a *plane field* (set of ∞^2 points with the structure of an S_2) and the line as a *range of points*. An important special case of the above hierarchy of projective geometries (which, in virtue of the freedom of choice of $\mathscr{R}$, is not really special at all) is the following. The vertices X_1, X_2, X_3 of the tetrahedron of reference determine a plane, the face of the tetrahedron opposite to X_0. A general point of this plane has coordinates $(0, x_1, x_2, x_3)$, and x_1, x_2, x_3 may be identified with the λ_1, λ_2, λ_3 of the preceding theory. We then have an allowable representation of the plane field determined by the plane $X_1X_2X_3$, the reference points being X_1, X_2, X_3 and the unit point the point $(0, 1, 1, 1)$ common to the plane and the line X_0E. This point is the *subordinate unit point* E_0 of the face $X_1X_2X_3$ of the tetrahedron of reference.

A general point of the line X_2X_3 is $(0, 0, x_2, x_3)$; and x_2, x_3 are coordinates in an allowable representation of this line, the reference points being X_2, X_3, and the unit point the point $(0, 0, 1, 1)$ in which X_2X_3 is met by X_1E_0. This point is the *subordinate unit point* E_{01} of the edge X_2X_3 of the tetrahedron of reference.

We can treat the one-dimensional projective geometry of X_2X_3 either as subordinate directly to the geometry of S_3 or as subordinate to the two-dimensional geometry of $X_1X_2X_3$ which is itself subordinate to the geometry of S_3, but in the end the choice of point of view is seen to make no difference. We do not need, therefore, to keep the original three-dimensional geometry separate from its various subordinate geometries, and we shall think of the full three-dimensional geometry of S_3 as comprising all the projective properties of S_3 itself and of its linear subspaces.

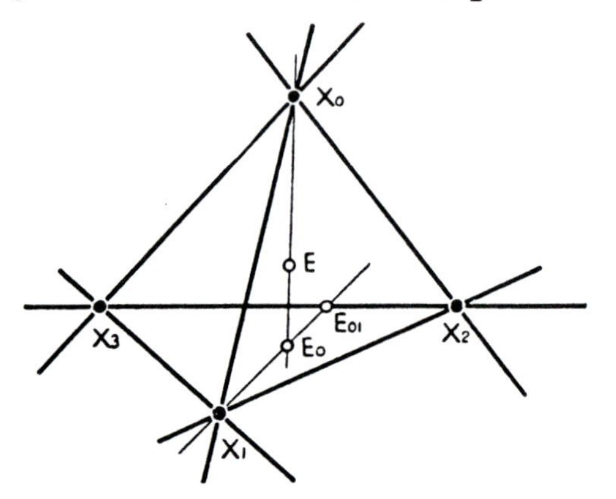

Consider, for example, the line determined by two points P_1, P_2. A general point of the line may be represented, using a non-homogeneous parameter θ, by the vector $\mathbf{x} = \mathbf{x}^{(1)}+\theta\mathbf{x}^{(2)}$. Then θ is a projective parameter on the line, and the cross ratio of two pairs of points may be expressed as usual in terms of the parameters of the points. In this way the cross ratio of two pairs of points on a line enters into the projective geometry of S_3.

The principle of duality

The fundamental incidence relation between a point and a plane, expressed in the equation

$$\sum_{i=0}^{3} u_i x_i = 0,$$

involves point-coordinates and plane-coordinates symmetrically. This, together with the fact that a non-singular linear transformation of point-coordinates induces a transformation of plane-coordinates of the same type, gives rise to a principle of duality in space. Duality is rather more complicated in space than in the plane since, in addition to points and planes, which are dual to each other, we now have an intermediate class of self-dual entities—the lines of space. The principle of duality may be stated formally as follows:

If T is any theorem, valid in three-dimensional projective geometry, the theorem T′ that is obtained from T by changing the word 'point' into the word 'plane', and vice versa, throughout the enunciation, and making the appropriate linguistic adjustments, is also valid in the same geometry.

To every kind of submanifold of S_3 corresponds a dual manifold, and this is true in particular of the range of points and the plane field.

The range of points is the set of ∞^1 points which lie on a fixed line, the axis of the range, and the dual manifold consists of the ∞^1 planes which pass through a fixed line. These planes are said to form a *pencil of planes* with the given line as axis. Two planes π_1, π_2, with coordinate vectors $\mathbf{u}^{(1)}, \mathbf{u}^{(2)}$, determine a unique pencil, a general plane of which is given by $\mathbf{u} = \mathbf{u}^{(1)}+\theta\mathbf{u}^{(2)}$; and θ is a projective parameter for the pencil, connected in the usual way with cross ratio. The cross ratio $\{\pi_1, \pi_2; \pi_3, \pi_4\}$ of two ordered pairs of planes of the same pencil is defined as the dual of the cross ratio $\{P_1, P_2; P_3, P_4\}$ for points of a range. It may be verified directly, and we shall also prove below (corollary to Theorem 4) that the cross ratio $\{\pi_1, \pi_2; \pi_3, \pi_4\}$ is equal to the cross ratio of the pairs of points intercepted on any transversal. In this way we have a fundamental connexion between the range and the pencil.

A plane field consists in the first instance of the set of ∞^2 points of a given plane, and to these points we may add the ∞^2 lines of the plane in order to make up the complete two-dimensional projective domain. The space-dual of the plane field is the *star of planes and lines*, the system of ∞^2 planes and ∞^2 lines that pass through a fixed point, the vertex of the star. Three linearly independent planes determine a star, a general plane of which is represented by a coordinate vector of the form $\mathbf{u} = \lambda_1 \mathbf{u}^{(1)}+\lambda_2 \mathbf{u}^{(2)}+\lambda_3 \mathbf{u}^{(3)}$. Two planes of the star intersect in a unique line of the star, and all those planes which pass through the line form a pencil of planes in the star.

Some important incidence theorems

We are now in a position to prove a number of incidence theorems which hold in S_3.

THEOREM 1. *Let A_0, A_1, A_2, A_3 be four points which form the vertices of a tetrahedron, and let P be a point which does not lie on any of the faces of the tetrahedron. Let $A_i P$ meet the opposite face in P_i $(i = 0, 1, 2, 3)$ and let p_i be the harmonic polar of P_i with respect to the triangle whose vertices are the vertices of the tetrahedron other than A_i. Then the four lines p_i lie in a plane.*

Proof. Take the given tetrahedron as tetrahedron of reference, and P as the point $(\xi_0, \xi_1, \xi_2, \xi_3)$. Then P_0 is the point $(0, \xi_1, \xi_2, \xi_3)$; and in the subordinate geometry of the plane $A_1 A_2 A_3$ the equation of the harmonic polar p_0 of this point with respect to the triangle of reference is

$$\frac{x_1}{\xi_1}+\frac{x_2}{\xi_2}+\frac{x_3}{\xi_3} = 0.$$

Thus p_0, regarded as a line of S_3, has equations

$$\frac{x_1}{\xi_1}+\frac{x_2}{\xi_2}+\frac{x_3}{\xi_3} = 0 = x_0,$$

and every point of it lies in the plane

$$\frac{x_0}{\xi_0}+\frac{x_1}{\xi_1}+\frac{x_2}{\xi_2}+\frac{x_3}{\xi_3} = 0.$$

Since this plane has a symmetric equation, it contains all four lines p_i.

The plane π which contains all the lines p_i is called the *harmonic polar plane* of P with respect to the given tetrahedron, and P is called the *harmonic pole* of π. Since the harmonic polar plane of the point $(\xi_0, \xi_1, \xi_2, \xi_3)$ is the plane $(1/\xi_0, 1/\xi_1, 1/\xi_2, 1/\xi_3)$ we see that, in particular, the unit point and the unit plane are harmonic pole and polar with respect to the tetrahedron of reference.

EXERCISE. With the notation of Theorem 1, show that if P projects from the edge $A_0 A_1$ into the point P_{01} of the opposite edge $A_2 A_3$ (that is to say, the plane $A_0 A_1 P$ meets $A_2 A_3$ in P_{01}), and P'_{01} is the harmonic conjugate of P_{01} with respect to A_2 and A_3, then P'_{01} and the five similar points P'_{ij} all lie in π.

THEOREM 2. *If two tetrahedra correspond to each other in such a way that the lines joining corresponding vertices pass through a common point, then the lines of intersection of corresponding faces lie in a common plane, and conversely.*

Proof. Take one of the tetrahedra as tetrahedron of reference $X_0 X_1 X_2 X_3$, and the point of concurrence of the joins as unit point E. If the vertices Y_0, Y_1, Y_2, Y_3 of the other tetrahedron have coordinate vectors $\mathbf{y}^{(0)}$, $\mathbf{y}^{(1)}$, $\mathbf{y}^{(2)}$, $\mathbf{y}^{(3)}$, we have, using an obvious notation,

$$\mathbf{y}^{(p)} = \mathbf{e}+\lambda_p \mathbf{x}^{(p)} \quad (p = 0, 1, 2, 3)$$

and hence

$$\mathbf{y}^{(p)}-\mathbf{y}^{(q)} = \lambda_p \mathbf{x}^{(p)}-\lambda_q \mathbf{x}^{(q)}.$$

$Y_p Y_q$ therefore meets $X_p X_q$ in the point P_{pq} given by $\lambda_p \mathbf{x}^{(p)}-\lambda_q \mathbf{x}^{(q)}$, and this point lies in the plane π represented by the symmetric equation

$$\sum_{i=0}^{3} \frac{1}{\lambda_i} x_i = 0.$$

Since the planes $X_p X_q X_r$ and $Y_p Y_q Y_r$ both contain the three points P_{qr}, P_{rp}, P_{pq}, which all lie in π, they meet in a line of π; and this proves the direct theorem. The converse follows by duality.

Remark. Two tetrahedra related as in Theorem 2 are said to be in perspective from the vertex E, and being in perspective is accordingly a self-dual relationship. Theorem 2 may be compared with Theorem 7 of Chapter IV.

THEOREM 3 (*Desargues's Theorem*). *If two triangles in S_3 correspond in such a way that the lines joining corresponding vertices are concurrent, then corresponding pairs of sides intersect, and the three points of intersection are collinear. Conversely, if two triangles correspond in such a way that the three pairs of corresponding sides intersect in collinear points, then the lines joining corresponding vertices are concurrent.*

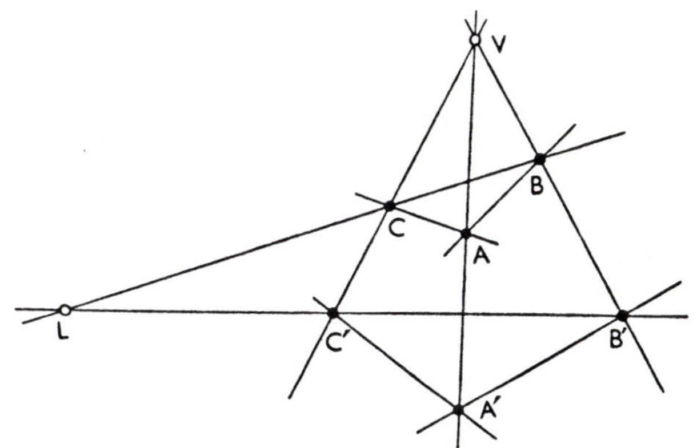

Proof. If the triangles lie in a common plane, the theorem reduces to Theorem 7 of Chapter IV; and we may therefore suppose that they lie in distinct planes π and π'.

Let the triangles be ABC and $A'B'C'$, with AA', BB', CC' all meeting at V. Then since BB' and CC' meet, the points B, B', C, C' are coplanar; and therefore BC and $B'C'$ meet in L, say. L then lies on the line $\pi\pi'$, and by symmetry the other two pairs of corresponding sides also meet in points of this same line.

If, conversely, it is given that the pairs of corresponding sides of the triangles ABC and $A'B'C'$ meet, the points B, C, B', C' are coplanar. Then the pair of lines BB', CC' meet, and similarly the pairs CC', AA' and AA', BB' also meet. The pairs of joins then lie in three planes, and since these planes clearly have not a line in common they have a unique common point V. V is then the common point of intersection of the three pairs of lines, and is common to AA', BB', and CC'.

Remarks

(i) Theorem 3 can be proved algebraically by an argument similar to that already used in proving Theorem 2. We have preferred to give a different proof in order to show how, as soon as a few basic incidence theorems are known, other theorems may be inferred by a more geometrical mode of argument.

(ii) Desargues's Theorem and its converse are not dual to each other in space, as they are in the plane. The proper three-dimensional analogue of Theorem 7 of Chapter IV is not Theorem 3 but Theorem 2.

(iii) Desargues's Theorem for coplanar triangles may be deduced from the theorem for non-coplanar triangles by constructing a third triangle, in a new plane, which is in perspective with each of the given triangles. The reader should do this.

Homographic ranges and pencils

In Chapter III we gave a detailed account of the formal theory of homographic correspondences, and in Chapter IV we saw how to apply this theory to concrete correspondences between ranges and pencils in S_2. Homographies are in fact possible whenever there are one-dimensional forms, i.e. systems of ∞^1 geometrical entities with the structure of an S_1. In three-dimensional space we already have three kinds of one-dimensional form—the range of points, the pencil of planes, and the plane pencil of lines—and any two forms, whether of the same kind or of different kinds, may be put in homographic correspondence. A $(1,1)$ correspondence between two one-dimensional forms is homographic if, when a projective parameter is introduced for each form, the parameters of corresponding elements are connected by a fixed bilinear equation

$$\theta' = \frac{\alpha\theta+\beta}{\gamma\theta+\delta} \quad (\alpha\delta-\beta\gamma \neq 0).$$

We now state the basic theorems on homographies in S_3, leaving the proofs to the reader whenever they are sufficiently simple.

THEOREM 4. *If a and b are fixed skew lines, then the correspondence between a variable point P of a and the plane bP of the pencil with axis b is homographic; in symbols, $(P) \barwedge b(P)$.*

COROLLARY. *A pencil of planes cuts homographic ranges on any two transversals.*

Remark. From this theorem we can deduce at once that the cross ratio of two ordered pairs of planes of a pencil is equal to the cross ratio of the pairs of points intercepted on an arbitrary transversal. This is the important result alluded to on p. 249.

THEOREM 5. *If P is a variable point of a fixed line a and B is a fixed point that does not lie on a, then the lines BP form a pencil of lines, and $B(P) \barwedge (P)$. Dually, if π is a variable plane through a fixed line a and β is a fixed plane that does not pass through a, then the lines $\beta\pi$ form a pencil of lines, and $\beta(\pi) \barwedge (\pi)$.*

THEOREM 6. *If P and Q are variable points of two fixed lines a and b, and P and Q correspond in a given homography ϖ, then it is possible to find two fixed points V, W and a fixed line c such that, for every corresponding pair (P, Q), VP and WQ meet in a point R of c.*

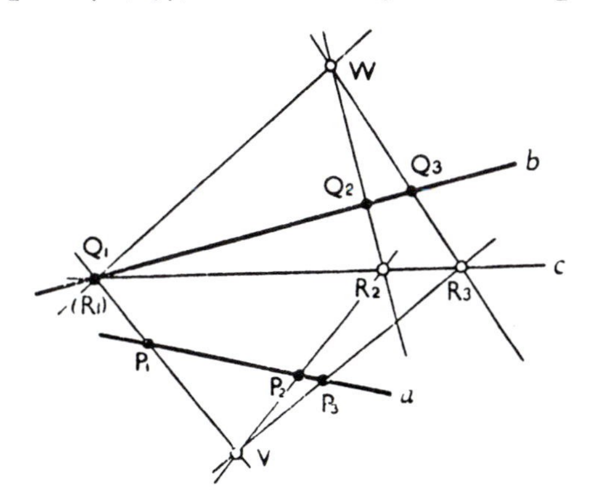

Proof. Select three corresponding pairs (P_1, Q_1), (P_2, Q_2), and (P_3, Q_3), and take a point V arbitrarily on $P_1 Q_1$. Then VP_2 and VP_3 are coplanar with Q_1, and a transversal $R_1 R_2 R_3$ can be found as shown, with R_1 coincident with Q_1. Since $Q_2 Q_3$ and $R_2 R_3$ are coplanar, $Q_2 R_2$ and $Q_3 R_3$ are also coplanar and meet in W, say. If we now take $R_2 R_3$ as the line c, the requirements of the theorem are satisfied; for the product of the perspectivity between (P) and (R) with vertex V and the perspectivity between (R) and (Q) with vertex W is a homography with three pairs in common with ϖ.

THEOREM 7. *Every homographic correspondence between two one-dimensional forms in S_3 can be generated by a finite sequence of operations of projection and section.*

Collinear plane fields and stars

In addition to the three kinds of one-dimensional form just discussed, S_3 also contains two important kinds of two-dimensional form, or system of ∞^2 geometrical entities with the structure of an S_2—namely the plane field and the star—and the analogue for such systems of the homographic correspondence between one-dimensional forms is, of course, the collineation. A $(1, 1)$ correspondence between two two-dimensional forms is a collineation if, when an allowable representation is chosen for each form, the parameters of corresponding elements are connected by fixed equations of the form

$$\lambda'_i = \sum_{k=1}^{3} \alpha_{ik} \lambda_k \quad (i = 1, 2, 3),$$

where $|\alpha_{rs}| \neq 0$. When two two-dimensional forms correspond in this way they are said to be *collinear*.

We shall meet important applications of collinear stars in Chapter XII in connexion with the twisted cubic curve and cubic surface.

§ 2. AFFINE AND EUCLIDEAN GEOMETRY OF THREE DIMENSIONS

Just as we were able to construct abstract models of two-dimensional affine and euclidean geometry by singling out certain geometrical entities in the projective plane and endowing them with recognizable individuality, so in a similar manner we can arrive at abstract models of affine and euclidean geometry of three dimensions.

We begin with real projective space $S_3(R)$ and adjoin to it ideal complex points, and then we single out a real plane ι, to be called the plane at infinity. To this plane, which is now regarded as an ideal plane, not properly belonging to the space, is assigned the invariable equation $x_0 = 0$. Thus we obtain affine space with its associated class $(\mathscr{R}_A)$ of allowable coordinate representations. In each of these representations the actual points of the space are in one–one correspondence with the triads (X_1, X_2, X_3) of real numbers.

Two planes of affine space are said to be parallel if and only if they meet in a line of ι; and a line is said to be parallel to a plane or to another line if it meets it in a point of ι. We can also define the mid-point of a finite segment AB in the usual way, as the harmonic conjugate with respect to A and B of the point at infinity on the line AB.

The affine geometry of space induces a subordinate affine geometry in every plane and on every line. If π is a given plane, the line at infinity in π is the line $\pi\iota$, and if l is a given line the point at infinity on l is the point $l\iota$.

By selecting a virtual conic Ω in ι and calling it the *absolute conic*, we can pass from affine space to (similarity) euclidean space. If we require Ω to have the equations

$$x_1^2+x_2^2+x_3^2 = 0 = x_0$$

in every allowable representation, we obtain the class of euclidean representations $(\mathscr{R}_E)$.

The tangent planes from any line l to Ω are called the isotropic planes through l, and the cone which joins any point P to Ω is called the isotropic cone at P.

Two planes are perpendicular if they are harmonic with respect to the isotropic planes through their line of intersection—in other words, if they meet ι in a pair of lines conjugate for Ω. Two lines (whether they intersect or not) are perpendicular if they meet ι in a pair of points conjugate for Ω.

The euclidean geometry of space induces a subordinate euclidean geometry in every (actual) plane. If π is a given plane, the absolute points in π are the conjugate complex points in which the line at infinity in the plane meets the virtual conic Ω.

Ω may be regarded as a disk quadric (see p. 265), and this suggests the possibility of passing from euclidean to non-euclidean geometry by replacing this disk quadric by a proper absolute quadric. This topic will be taken up in Chapter XIV.

THEOREM 8. *If $\mathscr{R}_A$ is any one allowable representation of affine space, then the whole class $(\mathscr{R}_A)$ of allowable representations consists of all those representations which can be derived from $\mathscr{R}_A$ by applying a transformation of the form*

$$X_i' = \sum_{k=1}^{3} b_{ik} X_k + c_i \quad (i = 1, 2, 3),$$

where the coefficients are arbitrary real numbers, subject to the condition $|b_{rs}| \neq 0$.

THEOREM 9. *If $\mathscr{R}_E$ is any one allowable representation of euclidean space, then the whole class $(\mathscr{R}_E)$ of allowable representations consists of all those representations which can be derived from $\mathscr{R}_E$ by applying a transformation of the form*

$$X'_i = c\sum_{k=1}^{3} e_{ik}X_k + a_i \quad (i = 1,2,3),$$

where a_1, a_2, a_3, c are arbitrary real numbers, with $c > 0$, and (e_{rs}) is any orthogonal matrix.

Proof. If (X_1, X_2, X_3) and (X'_1, X'_2, X'_3) are allowable coordinates in two representations $\mathscr{R}_E$ and $\mathscr{R}'_E$, then

$$X'_i = \sum_{k=1}^{3} b_{ik}X_k + c_i \quad (i = 1,2,3).$$

But, since Ω has the same equations in $\mathscr{R}_E$ and $\mathscr{R}'_E$,

$$\sum_{i=1}^{3} X_i'^2 \equiv h\sum_{i=1}^{3} X_i^2 + R,$$

where R involves only terms of degree lower than the second. By direct substitution, therefore,

$$h\sum_{i=1}^{3} X_i^2 \equiv \sum_{i=1}^{3}\Big(\sum_{k=1}^{3} b_{ik}X_k\Big)^2, \tag{1}$$

i.e.

$$\begin{aligned} h\sum_i\sum_j \delta_{ij}X_iX_j &\equiv \sum_i\Big(\sum_k b_{ik}X_k\Big)\Big(\sum_m b_{im}X_m\Big) \\ &\equiv \sum_k\sum_m\sum_i b_{ik}b_{im}X_kX_m \\ &\equiv \sum_i\sum_j\Big(\sum_k b_{ki}b_{kj}\Big)X_iX_j, \end{aligned}$$

by changing the dummy suffixes. Therefore

$$\sum_{k=1}^{3} b_{ki}b_{kj} = h\delta_{ij} \quad (i,j = 1,2,3).$$

But in virtue of (1), h is positive, and we may write $h = c^2$, where c is real. Then the last equation may be written in matrix form as

$$\mathbf{B}^T\mathbf{B} = c^2\mathbf{I},$$

and therefore $\mathbf{B} = c\mathbf{E}$, where $\mathbf{E}^T\mathbf{E} = \mathbf{I}$, i.e. where $\mathbf{E}$ is an orthogonal matrix.

Thus the transformation from $\mathscr{R}_E$ to $\mathscr{R}'_E$ is of the required form. The constants a_1, a_2, a_3, c may be any real numbers, but we must restrict the sign of c if each transformation is to appear only once in the whole set.

Remarks

(i) The euclidean transformations

$$X_i' = c\sum_{k=1}^{3} e_{ik} X_k + a_i \quad (i = 1, 2, 3)$$

form a group, the euclidean group in three dimensions.

(ii) In ordinary euclidean space, an allowable frame of reference is simply a rectangular coordinate system. A euclidean transformation can be interpreted as the transformation from one such coordinate system to another.

(iii) In the interpretation just mentioned, the columns of $\mathbf{E}$ are the direction cosines of the new axes with respect to the old. The relation $\mathbf{E}^T\mathbf{E} = \mathbf{I}$ gives the familiar orthogonality relations

$$\sum_{k=1}^{3} e_{ki} e_{kj} = \delta_{ij} \quad (i,j = 1, 2, 3)$$

for the direction cosines of three mutually perpendicular lines.

(iv) Since $\mathbf{E}^T\mathbf{E} = \mathbf{I}$ we have, on taking determinants,

$$|\mathbf{E}|^2 = 1.$$

Thus $|\mathbf{E}| = 1$ or -1. $|\mathbf{E}| = 1$ gives a proper orthogonal transformation and $|\mathbf{E}| = -1$ an improper one. In the second case there is a reflection, and the axes are changed from right-handed to left-handed or vice versa.

EXERCISES ON CHAPTER X

1. Given a coordinate representation $\mathscr{R}$ of S_3, for which the tetrahedron of reference is $X_0X_1X_2X_3$, write down (i) the point-equations of the four faces and the plane-equations of the four vertices of the tetrahedron, (ii) pairs of point-equations and pairs of plane-equations for the six edges, (iii) the equation of a general plane through the vertex X_0 and that of a general plane through the edge X_0X_1, and (iv) pairs of equations which represent a general line through X_0 and a general line in the face $X_1X_2X_3$.

Find also the line-coordinates of the six edges of the tetrahedron of reference.

2. If $XYZT$ is the tetrahedron of reference, E is the unit point $(1, 1, 1, 1)$, and P is the point (a, b, c, d), prove that:

(i) the equations of the line TP are $x/a = y/b = z/c$;

(ii) the equation of the plane ZTP is $x/a = y/b$;

(iii) the projection of PE from T on to the plane XYZ is the line

$$(b-c)x+(c-a)y+(a-b)z = 0 = t;$$

(iv) a general point of the transversal from P to XY and ZT is $(a, b, \lambda c, \lambda d)$;

(v) if EP meets the unit plane $(1, 1, 1, 1)$ in M, and the harmonic conjugate of P with respect to E and M is P', then the coordinates of P' are $(-a+b+c+d,\ a-b+c+d,\ a+b-c+d,\ a+b+c-d)$;

(vi) the equation of the locus of a variable point which is such that the transversal from it to XY and ZT meets EP is

$$(z-t)(bx-ay)-(x-y)(dz-ct) = 0.$$

3. If $f(x_1, x_2, x_3)$ and $g(x_2, x_3)$ are homogeneous polynomials, show that the equation $f(x_1, x_2, x_3) = 0$ represents a cone with vertex X_0 and the equation $g(x_2, x_3) = 0$ represents a set of planes through $X_0 X_1$.

Show also that the equation of any cone with vertex at the point of intersection of the three general planes $\pi_i = 0$ $(i = 1, 2, 3)$ is of the form

$$F(\pi_1, \pi_2, \pi_3) = 0.$$

4. A surface ϕ, whose equation is $f(x_0, x_1, x_2, x_3) = 0$, is met by the plane $\pi \equiv u_0 x_0 + u_1 x_1 + u_2 x_2 + u_3 x_3 = 0$ in a curve c. Show that the equation of the cone which projects c from the point (y_0, y_1, y_2, y_3) is

$$f(x_0 \pi_y - y_0 \pi,\ x_1 \pi_y - y_1 \pi,\ x_2 \pi_y - y_2 \pi,\ x_3 \pi_y - y_3 \pi) = 0,$$

where $$\pi_y = u_0 y_0 + u_1 y_1 + u_2 y_2 + u_3 y_3.$$

5. Show that the three planes

$$3x-3y+z-5t = 0,$$
$$5x+3y-5z+t = 0,$$
$$3x+3y-4z+2t = 0$$

meet in a line.

Find the harmonic conjugate of the third plane with respect to the first two.

Find also, in parametric form, the coordinates of a general point of the common line of intersection of the planes.

6. If $XYZT$ is the tetrahedron of reference, and X', Y', Z', T' are four points whose coordinate vectors are the columns of a non-singular skew-symmetric matrix, show that the tetrahedron $X'Y'Z'T'$ is both inscribed in and circumscribed about $XYZT$.

7. Show that the curve whose parametric equations are

$$x : y : z : t = a/(\theta - a) : b/(\theta - b) : c/(\theta - c) : d/(\theta - d)$$

passes through the reference points X, Y, Z, T, the unit point, and the point (a, b, c, d); and that it is met by any general plane in three points.

Find the equations of the four cones which project the curve from the points X, Y, Z, T, and show that the complete intersection of any two of these cones consists of the curve and one edge of the tetrahedron of reference.

8. Show that the curve whose parametric equations are

$$x : y : z : t = \theta^2 : \theta : 1 : \theta^2 - \theta + 1$$

is a conic, namely the section of the cone $y^2 = zx$ by the plane

$$x-y+z-t = 0.$$

Exhibit in a similar way, as the section of a cone by a plane, the conic whose parametric equations are

$$x:y:z:t = \theta^2+1:\theta-1:\theta^2-1:\theta^2+3\theta.$$

9. Three points A, B, C lie on a line k, and D is any point not on k. A variable line l, which does not meet k, is such that the cross ratio $l\{A, B; C, D\}$ has a given constant value. Show that l meets a fixed line through D.

10. $ABCD$ is a given tetrahedron and A', B', C' are three given points, none of which lies in a face of $ABCD$. Show that in general it is impossible to find a point D', not on a face of $ABCD$, such that $D'A'$ meets DA, $D'B'$ meets DB, and $D'C'$ meets DC; but that if A', B', C' are so related that one such point D' exists, then there is an infinity of such points, all on a straight line. Show also that in this case the triangles ABC and $A''B''C''$ are in perspective, A'', B'', C'' being the projections of A', B', C' from D' on to the plane ABC.

11. Two triads of points in space, (A, B, C) and (A', B', C'), are in perspective from a point D''; BC', $B'C$ meet in A'', CA', $C'A$ meet in B'', and AB', $A'B$ meet in C''. Prove that AA'', BB'', CC'' meet in a point D', and that $A'A''$, $B'B''$, $C'C''$ meet in a point D.

Show that any two of the tetrahedra $ABCD$, $A'B'C'D'$, $A''B''C''D''$ are in perspective in four ways, the centres of perspective being the vertices of the third tetrahedron.

12. Two given planes α, α' meet in a line l, and a collineation ϖ between P in α and P' in α' is such that no point of l is invariant. Show that a coordinate representation of space can be chosen in such a way that the coordinates of a general pair of corresponding points P, P' are $(\lambda, \mu, \nu, 0)$ and $(0, \lambda, \mu, \nu)$.

Show also that if ϖ is such that l is a self-corresponding line, and the two united points M, N of the homography induced on it by ϖ are distinct, then all the joins PP' of corresponding pairs for ϖ meet two fixed lines, one through M and the other through N.

13. Show that the points P, P' in which two given general planes α, α' are met by the transversal lines of two given skew lines correspond in a quadratic Cremona transformation of α into α'. Find simple equations for a transformation so generated, and investigate the fundamental points in α and α'. [*Hint.* Let the skew lines meet α in A, B and α' in A', B', and let AB, $A'B'$ meet the line $\alpha\beta$ in C', C. Take (A, B, C) as reference triad in α and (A', B', C') as reference triad in α'.]

14. The volume V of a tetrahedron in affine space, whose vertices are the points (X_i, Y_i, Z_i) $(i = 1, 2, 3, 4)$, is defined by the formula $6V = |X_i Y_i Z_i 1|$. Show that the value of V depends on the particular choice of the affine coordinate system, but that the ratio of the volumes of any two tetrahedra is an affine invariant.

15. Show that, in affine space, the equations of any two skew lines may be taken to be

$$X = 0 = Z-c, \qquad Y = 0 = Z+c$$

if the coordinate system is chosen suitably.

Investigate the locus of the mid-point of a variable transversal of two given skew lines which is parallel to a given plane.

16. If x, y, z, t are homogeneous cartesian coordinates, show that the line which contains the finite point $(a, b, c, 1)$ and the point at infinity $(l, m, n, 0)$ may be written in the form $(x-at)/l = (y-bt)/m = (z-ct)/n$.

Show that this line and the line given by

$$(x-a't)/l' = (y-b't)/m' = (z-c't)/n'$$

intersect if and only if $\sum (a-a')(mn'-m'n) = 0$, and that the lines are parallel if $l:m:n = l':m':n'$.

Show also that if the coordinates are rectangular, so that the absolute conic Ω is given by $x^2+y^2+z^2 = 0 = t$, the condition for the lines to be perpendicular is $ll'+mm'+nn' = 0$.

If the lines meet the plane at infinity in A, A', and the pole of AA' for Ω is B, show that the transversal of the lines which passes through B is their common perpendicular; and find the equations of this transversal.

17. The common perpendicular transversal of a variable generator of the right circular cone $X^2+Y^2 = Z^2$ and the line $X-1 = 0 = Y$ meets the generator in P. Investigate the locus of P on the cone.

18. If l, l' are given skew lines in euclidean space, show how it is possible, by taking their common perpendicular transversal as Z-axis, to define a rectangular coordinate system in which the equations of the two lines are $Y-mX = 0 = Z-c$ and $Y+mX = 0 = Z+c$ respectively.

A point moves in such a way that the line joining the feet of the perpendiculars from it to two given non-intersecting lines subtends a right angle at a fixed point. Prove that its locus is a hyperbolic cylinder whose generating lines are perpendicular to each of the given lines.

19. If two pairs of opposite edges of a tetrahedron are perpendicular, then so are the third pair. Show that this result can be regarded as a corollary to Hesse's Theorem (Chapter VI, Theorem 24).

20. Show that the altitudes of a tetrahedron $ABCD$ from the vertices A and B intersect if and only if the edges AB and CD are perpendicular.

21. If Θ, Θ_1, Θ_2, Θ_3 are four real quadratic polynomials in θ, of which three are linearly independent, show that the curve whose parametric equations in rectangular cartesian coordinates are $X = \Theta_1/\Theta$, $Y = \Theta_2/\Theta$, $Z = \Theta_3/\Theta$ is a conic; and find conditions for the conic to be (a) an ellipse, or a hyperbola, or a parabola, (b) a rectangular hyperbola, and (c) a circle.

Show how to find the plane of the conic, its asymptotes (if real), and its centre.

22. A system of homogeneous coordinates x, y, z, t in euclidean space is said to be *tetrahedral* if the unit point is the centroid of the tetrahedron of reference. Show that for such a system the plane at infinity is the unit plane.

Show that the tetrahedral coordinates x, y, z, t of a point P are proportional to the volumes of the tetrahedra $PYZT$, $PXZT$, $PXYT$, $PXYZ$, the volumes being given suitable signs.

23. A point P moves in such a way that the feet of the perpendiculars from it on to the faces of a tetrahedron $ABCD$ are coplanar. Show that it describes a cubic surface whose equation, in any system of homogeneous coordinates for which $ABCD$ is the tetrahedron of reference, is of the form $a/x+b/y+c/z+d/t = 0$, where a, b, c, d are constants.

24. A wire bent in the form of an ellipse rests on a horizontal plane π, with its own plane vertical and its axes respectively horizontal and vertical; and a point-source of light is placed at a point P. Find the locus of P when the shadow cast by the wire on π is (a) a parabola, and (b) a rectangular hyperbola.

Show that in every horizontal plane above the wire there are two positions of P for which the shadow is circular, and find the locus of these points as the horizontal plane varies.

CHAPTER XI

THE QUADRIC

IN the projective plane the simplest locus that we can consider, apart from the straight line, is the conic locus, and we have already seen how to establish its projective properties. In three-dimensional projective space, on the other hand, where a point has three degrees of freedom, there are two essentially different kinds of locus to be considered; and we have two particular loci, each of which is in its way a proper space-analogue of the conic, namely the quadric surface and the twisted cubic curve.

A *surface* is the locus of a variable point of space which has two degrees of freedom, and it may be defined by imposing a single analytical condition on a general point. If this condition takes the form of a quadratic equation $S(x_0, x_1, x_2, x_3) = 0$, the surface is a *quadric* surface; and it is then clearly analogous to the conic defined by the equation $S(x_0, x_1, x_2) = 0$ in S_2.

A *curve*, on the other hand, is the locus of a variable point of space which has one degree of freedom, and it may be defined analytically by taking as coordinates of the variable point four functions of a single parameter. In particular, the equations

$$x_0:x_1:x_2:x_3 = \theta^3:\theta^2:\theta:1$$

define a curve known as the *twisted cubic*; and this space curve is analogous to the conic given by the canonical representation

$$x_0:x_1:x_2 = \theta^2:\theta:1.$$

In the next three chapters we shall develop the projective geometry of the quadric and twisted cubic, discussing the quadric in this chapter and the next but one, and the twisted cubic in Chapter XII. The two manifolds are, however, closely connected with each other, and we shall not attempt to maintain a rigid separation.

§ 1. THE QUADRIC LOCUS AND QUADRIC ENVELOPE

DEFINITION. An *algebraic surface of order n* in S_3 is the totality of points whose coordinates in some assigned allowable representation $\mathcal{R}$ satisfy a fixed homogeneous equation of the nth degree:

$$f(x_0, x_1, x_2, x_3) = 0.$$

Dually, an *algebraic envelope of class n* in S_3 is the totality of planes whose coordinates satisfy a fixed homogeneous equation of the nth degree: $\phi(u_0, u_1, u_2, u_3) = 0$.

The surface or envelope is said to be *irreducible* if the polynomial f or ϕ is irreducible over the ground field K.

An algebraic surface of order 2 is called a quadric surface or quadric locus, and an algebraic envelope of class 2 is called a quadric envelope.

The equation of the general quadric locus may be written in either of the forms

$$S(x_0, x_1, x_2, x_3) \equiv \sum_{i=0}^{3} \sum_{k=0}^{3} a_{ik} x_i x_k = 0$$

and

$$S(\mathbf{x}) \equiv \mathbf{x}^T \mathbf{A} \mathbf{x} = 0,$$

where the matrix $\mathbf{A} = (a_{rs})$ may be taken, without loss of generality, to be symmetric.

The equation of the general quadric envelope may be written in the corresponding forms

$$\Sigma(u_0, u_1, u_2, u_3) \equiv \sum_{i=0}^{3} \sum_{k=0}^{3} A_{ik} u_i u_k = 0$$

and

$$\Sigma(\mathbf{u}) \equiv \mathbf{u}^T \mathfrak{A} \mathbf{u} = 0.$$

The equation $S = 0$ contains ten essentially different terms, four involving the squares x_i^2 and six involving the products $x_i x_j$, and there are accordingly nine effective coefficients. This gives us the important result:

THEOREM 1. *There is a unique quadric locus which passes through nine given points of general position; and dually, there is a unique quadric envelope which contains nine given planes of general position.*

Since the equation of the quadric locus is of similar algebraic form to that of the conic locus, many of the theorems established in Chapter V remain valid, possibly with trifling modification, for quadrics, and in many cases it will be sufficient to restate the results without proof. Thus, for instance, we have the following theorem on the projective invariance of the rank of the matrix $\mathbf{A}$.

THEOREM 2. *If the equation of a quadric locus S, referred to some allowable representation $\mathscr{R}$, is $\mathbf{x}^T \mathbf{A} \mathbf{x} = 0$, the rank $\rho[\mathbf{A}]$ is a projective characteristic of S.*

The number $\rho[\mathbf{A}]$ is called the rank of the quadric locus, and we denote it by r. The introduction of the concept of rank makes possible a projective classification of the various types of quadric locus, with a dual classification of quadric envelopes.

Rank	*Quadric locus*	*Quadric envelope*
4	Proper quadric locus	Proper quadric envelope
3	Proper quadric cone	Disk quadric
2	Pair of distinct planes	Pair of distinct points
1	Repeated plane	Repeated point

This classification may be derived, in the following way, from Theorem 2 of the Appendix. According to this theorem it is possible, by selecting the representation $\mathscr{R}$ suitably, to put the equation of any given quadric locus S into the form

$$\sum_{i=0}^{r-1} d_i x_i^2 = 0,$$

where r is the rank of S and each of the coefficients d_i is 1 or -1. If the ground field K is the complex field, all the d_i may be made equal to 1.

Let us now consider the possibilities that arise.

Case 1: $r = 4$.

The equation of S is

$$d_0 x_0^2 + d_1 x_1^2 + d_2 x_2^2 + d_3 x_3^2 = 0.$$

This is the general case, and we shall not discuss it at this stage.

Case 2: $r = 3$.

The equation of S is

$$d_0 x_0^2 + d_1 x_1^2 + d_2 x_2^2 = 0.$$

The point X_3, given by $(0, 0, 0, 1)$, lies on the surface; and if P is any other point of S, every point of the line $X_3 P$ is a point of S. The surface is accordingly generated by lines through X_3, and we call it a *quadric cone* with vertex X_3. It is sometimes called a proper cone, to distinguish it from the more special types of quadric which follow. The section of S by the plane $X_0 X_1 X_2$ is the conic

$$d_0 x_0^2 + d_1 x_1^2 + d_2 x_2^2 = 0 = x_3,$$

and the cone is generated by the lines which join X_3 to the points of this conic.

Case 3: $r = 2$.

The equation of S is $d_0 x_0^2 + d_1 x_1^2 = 0$, and this may be factorized as follows:

$$\{\surd(d_0)x_0 + \surd(-d_1)x_1\}\{\surd(d_0)x_0 - \surd(-d_1)x_1\} = 0.$$

The surface is therefore a pair of planes which meet in the line $X_2 X_3$.

Case 4: $r = 1$.

The equation is $d_0 x_0^2 = 0$, and this represents the plane $X_1 X_2 X_3$ taken twice.

We say the quadric locus is *proper* when $r = 4$ and *degenerate* when $r < 4$. The reasons for this terminology will become clear below; but it should be noted that the terms 'proper' and 'irreducible' are not equivalent for quadrics, as they are for conics. The cone, corresponding to $r = 3$, is irreducible, but it is not a proper quadric locus.

The classification of quadric envelopes is dual to the classification just considered, and the only type that requires separate discussion is the *disk quadric*, the dual of the proper cone. The reduced equation of this envelope is

$$d_0 u_0^2 + d_1 u_1^2 + d_2 u_2^2 = 0.$$

The plane ξ_3, given by $(0, 0, 0, 1)$, belongs to the envelope; and if π is any other plane of the envelope, every plane through the line $\xi_3 \pi$ also belongs to it. The planes of the envelope thus form a system of pencils of planes with axes in the fixed plane ξ_3.

If, now, π has plane-coordinates (u_0, u_1, u_2, u_3), the line-coordinates in ξ_3 of the line $\xi_3 \pi$ are (u_0, u_1, u_2) and the envelope of the lines $\xi_3 \pi$ is therefore the conic $d_0 u_0^2 + d_1 u_1^2 + d_2 u_2^2 = 0$ in ξ_3. The planes of the quadric envelope are thus all the planes through all the tangents to this conic—∞^1 pencils each of ∞^1 planes, giving ∞^2 planes in all. This is the reason for the name 'disk quadric'. The reader may visualize the system by thinking of the tangent planes to the ellipsoid $\dfrac{X^2}{a^2} + \dfrac{Y^2}{b^2} + \dfrac{Z^2}{c^2} = 1$ and letting c tend to zero.

The Joachimsthal theory

The polar properties of the quadric, like those of the conic, follow readily from Joachimsthal's equation

$$S_{yy} + 2\theta S_{yz} + \theta^2 S_{zz} = 0,$$

and since the formal algebra is identical with that in § 4 of Chapter

V, except that the summations are now from 0 to 3 instead of from 0 to 2, we shall not repeat the details.

Two points Q and R, with coordinate vectors $\mathbf{y}$ and $\mathbf{z}$ respectively, are said to be *conjugate points* with respect to the quadric locus S if they are harmonic with respect to the two points A_1, A_2 in which QR meets S, and the analytical condition for conjugacy is therefore $S_{yz} = 0$.

If Q is a fixed point, the locus of a variable point of space that is conjugate to Q is the plane $S_y = 0$, and this is called the *polar plane* of Q with respect to S.

Now suppose Q is a point of S, and R is a point, distinct from Q, that is conjugate to Q. Then one of the points A_1, A_2 coincides with Q; and, since the pairs (A_1, A_2) and (Q, R) are harmonic, the second of these points also coincides with Q. Thus any line which joins a point Q of S to a conjugate point meets S in points which coincide at Q. It is therefore a *tangent line* to S at Q, and the polar plane of Q is the *tangent plane* at Q, generated by all the tangent lines at this point.

If Q does not lie on S, the tangent lines through Q generate a quadric cone, the *enveloping cone* or *tangent cone* with vertex Q; and the equation of this cone is

$$S_y^2 - SS_{yy} = 0.$$

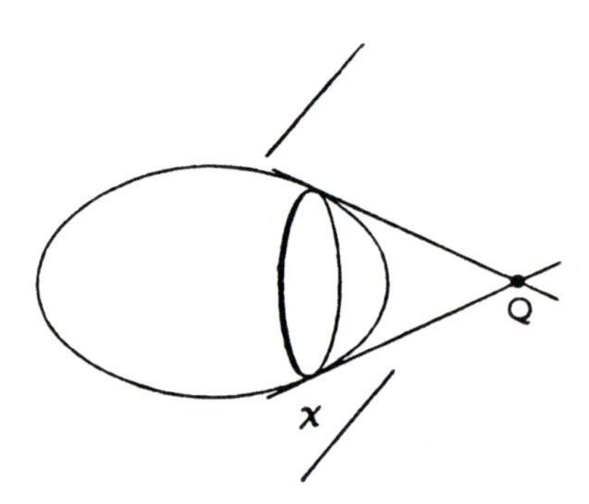

The tangent lines which generate the cone all touch S at points lying in the polar plane χ of Q. The plane χ cuts S in a conic, and the enveloping cone has ring contact with S along this conic.

THEOREM 3. *If the rank of S is 4, then the totality of tangent planes of S is a quadric envelope Σ. If the equation of S is*

$$\sum_i \sum_k a_{ik} x_i x_k = 0,$$

the equation of Σ is

$$\sum_i \sum_k A_{ik} u_i u_k = 0,$$

where A_{ik} is the cofactor of a_{ik} in the determinant $|a_{rs}|$.

Proof. As for Theorem 9 of Chapter V.

The properties of the general quadric envelope may be inferred from the properties of the general quadric locus by the principle of duality.

Let Σ be a general quadric envelope. Two planes χ and ρ are said to be *conjugate planes* with respect to Σ if they are harmonic with respect to the two planes of Σ which pass through the line $\chi\rho$. The condition for conjugacy is accordingly $\Sigma_{vw} = 0$.

Every plane χ has a *pole*, a point which is common to all planes conjugate to χ, and the pole of the plane $\mathbf{v}$ is the point $\Sigma_v = 0$.

A line t is called a *tangent line* of Σ if the two planes of the envelope which pass through it coincide. The line of intersection of a plane χ of Σ with any conjugate plane is necessarily a tangent line of Σ, and the pole of χ is the point of concurrence of all those tangent lines of Σ which lie in χ. Such a point is said to be a *point of the envelope* Σ (dual to a tangent plane of S).

If χ does not belong to Σ, the tangent lines of Σ which lie in χ envelop a conic in χ. The equation of this conic, regarded as a disk quadric, is

$$\Sigma_v^2 - \Sigma\,\Sigma_{vv} = 0.$$

If finally the rank of Σ is 4, the points of Σ, as just defined, are the points of a quadric locus S; and if the equation of Σ is

$$\sum_i \sum_k A_{ik} u_i u_k = 0$$

that of S is

$$\sum_i \sum_k a_{ik} x_i x_k = 0,$$

where the a_{ik} are cofactors in $|A_{rs}|$.

Here, then, we have the fundamental polar properties of the quadric locus and quadric envelope. They are valid without restriction for proper quadrics, but in the case of degenerate quadrics they may need suitable modification. We leave it to the reader to supply the necessary qualifications, guided by what he already knows of the geometry of quadrics from a more elementary treatment and also by the analogy with Chapter V.

Like the proper conic, the proper quadric is best looked upon as a single self-dual figure, comprising both a quadric locus and a quadric envelope. This is made clear by the following theorems, which correspond to Theorems 15 and 16 of Chapter V.

THEOREM 4. *If S is a proper quadric locus and Σ is the associated envelope, then Σ is a proper quadric envelope. Furthermore, the quadric locus associated with the envelope Σ is the original locus S.*

THEOREM 5. *If S is a proper quadric locus and Σ is the associated envelope, then a plane π is the polar plane of a point P with respect to S if and only if P is the pole of π with respect to Σ.*

In virtue of these theorems, we may take S and Σ together as a *proper quadric* ψ. This quadric determines a single polarity of space, the (1, 1) correspondence between points and planes given by $\mathbf{u} = \mathbf{A}\mathbf{x}$, $\mathbf{x} = \mathbf{A}^{-1}\mathbf{u}$. The quadric has both a point-equation

$$S \equiv \sum_i \sum_k a_{ik} x_i x_k = 0$$

and a plane-equation

$$\Sigma \equiv \sum_i \sum_k A_{ik} u_i u_k = 0,$$

each of which determines the other.

EXERCISE. Show that, with the two definitions of tangent given above, t is a tangent line of S if and only if it is a tangent line of Σ.

Theorem 4 holds only for proper quadrics, and a degenerate quadric locus does not necessarily define an associated quadric envelope at all. Let us consider first the case of a proper cone S, with vertex V. At any point of S, other than V, we have a uniquely defined tangent plane, generated by the plane pencil of tangent lines through the point; but every such tangent plane touches the cone at all points of a generator, instead of at a unique point as in the case of the proper quadric. Every line through V is, in a sense, to be regarded as a tangent line to S at V, since it cuts S in two coincident points there, and every plane through V is made up of tangent lines and counts as a tangent plane. Thus the tangent planes to S are all the planes through the vertex V. This fits in with the fact that if we try to derive a plane-equation from

$$d_0 x_0^2 + d_1 x_1^2 + d_2 x_2^2 = 0$$

by taking cofactors, we get simply

$$d_0 d_1 d_2 u_3^2 = 0,$$

i.e. the plane-equation of the point V taken twice.

Nevertheless, we often find it convenient to distinguish between an arbitrary plane through V and a *proper* tangent plane to the cone, which touches it along a generator. The proper tangent planes form an ∞^1 system only, and this sytem is not, strictly speaking, a surface (regarded as the envelope of its ∞^2 tangent planes) but a new geometrical entity known as a *developable*. A developable may be defined as the dual of a curve in space.

Consider, in fact, any space curve c, other than a straight line.

At every point of c (with the exception of singular points, which we need not consider in this connexion) there is a unique tangent line, or line which meets the curve in two coincident points; and every plane through the tangent line is a tangent plane. Thus c has ∞^1 tangent lines and ∞^2 tangent planes, and the tangent planes form ∞^1 pencils of planes with the tangent lines as axes. The dual of the curve c is a developable δ, which has ∞^1 planes, ∞^1 generating lines (one in each plane) and ∞^2 points, which form ∞^1 ranges on the generating lines as axes. If, now, c is a plane curve, all its tangent lines lie in a plane; and the dual developable therefore has the property that all its generating lines pass through a point. Thus *the space dual of a plane curve is a cone.*

The quadric cone and the conic are dual to each other in space in two distinct ways.

(i) The cone is a quadric locus of rank 3, consisting of ∞^2 points lying on ∞^1 generating lines. Dually, the conic is a quadric envelope of rank 3 (disk quadric) consisting of ∞^2 planes passing through ∞^1 tangent lines.

(ii) The cone is a developable of class 2. It consists of ∞^1 (proper) tangent planes, two of which pass through a general line through the vertex. Dually, the conic is a plane curve of order 2. It consists of ∞^1 points, two of which lie on a general line in its plane. The conic is a quadratic locus in a plane field, whereas the cone is a quadratic envelope in a star; and they are thus dual to each other.

If a quadric locus is of rank 2, it breaks up into a pair of planes. The equation of the locus may be put in the form $d_0 x_0^2+d_1 x_1^2 = 0$; and since all the cofactors in the matrix of the quadric are zero we cannot obtain any envelope at all by the usual algebraic procedure. We can show, however, by a different kind of argument (p. 278 below) that the quadric breaks up, as an envelope, into two points which lie on the line of intersection of the two planes. This type of degenerate quadric may accordingly be called a *bifocal plane-pair.* It can degenerate still further, since either the two planes, or the two points, or both, may be coincident.

Example. To find the point-equations of a conic when the plane-equation is given, and vice versa.

Suppose we are given the plane equation

$$2u_1^2+5u_3^2+4u_2 u_0+2u_0 u_3-4u_1 u_3+6u_2 u_3 = 0. \tag{1}$$

The matrix of coefficients and its matrix of cofactors are as follows:

$$\begin{pmatrix} 0 & 0 & 2 & 1 \\ 0 & 2 & 0 & -2 \\ 2 & 0 & 0 & 3 \\ 1 & -2 & 3 & 5 \end{pmatrix}, \qquad \begin{pmatrix} -18 & 12 & -6 & 12 \\ 12 & -8 & 4 & -8 \\ -6 & 4 & -2 & 4 \\ 12 & -8 & 4 & -8 \end{pmatrix}.$$

Since the original matrix is of rank 3, equation (1) represents a disk quadric or conic. We require to find two point-equations which represent surfaces whose complete intersection is the conic, and a convenient pair of surfaces is the plane of the conic and the cone which joins it to a vertex of the tetrahedron of reference.

The plane of the conic is the polar plane of any general point of space, X_0 for example. It is therefore the plane $(-18, 12, -6, 12)$, i.e. $(-3, 2, -1, 2)$.

In the star with vertex X_3, the plane-equation of the enveloping cone with vertex X_3 is obtained by putting $u_3 = 0$ in (1), and is therefore

$$2u_1^2+4u_2u_0 = 0,$$

i.e.
$$u_1^2+2u_2u_0 = 0;$$

and the associated point-equation is

$$x_1^2+2x_2x_0 = 0.$$

Thus a pair of point-equations of the given conic is

$$3x_0-2x_1+x_2-2x_3 = 0 = x_1^2+2x_2x_0. \tag{2}$$

We may note, incidentally, that the quadratic equation whose matrix is the matrix of cofactors is

$$9x_0^2+4x_1^2+x_2^2+4x_3^2-4x_1x_2+6x_2x_0-12x_0x_1-12x_0x_3+8x_1x_3-4x_2x_3 = 0,$$

i.e.
$$(3x_0-2x_1+x_2-2x_3)^2 = 0.$$

This represents the plane of the conic taken twice, in accordance with what we said above in connexion with the plane-equation of the proper cone.

Now suppose, conversely, that we are given a pair of point-equations (2), representing a conic, and we wish to find the plane-equation of the conic, regarded as a disk quadric.

If π is any tangent plane, it meets the plane of the conic in a tangent line to the conic; and the plane joining this line to X_3 must be a proper tangent plane to the cone $x_1^2+2x_2x_0 = 0$. Its coordinates must therefore satisfy the equation $u_1^2+2u_2u_0 = 0$. Now if π is the plane (u_0, u_1, u_2, u_3), a general plane through the line in which it meets the plane of the conic is

$$3x_0-2x_1+x_2-2x_3+\lambda(u_0x_0+u_1x_1+u_2x_2+u_3x_3) = 0.$$

For this plane to pass through X_3, $-2+\lambda u_3 = 0$. The value of λ is therefore $2/u_3$, and the equation of the plane is

$$(2u_0+3u_3)x_0+(2u_1-2u_3)x_1+(2u_2+u_3)x_2 = 0.$$

The coefficients, v_0, v_1, v_2 say, in this equation must satisfy $v_1^2+2v_2v_0 = 0$, and hence

$$4(u_1-u_3)^2+2(2u_0+3u_3)(2u_2+u_3) = 0,$$

i.e.
$$4u_1^2+10u_3^2+8u_2u_0+4u_0u_3-8u_1u_3+12u_2u_3 = 0,$$

i.e.
$$2u_1^2+5u_3^2+4u_2u_0+2u_0u_3-4u_1u_3+6u_2u_3 = 0;$$

and this is the original equation (1).

Polar lines with respect to a quadric

Every proper quadric determines a polarity of space, which transforms points into planes and planes into points, and we may ask what effect this transformation has upon the lines of space. A line may be treated either as the axis of a range of points or as the axis of a pencil of planes; and with either interpretation it is found to transform into another line—its *polar line* with respect to the quadric.

THEOREM 6. *If ψ is a proper quadric and l is a given line, the polar planes of all points of l pass through a second line l' and the poles of all planes through l lie on l'. The relation between l and l' is symmetrical.*

Proof. Let $\mathbf{x}^{(1)}$, $\mathbf{x}^{(2)}$ represent two points Q_1, Q_2 of l, with polar planes $\mathbf{u}^{(1)}$, $\mathbf{u}^{(2)}$ respectively. Then the polar plane of the general point $\mathbf{x}^{(1)}+\lambda\mathbf{x}^{(2)}$ of l is $\mathbf{u}^{(1)}+\lambda\mathbf{u}^{(2)}$, and the polar planes therefore form a pencil of planes, with axis l', say.

If π is any plane through l, it may be fixed by Q_1, Q_2 and a point R not on l. Then the pole of π with respect to ψ is the common point of the polar planes of Q_1, Q_2, R; and since two of these polar planes pass through l, the pole of π lies on l.

The relation between l and l' is symmetrical because it amounts simply to the conjugacy of every point of l with every point of l' with respect to ψ.

COROLLARY. *The polar line l' is the line of intersection of the tangent planes at the two points in which ψ is cut by l, and it is also the line which joins the points of contact of the two tangent planes of ψ which pass through l.*

In general, a line and its polar line with respect to a quadric ψ are skew. If l and l' meet, in a point P, say, then the polar plane of P passes through both lines, and therefore through P. This means that P is a point of ψ, the plane of l and l' is the tangent plane at P, and l and l' are tangent lines at P.

If l coincides with its polar line, then the polar plane of every point of l passes through the point, and l therefore lies wholly on the quadric.

If l and m are two lines, so related that l meets m', the polar line of m, then m meets l', the polar line of l; and in this case l and m are said to be *conjugate lines* with respect to ψ. Their two polar lines l', m' are then also conjugate lines.

Self-polar tetrahedra

Consider a proper quadric ψ. If A is a general point of space, it has a polar plane α with respect to ψ; and if B is a general point of α, its polar plane β passes through A. If C is a general point of the line $\alpha\beta$, its polar plane γ passes through A and B, and cuts $\alpha\beta$ in a unique point D. Then $ABCD$ is a *self-polar tetrahedron* for ψ, and it has the following properties: (i) every vertex is the pole of the opposite face; (ii) any two vertices are conjugate points for ψ, and any two faces are conjugate planes; (iii) any pair of opposite edges are polar lines. There are ∞^6 self-polar tetrahedra for any proper quadric ψ.

EXERCISES

(i) What becomes of the relation between a point and its polar plane or a plane and its pole if the quadric is degenerate? Discuss all possible cases.

(ii) What self-polar tetrahedra exist for each of the types of degenerate quadric?

(iii) How must the tetrahedron of reference be related to a quadric locus in order that the equation of the locus may reduce to the form

$$\sum_{i=1}^{r-1} d_i x_i^2 = 0?$$

§ 2. PLANE SECTIONS AND GENERATORS

Since every quadric is an algebraic surface of the second order, it is met by any plane in an algebraic plane curve of the second order, i.e. a conic.

Consider a proper quadric ψ and a plane π. We may suppose that the coordinate system is chosen in such a way that the equations of π and ψ are

$$x_0 = 0$$

and

$$\sum_{i=0}^{3} \sum_{k=0}^{3} a_{ik} x_i x_k = 0.$$

Then the curve of section is given by

$$\sum_{i=1}^{3} \sum_{k=1}^{3} a_{ik} x_i x_k = 0 = x_0,$$

and it is therefore a conic in π which is degenerate if and only if

$$\begin{vmatrix} a_{11} & a_{12} & a_{13} \\ a_{21} & a_{22} & a_{23} \\ a_{31} & a_{32} & a_{33} \end{vmatrix} = 0,$$

i.e. $A_{00} = 0$. We thus have the theorem:

THEOREM 7. *Every plane section of a quadric locus is a conic locus. If the quadric is proper, the conic is degenerate if and only if the plane is a tangent plane of the quadric.*

THEOREM 8. *Through every point of a proper quadric there pass two distinct lines which lie wholly on the surface.*

Proof. Let P be a point of a proper quadric ψ. Then the tangent plane π at P cuts ψ in a line-pair, and since every line in π which passes through P cuts ψ in coincident points at P, P is a double point of the line-pair. Further, the lines of the pair are distinct; for if they were coincident, π would be the tangent plane to ψ at every point of the repeated line, and this is impossible since π has a unique pole with respect to the proper quadric ψ.

Remarks

(i) The two lines in which ψ is met by the tangent plane at P are called the *generating lines* or *generators* through P.

(ii) If ψ is a proper cone, a plane π meets it in a proper conic, a pair of distinct lines, or a repeated line according as it is a plane which does not pass through the vertex of ψ, a plane which passes through the vertex but is not a proper tangent plane, or a proper tangent plane.

THEOREM 9. *The generators of a proper quadric ψ form two systems, and one generator of each system passes through any assigned point of ψ. Any two generators of the same system are skew to each other, whereas two generators which are not of the same system necessarily intersect.*

Proof. Choose a fixed point P_0 arbitrarily on ψ. Then there are two generators through P_0, which we may call u_0 and v_0. If Q is any point of u_0 there are two generators through Q, u_0 itself and another generator v. The lines v and v_0 are skew, since otherwise their plane would meet ψ in a cubic curve made up of v, v_0, and u_0. Thus v is a generator of ψ which meets u_0 but not v_0, and we call it the v-generator through Q. In just the same way we can define a u-generator, which meets v_0 but not u_0, through every point of v_0. There are thus two systems of generators on ψ—the u-generators and the v-generators.

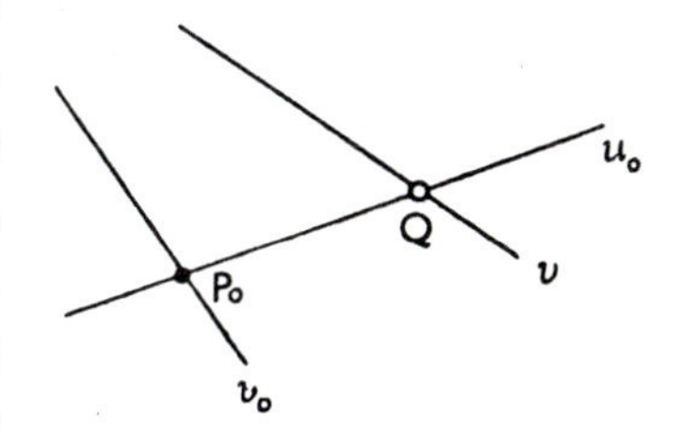

If P is any point of ψ which does not lie on u_0 or v_0, the tangent plane at P cuts ψ in the pair of generators through P; and since this plane cuts u_0 and v_0 each in a point of ψ, one of the generators through P is a u-generator and the other is a v-generator. The two systems therefore include all the generators of ψ.

Two generators of the same system are necessarily skew, since if they met we would have a cubic section of ψ as above.

Finally, let u_1, v_1 be any two generators, one of which belongs to each system, and let R be any point of v_1. Then the plane Ru_1 cuts ψ in a conic, which must break up into u_1 and another line. This residual line is a generator; and since it meets u_1 it is the v-generator through R, i.e. the generator v_1. The conic of section is therefore the line-pair (u_1, v_1), and hence u_1, v_1 are intersecting lines.

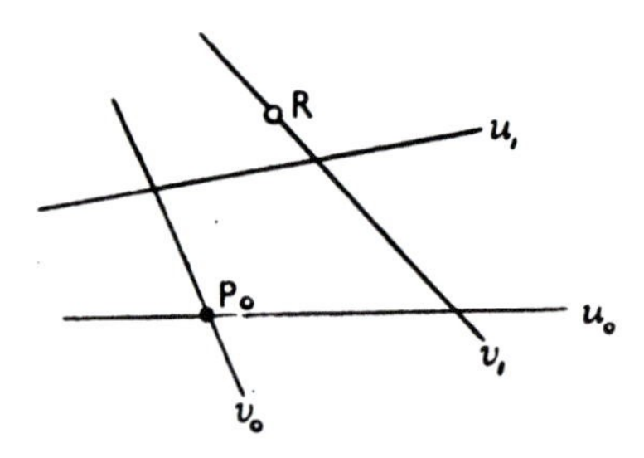

COROLLARY. *A generator of a proper quadric ψ is a self-dual entity. Its points all lie on ψ, and planes through it all touch ψ.*

We remarked on p. 267 that a proper quadric ψ may be thought of indifferently as the locus of a variable point or as the envelope of a variable plane; and we see now that, by virtue of Theorem 9, the same manifold may equally well be treated as a *ruled surface*, swept out by a variable line. The proper quadric may thus be regarded as (i) a *quadric locus*, or assemblage of ∞^2 points; or (ii) a *quadric envelope*, or assemblage of ∞^2 planes; or (iii) a *regulus*, or assemblage of ∞^1 lines.

The same quadric can be generated as a regulus in two distinct ways. Alternatively we can say that it gives rise to two *complementary reguli*, each of which determines the other as the set of all its transversals. Since each regulus comprises ∞^1 generating lines, we obtain ∞^2 line-pairs by taking one generator from each regulus; and in this way we obtain once again the ∞^2 points and the ∞^2 planes of ψ.

A quadric is uniquely determined by nine general points (or nine general planes), and this suggests that three generators may suffice to determine a quadric. More precisely, *there is a unique quadric which has three given mutually skew lines as generators of the same system*. For, if we take three points on each of the lines, there is a unique quadric ψ which contains the nine points; and sin each of the given lines meets this quadric in three points it lies wholly on it. The quadric must be proper, since neither the cone nor the plane-pair can contain three mutually skew lines.

If the given lines are v_1, v_2, v_3, a unique transversal can be drawn to v_1 and v_2 from any point of v_3. In this way we obtain ∞^1 lines, which each meet ψ in three points, i.e. the lines of the regulus complementary to the regulus which contains v_1, v_2, and v_3; and we can now construct this first regulus by selecting three lines of the second one and drawing transversals to them. In this way we are able to construct the generating lines, and hence the points and planes, of the quadric determined by v_1, v_2, and v_3.

The canonical equation of the quadric

By using two pairs of generators of a proper quadric ψ to define the tetrahedron of reference, we obtain a simple representation of the quadric, analogous in many respects to the canonical representation $x_1^2 = x_2 x_0$ of the proper conic.

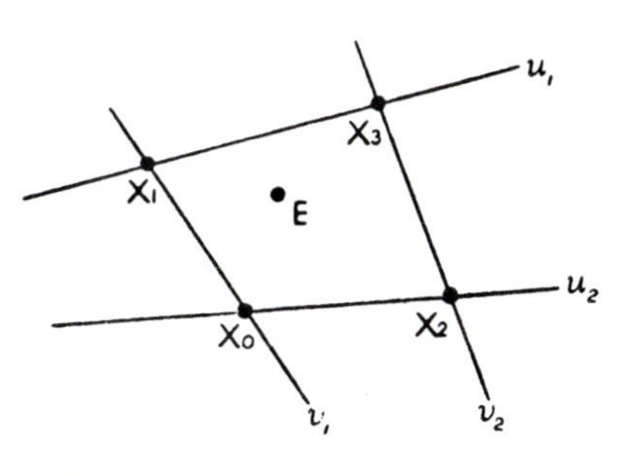

If we take two generators u_1, u_2 and two other generators v_1, v_2 as pairs of opposite edges of the tetrahedron of reference, as indicated, and a general point of ψ as unit point E, the point equation of ψ becomes

$$x_1 x_2 - x_0 x_3 = 0.$$

The associated plane-equation is

$$u_1 u_2 - u_0 u_3 = 0.$$

If we now put $x_1 = \theta x_3$, $x_2 = \phi x_3$, then

$$x_0 x_3 = x_1 x_2 = \theta\phi x_3^2,$$

i.e.

$$x_0 = \theta\phi x_3,$$

and so we have the parametric representation

$$x_0 : x_1 : x_2 : x_3 = \theta\phi : \theta : \phi : 1.$$

This may be called the *canonical parametric representation* of the proper quadric locus.

The points of ψ for which θ has the fixed value θ_0 are the points whose coordinates satisfy simultaneously the two equations

$$x_0 - \theta_0 x_2 = 0 \quad \text{and} \quad x_1 - \theta_0 x_3 = 0.$$

They are therefore the points of a certain line; and since the coordinates of every point of this line satisfy the equation

$$x_1 x_2 - x_0 x_3 = 0,$$

the line is a generator of ψ. Thus *the equations* $\theta =$ *constant and* $\phi =$ *constant give the two systems of generators of* ψ. The u-generators are given by $\phi =$ constant, and the v-generators by $\theta =$ constant.

As θ_0 takes different values, the planes $x_0 - \theta_0 x_2 = 0$ and $x_1 - \theta_0 x_3 = 0$ describe homographic pencils with axes $X_1 X_3$ and $X_0 X_2$ respectively. Thus the v-generators of ψ are the lines of intersection of two homographic pencils of planes with axes u_1 and u_2. Since u_1 and u_2 can be any two u-generators, we have proved the theorem:

THEOREM 10. *Every proper quadric* ψ *may be generated as the locus of the line of intersection of corresponding planes of two related pencils, and the axes of the pencils can be any two non-intersecting generators of* ψ. *Dually,* ψ *may be generated as the locus of the line which joins corresponding points of two related ranges, and the axes of the ranges can be any two non-intersecting generators of* ψ.

COROLLARY. *The* u*-generators of* ψ *cut homographic ranges on any two* v*-generators.*

The corollary may be proved independently as follows. Let v_1, v_2 be fixed v-generators, cut by a variable u-generator u in P_1, P_2. Then the polar plane of P_1 is the plane $v_1 u$, i.e. the plane $v_1 P_2$, and hence $(P_1) \barwedge v_1(P_2) \barwedge (P_2)$.

We now give the converse of Theorem 10.

THEOREM 11. *The ruled surface generated by the line of intersection of corresponding planes of two homographically related pencils, whose axes are skew lines* v_1, v_2, *is a proper quadric; and* v_1, v_2 *belong to the complementary regulus. Dually, the ruled surface generated by the line which joins corresponding points of two homographically related ranges, whose axes are skew lines* v_1, v_2, *is a proper quadric; and* v_1, v_2 *belong to the complementary regulus.*

Proof. As for Theorem 12 of Chapter VI.

We now have a projective generation of the quadric determined by three mutually skew lines (p. 274). Let v_1, v_2, v_3 be three such lines. Then, if P is a variable point of v_3,

$$v_1(P) \barwedge (P) \barwedge v_2(P);$$

and since the planes $v_1 P$ and $v_2 P$ intersect in a line that meets v_1, v_2, and v_3, the quadric generated by these two homographic pencils is the quadric defined by v_1, v_2, and v_3.

The degenerate quadrics, as well as the proper quadric, can be generated projectively by means of homographically related pencils of planes. Suppose, first of all, that we take two homographically related pencils of planes with intersecting axes a, b. If the axes meet in V, every two corresponding planes intersect in a line through V, and the lines of intersection generate a cone with vertex V. Since the homographic pencils of planes cut a general fixed plane σ in homographic pencils of lines $A(P)$, $B(P)$, the section of the cone by this fixed plane is a conic; and the cone is therefore a quadric cone.

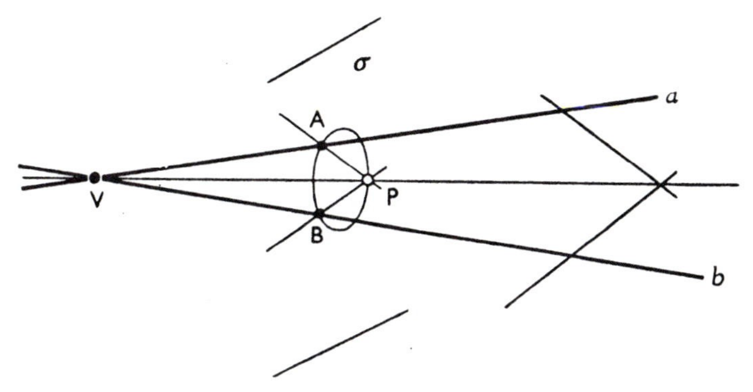

Now suppose we take two pencils of planes whose axes are skew but which are related by a degenerate homography, that is to say a correspondence

$$\theta' = \frac{\alpha\theta+\beta}{\gamma\theta+\delta},$$

for which $\alpha\delta-\beta\gamma = 0$, i.e. a correspondence $(\theta-k)(\theta'-k') = 0$. Then to all the planes of the first pencil correspond the same fixed plane π_0' of the second, and to all the planes of the second pencil correspond the same fixed plane π_0 of the first. The assemblage of lines of intersection thus becomes two plane pencils of lines, one in π_0 and one in π_0'. The vertices A, B of these pencils lie on the line of intersection of π_0 and π_0'; and the quadric generated by the two homographic pencils is a bifocal plane-pair (p. 269). The second system of generators is derived in a similar way from a degenerate homographic correspondence between pencils of planes with two lines of the first system as axes, π_0 and π'_0 again being the special planes. These generators also form two plane pencils, but they lie in the opposite planes to the first pair of pencils, thus:

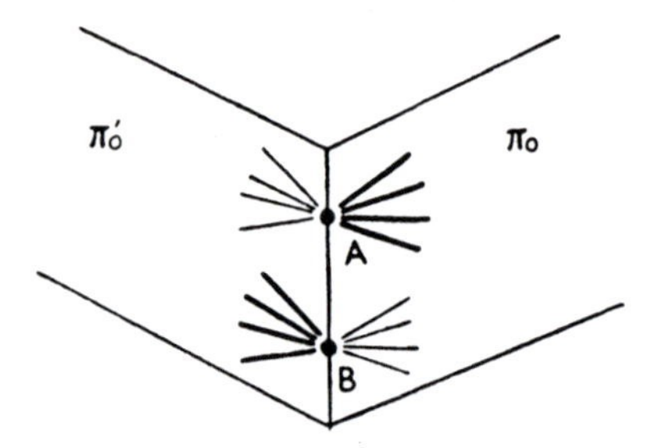

regulus I: a pencil in π_0 with vertex A + a pencil in π'_0 with vertex B;

regulus II: a pencil in π_0 with vertex B + a pencil in π'_0 with vertex A.

If the axes of the pencils are coplanar, and their plane assumes the roles of both π_0 and π'_0, the quadric is still more special, as its two planes now coincide.

§ 3. Projection of a Quadric on to a Plane

In the geometry of the conic, as the reader will remember, many remarkable properties follow from one theorem of fundamental importance, which states that the projective geometry of the plane induces a one-dimensional projective geometry on any proper conic; and this theorem depends essentially on the fact that the points of the conic subtend homographic pencils at any two fixed points of the curve. Unfortunately, as we can easily show, the points of a proper quadric do not subtend collinear stars at two fixed points V_1, V_2 of the surface. For, if P is a variable point of ψ, the ray V_2P describes a quadric cone when V_1P describes a general plane through V_1, so that the stars described by the rays V_1P and V_2P cannot be collinearly related. It follows that the projective geometry of S_3 does not induce a two-dimensional projective geometry on ψ.

We can, however, project a quadric ψ from any fixed point V of itself on to a plane π which does not pass through V; and in this projection the correspondence between a point P of ψ and its projection P' on π is (1, 1) in general. More precisely, if P does not fall at V, then it projects into a unique point P', and if P' is any point of π which does not lie in the tangent plane to ψ at V, then P'

is the projection of a unique point P of ψ. The plane representation of ψ obtained in this way is of great importance, since it enables us to study all the curves which lie on ψ, and the relations which exist between them, by studying the plane curves into which they project. We propose therefore to examine the representation in sufficient detail to show how it may be used for this purpose.

Before discussing the geometrical properties of the projection we shall first obtain its equations in a simple form. Let the equation of ψ be taken in canonical form

$$x_0 x_3 - x_1 x_2 = 0,$$

with V as the reference point $X_3 \equiv (0,0,0,1)$ and π as the plane $x_3 = 0$. The projection of any point $P \equiv (x_0, x_1, x_2, x_3)$, other than V itself, is then the point $P' \equiv (x_0, x_1, x_2, 0)$. It is evident also that any point $(x_0, x_1, x_2, 0)$, for which $x_0 \neq 0$, is the projection of the unique point $(x_0, x_1, x_2, x_1 x_2/x_0)$ of ψ. We have here the required algebraic representation of the transformation.

In order to use the projection, it is essential to realize that it has exceptional elements, and to understand the role which they play. These elements are related in a simple way to the vertex V, as we shall now show.

Let α be the tangent plane to ψ at V, meeting π in a line a, and let u_0, v_0 be the two generators of ψ through V. Then u_0, v_0 meet π in two points U_0, V_0, which lie on a.

The point V of ψ has itself no single corresponding point in π, but points of ψ in its 'first neighbourhood'—by which we understand directions through V in the tangent plane α—project into the individual points of a. This means that if P describes a path on ψ, which passes through V in the direction of the tangent line t, then the path described by P' passes through that point of a which corresponds to t; and conversely, if T is any point of a, other than U_0 and V_0, then as P' describes a path in π, distinct from a, passing through T, P describes a path on ψ that passes through V in the direction t corresponding to T.

Furthermore, all the points of u_0 project into the same point U_0, and all the points of v_0 into V_0; but here again we have a state of affairs similar to that just described, in that directions through U_0 correspond to the individual points of u_0, and directions through V_0 to the individual points of v_0. To see this, we observe that a variable point K of u_0 is homographically related to the tangent plane κ to ψ at this point, and hence to the line k through U_0 in which π is met by κ; and if P describes a general path passing through K, then P' describes a path passing through U_0 in the direction of k.

The above results may easily be verified algebraically by the method used in Chapter IX in the discussion of the reciprocal transformation. We may accordingly formulate the following theorem.

THEOREM 12. *Any proper quadric ψ may be represented birationally on a plane π by projecting it from a point V of itself. This representation is* (1, 1) *in general; but the neighbourhood of V on ψ is represented homographically on a line a of π, and the generators of ψ through V project respectively into two points U_0, V_0 of a, whose neighbourhoods in π correspond homographically to the generators in question.*

We note particularly the following facts.

(i) If c is a curve on ψ and c' is its projection in π, then every intersection of c' with a which is not at U_0 or V_0 corresponds to a branch of c at V. Intersections of c' with a at U_0 and V_0 correspond to intersections of c with u_0 and v_0 respectively.

(ii) If c does not pass through V, then c' meets a only at U_0 and V_0; and if c meets u_0 and v_0 respectively in sets of α and β distinct points, remote from V, then c' passes α times through U_0 and β times through V_0.

(iii) If c'_1, c'_2 are two curves in π, passing through U_0 (or V_0), then intersections of c'_1 with c'_2 at U_0 (or V_0) do not represent intersections of the corresponding curves c_1 and c_2 on ψ. The only exception is when a branch of c'_1 touches a branch of c'_2 at U_0 (or V_0).

Properties of the plane representation of the proper quadric

Since the u-generators of ψ all meet v_0, their projections all pass through V_0. Thus the two systems of generators of ψ project into two pencils of lines in π, with vertices V_0, U_0 respectively.

There are ∞^2 planes through V; and the conics in which ψ is cut by these planes project into the lines of π.

The sections of ψ by planes which do not pass through V project into conics through U_0 and V_0; and the residual points of intersection of two such conics arise from the two points in which ψ is met by the common line of the two corresponding planes of section.

Let us now consider an arbitrary curve c on ψ, which does not pass through V. Such a curve may for convenience be called a 'general' curve on ψ. If c meets u_0 in α distinct points† and v_0 in β distinct points, its projection c' passes α times through U_0 and β times through V_0; and since it does not meet a in any point other than U_0 and V_0 its order, as we shall here assume, is $\alpha+\beta$. We may say, then, that c' is a plane curve of the type $C^{\alpha+\beta}(U_0^\alpha, V_0^\beta)$.

Now suppose u is an arbitrary u-generator of ψ, represented in π by a line u' through V_0. Since u' meets c' β times at V_0, and c' is of order $\alpha+\beta$, u' has α free intersections with c'. It follows that c is met by u in α points; and, by the same argument, it is met by an arbitrary v-generator in β points. We thus have the theorem:

THEOREM 13. *If a curve c on ψ meets a u-generator u_0 in α distinct points and a v-generator v_0 in β distinct points, then it meets every u-generator in α points and every v-generator in β points. If the quadric is projected on to a plane from the point of intersection of u_0 and v_0, the projection of c is a curve of the type $C^{\alpha+\beta}(U_0^\alpha, V_0^\beta)$.*

A curve c of the kind just referred to is called an (α, β)-*curve* on ψ. Its order—i.e. the number of points in which it is met by a general plane—is $\alpha+\beta$; for since c does not pass through V, its order is equal to the order of its projection on π. We may observe also that any general tangent plane of ψ evidently meets c in $\alpha+\beta$ points.

The u-generators of ψ are, of course, $(0, 1)$-curves, and the v-generators are $(1, 0)$-curves. Any conic on ψ, being a plane section of the quadric, is a $(1, 1)$-curve. We shall see later on (p. 308) that ψ contains two families of twisted cubics, which are $(1, 2)$-curves and $(2, 1)$-curves respectively. The quartic curve in which ψ is met by any other quadric ψ' is plainly a $(2, 2)$-curve on ψ.

The freedom of (α, β)-curves on ψ is the same as that of curves $C^{\alpha+\beta}(U_0^\alpha, V_0^\beta)$ in π, and this may be determined by examining the form of the equation of such a curve when U_0 and V_0 are taken as reference points. An unrestricted plane curve of order $\alpha+\beta$ has $\frac{1}{2}(\alpha+\beta)(\alpha+\beta+3)$ degrees of freedom; and since the conditions U_0^α, V_0^β reduce this number by $\frac{1}{2}\alpha(\alpha+1)$ and $\frac{1}{2}\beta(\beta+1)$ respectively, the number of degrees of freedom remaining is $\alpha\beta+\alpha+\beta$. In other words, there are $\infty^{\alpha\beta+\alpha+\beta}$ (α, β)-curves on ψ.

Lastly, in order to find the number of intersections of a general (α, β)-curve with a general (α', β')-curve we have to find the number

† I.e., c has only simple points on u_0 and does not touch it.

of intersections, not at U_0 or V_0, of two general curves

$$C^{\alpha+\beta}(U_0^\alpha, V_0^\beta) \quad \text{and} \quad C^{\alpha'+\beta'}(U_0^{\alpha'}, V_0^{\beta'}).$$

If we assume that, in general, the number of common points of two curves is the product of the orders of the curves, and that, subject to suitable restrictions, the number of these intersections absorbed at a common point is the product of the multiplicities of this point—as is in fact the case if there are no contacts between the branches—the required number is

$$(\alpha+\beta)(\alpha'+\beta')-\alpha\alpha'-\beta\beta' = \alpha\beta'+\alpha'\beta.$$

§ 4. Affine and Euclidean Specializations

We have been dealing so far with quadrics in projective space S_3, for the most part with the complex field as ground field, but now we shall turn to the affine and euclidean specializations of the theory, which are of considerable geometrical interest. The space that we consider is real affine or euclidean space, with the plane at infinity ι adjoined as an ideal plane. The specializations which arise are analogous to those considered at the end of Chapter V, and we shall not need to discuss them in much detail.

We shall confine our attention now to real quadrics, i.e. quadrics whose equations, referred to any allowable coordinate representation, involve only real coefficients. A quadric that is real in this sense may, of course, be virtual, and have no real points; for example, the quadric given by $x_0^2+x_1^2+x_2^2+x_3^2 = 0$.

Affine properties of a quadric ψ are projective relations between ψ and the plane at infinity ι, and they are therefore closely connected with the section of ψ by ι—the conic at infinity h on ψ. This conic may be of several different kinds, distinguishable from one another in the projective geometry of the real plane, and we have a corresponding affine classification of quadrics:

Case 1: h is a non-virtual proper conic.
In this case ψ is said to be a hyperboloid.

Case 2: h is a virtual proper conic.
ψ is said to be an ellipsoid (real or virtual).

Case 3: h is a pair of distinct real lines.
ψ is said to be a hyperbolic paraboloid.

Case 4: h is a pair of conjugate complex lines.

ψ is said to be an elliptic paraboloid.

Case 5: h is a repeated (real) line.

ψ is said to be a parabolic cylinder.

This classification refers, with the exception of case 5, to proper quadrics. The modifications and additions which are needed in the case of degenerate quadrics may be left to the reader, since they are familiar from the elementary geometry of quadrics. Thus, for example, we call a cone whose vertex is at infinity a cylinder.

The pole of the plane at infinity is called the *centre* of ψ. In cases 1 and 2 the centre is a finite point, and is therefore a centre of symmetry. The hyperboloid and ellipsoid are referred to as central quadrics. Since the plane at infinity cuts a paraboloid in a pair of lines, it is a tangent plane, and the centre of a paraboloid is therefore at infinity. For this reason paraboloids have no centre of symmetry. Their affine geometry is rather special, just as in the plane the affine geometry of the parabola is special.

A line which passes through the centre of a quadric is called a *diameter*, and a plane which passes through the centre is called a *diametral plane*. A diameter and a diametral plane of a central quadric ψ are said to be *conjugate* if they meet ι in a point and a line which are pole and polar for the conic at infinity h on ψ—i.e. if the diametral plane is the polar plane with respect to ψ of the point at infinity on the diameter. It may be verified that a diametral plane bisects all chords parallel to the conjugate diameter, and a diameter contains the centres of the conics in which ψ is cut by planes parallel to the conjugate diametral plane.

Three diameters of ψ are said to be *mutually conjugate* if they meet ι in the vertices of a triangle which is self-polar for h. Each is then conjugate to the diametral plane determined by the other two; and the three diametral planes which contain the three pairs of diameters are said to be mutually conjugate diametral planes.

Exercises

(i) If d is a diameter of a central quadric ψ, show that its polar line d' lies in ι, and that planes through d' are parallel to the diametral plane conjugate to d.

(ii) If ψ is a proper central quadric, and three mutually conjugate diameters of ψ are taken as axes of reference for a cartesian coordinate system, show that the equations of ψ take the forms $aX^2+bY^2+cZ^2 = 1$ and $bcU^2+caV^2+abW^2 = abc$ respectively. How must the equations be modified if ψ is a proper cone with a finite vertex?

In order to fit the euclidean properties of the quadric into our projective scheme, we need to introduce the absolute conic Ω, as explained in Chapter X, § 2. Ω is a virtual conic, lying in the plane ι, whose equations in every allowable representation $\mathscr{R}_E$ are

$$x_1^2+x_2^2+x_3^2 = 0 = x_0.$$

When we regard the absolute conic as a disk quadric, its plane-equation takes the simple form

$$\Omega \equiv u_1^2+u_2^2+u_3^2 = 0.$$

If, now, we take any quadric ψ in euclidean space, we have a pair of conics in the plane at infinity, namely Ω and the conic at infinity h on ψ, and many euclidean properties of ψ can be interpreted as projective relations between these two conics.

Suppose, first of all, that h coincides with Ω. Then ψ is a sphere, and it possesses the properties (e.g. complete symmetry with respect to its centre) that are commonly associated with this special type of quadric.

In general, h and Ω have four distinct points in common, and they then have a unique common self-polar triangle. If ψ is a central quadric, the three diameters whose points at infinity are the vertices of this triangle are both mutually conjugate and also mutually perpendicular, and they are called the *principal axes* of ψ. It may easily be verified that ψ is symmetrical with respect to the three diametral planes determined by the axes, its *principal planes*. The axes of ψ define a rectangular coordinate system (i.e. an allowable representation $\mathscr{R}_E$) with respect to which the point-equation of ψ is of the form

$$aX^2+bY^2+cZ^2 = 1.$$

EXERCISE. Consider the various special cases which arise when (a) the points of intersection of h and Ω are not distinct, and (b) the centre of ψ is at infinity.

Confining our attention to the general case, in which ψ is a central quadric and the common points of h and Ω form a proper quadrangle, we now turn to the problem of finding all the circular sections of ψ. A plane π will cut ψ in a circle if and only if the line at infinity in π cuts h and Ω in the same two points, i.e. if this line is a side of the quadrangle already referred to. We see, then, that *a general quadric ψ has six families of circular sections, the circles of any one family being cut by a pencil of (parallel) planes whose axis* *is one of the common chords of h and* Ω. Each of the six pencils contains one diametral plane and two tangent planes. The centres of the circles lie on the diameter conjugate to the diametral plane, and the extremities of this diameter are the point-circles cut by the two tangent planes. A point on a quadric which is the centre of a point-circle lying on the surface (i.e. the point of intersection of a pair of isotropic generators) is called an *umbilic*; and we may therefore say that a general quadric possesses twelve umbilics.

EXERCISES

(i) Show that the four common points of h and Ω and the twelve umbilics of ψ form a set of sixteen points of ψ which lie in sets of four on eight generators.

(ii) If the equation of ψ, referred to its principal axes, is

$$aX^2+bY^2+cZ^2 = 1 \quad (a < b < c),$$

obtain the equations of the planes of circular section, and show that only two of the six pencils contain real planes.

Various special cases arise when the common points of h and Ω coincide in different ways. The most interesting of these occurs when h has double contact with Ω. In this case there is only one pencil of planes of circular section, the axis of this pencil being the chord of contact of h and Ω. The diameter of ψ which contains the centres of the circles is perpendicular to the planes of the circles, and the quadric is therefore a *quadric of revolution*. Since h and Ω have ∞^1 common self-polar triangles in this case, the axes of a quadric of revolution are partly indeterminate.

Transformation to principal axes

The principal axes of a central quadric ψ define, as we have already seen, a rectangular cartesian system of reference, and in terms of this system the equation of the quadric assumes the simple form $aX^2+bY^2+cZ^2 = 1$. If, now, we are given the equation of ψ referred to an arbitrary triad of rectangular axes, we may wish to find the principal axes of ψ and introduce new coordinates for which these lines are the axes of reference. Not only is the method of doing this of interest in itself, but the algebra on which the solution depends is used in solving fundamental problems in various branches of mathematics.† We shall now discuss the problem of

† In the language of algebra, the problem is just that of reducing a quadratic form $\mathbf{x}^T\mathbf{A}\mathbf{x}$ to diagonal form by means of an *orthogonal* transformation $\mathbf{x} = \mathbf{P}\mathbf{x}'$. The solution indicated in the text is quite general, and there is no need for $\mathbf{x}$ to be a 3-vector.

transformation of the equation of a quadric to principal axes, first explaining the principles involved in the solution, and then illustrating the method by means of a numerical example. It will be sufficient to take a quadric which is already referred to rectangular axes through its centre, since if the origin is not at the centre we can begin with a preliminary translation.

PROBLEM. If the equation of a central quadric ψ, referred to arbitrary rectangular axes through its centre, is

$$\sum_{i=1}^{3}\sum_{k=1}^{3} a_{ik} X_i X_k = 1 \quad (a_{ki} = a_{ik}),$$

to find the axes of ψ and to transform the equation by taking these axes as new axes of reference.

Solution. The equation of ψ, written homogeneously, is

$$\sum_{i=1}^{3}\sum_{k=1}^{3} a_{ik} x_i x_k - x_0^2 = 0.$$

If the point at infinity on one of the axes of ψ is $(0, y_1, y_2, y_3)$, the conjugate diametral plane is

$$\sum_{i=1}^{3}\Big(\sum_{k=1}^{3} a_{ik} y_k\Big) x_i = 0;$$

and since this plane is perpendicular to the axis

$$\sum_{i=1}^{3} a_{ik} y_k = \lambda y_i \quad (i = 1, 2, 3),$$

i.e.

$$\mathbf{A}\mathbf{y} = \lambda \mathbf{y}.$$

Thus the directions of the axes are given by the characteristic vectors $\mathbf{y}$ of the 3×3 matrix $\mathbf{A} = (a_{rs})$. To find these vectors we have first to solve the characteristic equation

$$|\mathbf{A} - \lambda \mathbf{I}| = 0,$$

i.e.

$$\begin{vmatrix} a_{11}-\lambda & a_{12} & a_{13} \\ a_{21} & a_{22}-\lambda & a_{23} \\ a_{31} & a_{32} & a_{33}-\lambda \end{vmatrix} = 0.$$

Suppose the roots λ_1, λ_2, λ_3 of this equation are distinct. If λ is equal to any one of the characteristic roots λ_j, the three equations $\sum_{k=1}^{3} a_{ik} y_k = \lambda y_i$ are consistent, and we may solve them for the ratios $y_1 : y_2 : y_3$. In this way we obtain three non-zero vectors $\mathbf{y}^{(1)}$, $\mathbf{y}^{(2)}$,

$\mathbf{y}^{(3)}$, corresponding to λ_1, λ_2, λ_3 respectively; and from these we can derive three unit vectors $\mathbf{e}^{(i)} = \dfrac{\mathbf{y}^{(i)}}{|\mathbf{y}^{(i)}|}$, uniquely determined except for sign. The components of these vectors are then the direction cosines of the axes of ψ. That they are in fact the direction cosines of three mutually perpendicular lines can be seen algebraically as follows:

Since
$$\mathbf{A}\mathbf{e}^{(i)} = \lambda_i \mathbf{e}^{(i)}$$
and
$$\mathbf{A}\mathbf{e}^{(j)} = \lambda_j \mathbf{e}^{(j)},$$
we have
$$\mathbf{e}^{(j)T}\mathbf{A}\mathbf{e}^{(i)} - \mathbf{e}^{(i)T}\mathbf{A}\mathbf{e}^{(j)} = \lambda_i \mathbf{e}^{(j)T}\mathbf{e}^{(i)} - \lambda_j \mathbf{e}^{(i)T}\mathbf{e}^{(j)},$$
i.e.
$$0 = (\lambda_i - \lambda_j)(\mathbf{e}^{(i)}, \mathbf{e}^{(j)}).$$
But $\lambda_i \neq \lambda_j$, by hypothesis, and therefore $(\mathbf{e}^{(i)}, \mathbf{e}^{(j)}) = 0$.

If, now, $\mathbf{e}_i$ and $\mathbf{e}'_i$ $(i = 1, 2, 3)$ are the sets of unit vectors along the old axes and along the principal axes of ψ, we can write $\mathbf{e}'_i = \mathbf{e}^{(i)}$. If $\mathbf{e}'_i$ has components $(e'_{i1}, e'_{i2}, e'_{i3})$, and $\mathbf{r}$ is the position vector of a general point of space,
$$\begin{aligned}\mathbf{r} &= \sum_{i=1}^{3} X'_i \mathbf{e}'_i \\ &= \sum_i X'_i\Big(\sum_k e'_{ik}\mathbf{e}_k\Big) \\ &= \sum_k \Big(\sum_i e'_{ik} X'_i\Big)\mathbf{e}_k \\ &= \sum_k X_k \mathbf{e}_k,\end{aligned}$$
and hence
$$X_k = \sum_i e'_{ik} X'_i \quad (k = 1, 2, 3).$$
The equations of transformation are accordingly
$$\mathbf{r} = \mathbf{P}\mathbf{r}', \qquad \mathbf{r}' = \mathbf{P}^{-1}\mathbf{r},$$
where
$$\mathbf{P} = (e'_{sr}).$$

Remarks

(i) The orthogonality relations $(\mathbf{e}^{(i)}, \mathbf{e}^{(j)}) = 0$, obtained above, ensure that $\mathbf{P}$ is an orthogonal matrix, in conformity with Theorem 9 of Chapter X.

(ii) We have assumed that λ_1, λ_2, λ_3 are all distinct. If two of them coincide, ψ is a quadric of revolution, and if all three coincide it is a sphere. In these cases the unit vectors $\mathbf{e}^{(i)}$ are not uniquely determined by the equation $\mathbf{A}\mathbf{e} = \lambda\mathbf{e}$, and the axes are partly arbitrary.

(iii) It is a well-known algebraic theorem that, for a real symmetric matrix $\mathbf{A}$, the characteristic roots $\lambda_1, \lambda_2, \lambda_3$ are all real.

Let us now take a numerical example, transforming the quadric $2X^2+5Y^2+5Z^2-ZX = 1$ to principal axes.

The characteristic equation is

$$\begin{vmatrix} 2-\lambda & 0 & -\frac{1}{2} \\ 0 & 5-\lambda & 0 \\ -\frac{1}{2} & 0 & 5-\lambda \end{vmatrix} = 0,$$

i.e.
$$(5-\lambda)\{(2-\lambda)(5-\lambda)-\tfrac{1}{4}\} = 0,$$

i.e.
$$(5-\lambda)(39-28\lambda+4\lambda^2) = 0;$$

and its roots are 5, $\frac{1}{2}(7\pm\sqrt{10})$.

Three characteristic vectors are $(0,1,0)$, $(1,0,-3\mp\sqrt{10})$, and by normalizing them we obtain the vectors

$$\mathbf{e}_1' = \frac{1}{\sqrt{(20+6\sqrt{10})}}(1,0,-3-\sqrt{10}),$$

$$\mathbf{e}_2' = (0,1,0),$$

$$\mathbf{e}_3' = \frac{1}{\sqrt{(20-6\sqrt{10})}}(1,0,-3+\sqrt{10}),$$

corresponding to the characteristic roots in the order $\frac{1}{2}(7+\sqrt{10})$, 5, $\frac{1}{2}(7-\sqrt{10})$. They give the orthogonal transformation

$$X = \frac{1}{\sqrt{(20+6\sqrt{10})}}X' \qquad +\frac{1}{\sqrt{(20-6\sqrt{10})}}Z',$$

$$Y = \qquad Y',$$

$$Z = \frac{-3-\sqrt{10}}{\sqrt{(20+6\sqrt{10})}}X' \qquad +\frac{-3+\sqrt{10}}{\sqrt{(20-6\sqrt{10})}}Z',$$

and the transformed equation of the quadric is

$$\frac{7+\sqrt{10}}{2}X'^2+5Y'^2+\frac{7-\sqrt{10}}{2}Z'^2 = 1.$$

This example is especially simple, since one of the original axes OY is already a principal axis of the quadric, as can be seen from the form of the equation with which we started.

It will be observed that the coefficients in the transformed equation are the characteristic roots $\lambda_1, \lambda_2, \lambda_3$. This is quite general, and follows from standard theorems on the reduction of quadratic forms. Thus the characteristic roots of the matrix $\mathbf{A}$ may be interpreted in terms of the lengths of the axes of the quadric

$$\sum_{i=1}^{3}\sum_{k=1}^{3} a_{ik}X_iX_k = 1.$$

EXERCISE. If the axes are rectangular, show that a necessary and sufficient condition for a real proper quadric, represented by the equation

$$(a,b,c,d,f,g,h,p,q,r\between X,Y,Z,1)^2$$
$$\equiv aX^2+bY^2+cZ^2+2fYZ+2gZX+2hXY+2pX+2qY+2rZ+d = 0,$$

to be a quadric of revolution is that, for some value of λ, the matrix

$$\begin{pmatrix} a-\lambda & h & g \\ h & b-\lambda & f \\ g & f & c-\lambda \end{pmatrix}$$

is of rank 1. Examine the various ways in which this condition can be satisfied.

§ 5. STEREOGRAPHIC PROJECTION OF A SPHERE ON TO A PLANE

In § 3 we discussed in some detail the projection of a proper quadric ψ from a point V of itself on to a plane π, and we showed that plane sections of ψ project into conics through two fixed points U_0, V_0 of π or, if they pass through V, into lines of π. In extended euclidean space we can obtain some striking specializations of this representation by arranging matters so that U_0, V_0 are the absolute points I, J of π. If we do this, the projection will have the remarkable property that the totality of plane sections of ψ projects into the totality of circles (in the wide sense that includes straight lines) of π.

The necessary arrangement is quite simple: we take as V any umbilic of ψ, and as π any plane parallel to the tangent plane of ψ at V. Then the generators of ψ through V, which are isotropic lines lying in the tangent plane at V, meet π in points which lie both on the line at infinity in π and also on the absolute conic Ω, and these points are therefore the absolute points I, J in π as required.

If ψ is a sphere, the arrangement is even simpler. Since every point of a sphere is an umbilic, V may now be any point of ψ, and the only requirement is that π shall be parallel to the tangent plane at V. This kind of projection of a sphere on to a plane, in which all plane sections project into circles or straight lines, is known as *stereographic projection*. We shall now examine some of its properties.

If P, Q are any two points of ψ, the sections of ψ by planes through the line PQ evidently project into circles of the coaxal system whose common points are the projections P', Q' of P, Q. The two tangent planes of ψ which pass through PQ cut ψ in line-pairs, and these project into pairs of isotropic lines in π, i.e. into the point-circles of the coaxal system. The points of contact M, N of the tangent planes project into the two limiting points M', N' of the system.

The two lines PQ and MN are polar lines for ψ, and are therefore symmetrically related. It follows that the sections of ψ by planes through MN project into circles of the coaxal system whose common points are M', N' and whose limiting points are P', Q', i.e. the conjugate coaxal system.

If two planes are such that each passes through one of a given pair of polar lines, the planes are conjugate for ψ; and conversely, if two planes are conjugate for ψ, they contain an infinity of pairs of lines, one line in each plane, which are polar for ψ. We have therefore the result: *Two circles in π are orthogonal if and only if the corresponding sections of ψ are cut by planes which are conjugate for ψ.* The extension of this result to a circle and a line (or a pair of straight lines) cutting orthogonally is immediate.

If c is a circle on ψ, cut by a plane α, and A is the pole of α, the enveloping cone of ψ whose vertex is A has ring contact with ψ along c; and it is plain, from consideration of elementary geometry, that every plane through A cuts ψ in a circle that is orthogonal to c. Thus the result just established amounts to the fact that orthogonal circles on ψ project into orthogonal circles in π. It is, in fact, a well-known property of stereographic projection that it not only leaves orthogonality invariant but preserves the magnitudes of all angles, i.e. stereographic projection is a conformal mapping of the sphere on a plane.

From the result proved above we deduce that if c' is any circle in π, representing the section of ψ by a plane α, then all circles or straight lines which cut c' orthogonally represent sections of ψ by planes through the pole A of α. In particular, all diameters of c' represent sections of ψ by planes through VA, so that the centre of c' is simply the projection of A. Hence *if a circle c' in π represents the section of ψ by a plane α, its centre is the projection from V of the pole A of α.*

Keeping to the same notation, we may now ask what self-transformation of ψ corresponds in π to inversion with respect to c'. Two points P', Q' are inverse for c' if and only if every circle through them cuts c' orthogonally, and the corresponding points P, Q of ψ must therefore be such that every plane through them passes through the pole A of α. In other words they must be collinear with A, and so we have the result: *Inversion with respect to a circle c' in π corresponds to the self-transformation of ψ generated by chords through a fixed point A that does not lie on the surface; and conversely, chords through any point A, not on the surface, generate a self-transformation of ψ which corresponds in π to inversion with respect to the circle representing the section of ψ by the polar plane of A.*

These results may serve to illustrate the process of translating the geometry of circles in the plane into geometry on a quadric with special reference to a fixed point of itself.

EXERCISES ON CHAPTER XI

1. Show that the quadric locus $S(x, y, z, t) = 0$ is a cone if and only if there exists a triad of linearly independent linear forms π_1, π_2, π_3 in x, y, z, t such that S is expressible as a quadratic form in π_1, π_2, π_3. Show further that when this condition is satisfied the vertex of the cone is the point V which is common to the three planes $\pi_i = 0$, and that S can be expressed as a quadratic form in π_1', π_2', π_3', where $\pi_i' = 0$ $(i = 1, 2, 3)$ are any three linearly independent planes through V.

Show that the quadric

$$x^2+y^2+z^2+t^2+2yz+2zx+2xy+2xt+2yt-2zt = 0$$

is a cone with vertex $(-1, 1, 0, 0)$; and write down the general equation of a quadric cone with this point as vertex.

2. Find necessary and sufficient conditions for the plane (u, v, w, p) to touch the cone whose equation is given in Exercise 1. [*Hint.* The plane must pass through the vertex and also touch, for example, the section of the cone by the plane $x = 0$.]

3. Investigate the projective character and the singular points (if any) of each of the following quadric loci:

(i) $-y^2+z^2+zx+xy+xt-yt+zt = 0$,

(ii) $x^2+y^2+z^2-yz-zx-xy+t^2 = 0$,

(iii) $x^2+2y^2+t^2+2xy-2zt = 0$.

Find the equation of the quadric envelope defined by the locus (iii).

4. Investigate the projective character and the singular planes (if any) of each of the following quadric envelopes:

(i) $u^2-v^2-w^2+p^2-2vw-2up = 0$,

(ii) $p^2-vw-wu-up-vp = 0$,

(iii) $vw+wu+uv-p^2 = 0$.

Find the equation of the quadric locus defined by the envelope (iii).

5. Find the tangent cones from the point $(-1, 1, 1, 1)$ to the quadrics (ii) and (iii) in Exercise 3, and explain why the first is a plane-pair and the second is a repeated plane. Find the planes concerned.

Find also the plane-equations of the sections by the plane $x-t = 0$ of the quadrics (ii) and (iii) in Exercise 4. Explain why the first of these sections is a point-pair, and find the points.

6. Obtain the plane-equations of the cone (ii) in Exercise 3 and the point-equations of the conic (disk quadric) (ii) in Exercise 4.

7. Find the point-equation of the cone whose plane-equations are

$$w-p = 0 = u^2-4uv-v^2-p^2.$$

Find also the plane-equation of the conic (disk quadric) whose point-equations are $z+t = 0 = x^2-y^2-3t^2$.

8. If θ and ϕ are projective parameters for the two systems of generators on a quadric ψ, prove that the point of intersection of a variable pair of generators, one belonging to each system, describes a proper conic on ψ if and only if the parameters θ, ϕ of the generators are connected by a fixed non-singular bilinear relation.

9. Two lines l and m meet the quadric $(\theta\phi, \theta, \phi, 1)$ in the pairs of points (θ_1, ϕ_1), (θ_2, ϕ_2) and (θ_3, ϕ_3), (θ_4, ϕ_4). Show that l meets m if and only if

$$\{\theta_1, \theta_2; \theta_3, \theta_4\} = \{\phi_1, \phi_2; \phi_3, \phi_4\}.$$

If l meets both m and its polar line m', prove that each of the two cross ratios has the value -1.

10. Show that a line is its own polar line with respect to a quadric if and only if it is a generator of the quadric.

If $ABCD$ is a given tetrahedron and l is a given line of general position, prove that there exists a unique quadric for which $ABCD$ is a self-polar tetrahedron and l is a generator.

Deduce that the range of points in which l meets the faces of the tetrahedron is homographic with the pencil of planes joining l to the opposite vertices.

11. A plane π_0 meets four generators g_r $(r = 1, 2, 3, 4)$ of a quadric at their intersections $g_r g'_r$ with generators g'_r of the opposite system, and the intersections $g_r g'_{r+1}$ lie in a second plane π_1 (g'_{n+4} being defined as g'_n for all n). Prove that each set of generators is a harmonic set.

Prove that the intersections $g_r g'_{r+2}$ and $g_r g'_{r+3}$ lie in two planes π_2, π_3 coaxial with π_0 and π_1, and that these four planes form a harmonic pencil.

12. Show that the two systems of generators of a quadric ψ project from any point, not on ψ, into the tangents to a conic.

If g_1, g_2, g_3 are three generators of one system and g'_1, g'_2, g'_3 are three generators of the other system, show that the three lines $g_2 g'_3 \cdot g'_2 g_3$, $g_3 g'_1 \cdot g'_3 g_1$, $g_1 g'_2 \cdot g'_1 g_2$ are concurrent, and hence prove Brianchon's Theorem for a conic.

13. Find the coordinates of the poles X', Y', Z', T' of the faces of the tetrahedron of reference with respect to the quadric

$$\Sigma \equiv (A, B, C, D, F, G, H, L, M, N \between u, v, w, p)^2 = 0.$$

If $FL = GM = HN$, show that $XYZT$ and $X'Y'Z'T'$ are in perspective from a point V, and that in this case the form Σ can be expressed as a linear combination of the squares of the left-hand sides of the plane-equations of the points X, Y, Z, T, V. Show further that any four of the five points X, Y, Z, T, V form a tetrahedron which is in perspective with its polar tetrahedron from the fifth point as vertex.

14. Show that the point P, with coordinates (X_0, Y_0, Z_0), is the centre of the section of the quadric ψ whose cartesian equation is $aX^2+bY^2+cZ^2 = 1$ by the plane π whose equation is

$$aX_0(X-X_0)+bY_0(Y-Y_0)+cZ_0(Z-Z_0) = 0.$$

15. With the notation of Exercise 14, show that if P describes a plane whose pole for ψ is A then π envelops a quadric which is inscribed both in the asymptotic cone of ψ and also in the tangent cone to ψ from A.

Show also that if P describes a line then π envelops a cylinder.

16. Show that the normals to a quadric at the points of a generator lie on a hyperbolic paraboloid.

17. If $fgh \neq 0$, show that the conditions for the quadric ψ whose rectangular cartesian equation is $(a,b,c,d,f,g,h,l,m,n \between X,Y,Z,1)^2 = 0$ to be a quadric of revolution are

$$(gh-af)/f = (hf-bg)/g = (fg-ch)/h.$$

If $fgh = 0$, show that for ψ to be a quadric of revolution at least two of f, g, h must be zero, and that when $g = h = 0$ a necessary and sufficient condition is $f^2 = (b-a)(c-a)$.

A conic k has equations $aX^2+bY^2-1 = 0 = Z$. Prove that the quadrics of revolution through k form two families, and that the equation of a general member of one of these families is

$$aX^2+bY^2-1+\lambda Z^2+2\sqrt{\{(\lambda-b)(a-b)\}}XZ+2\mu Z = 0,$$

where λ and μ are arbitrary constants. Find the centre and axis of revolution of this quadric.

18. If a line l meets a given central quadric ψ in U and V, prove that the following three conditions on l are equivalent:

(i) l is perpendicular to its polar line for ψ;

(ii) the normals to ψ at U and V intersect;

(iii) l is a principal axis of some plane section of ψ.

19. If h is the conic at infinity on a quadric ψ, show that (i) if h is outpolar to the absolute conic Ω then ψ possesses an infinity of sets of three perpendicular generators of each system, and (ii) if ψ is a cone and h is inpolar to Ω then ψ possesses an infinity of sets of three mutually perpendicular tangent planes.

By means of (ii), or otherwise, show that the locus of a point from which three mutually perpendicular tangent planes can be drawn to a given quadric is in general a sphere, but that if the quadric is a paraboloid the locus is a plane.

20. If A is a fixed point of a quadric ψ, and P, Q, R are three variable points of ψ such that the lines AP, AQ, AR are mutually perpendicular, show that the plane PQR passes through a fixed point on the normal to ψ at A.

21. If ψ is a given quadric in euclidean space, and P is a given general point, show that there are six planes through P each of which cuts ψ in a conic with P as one focus.

22. If two quadrics ψ, ψ' touch along a conic k, prove that the tangent plane π to ψ' at an umbilic U meets ψ in a conic for which U is a focus and the line of intersection of π with the plane of k is the corresponding directrix. Use this result to obtain a construction for the foci of any plane section of a right circular cone.

23. Show that the sphere $X^2+Y^2+Z^2 = a^2$ admits of the parametric representation

$$X = a\frac{\lambda+\mu}{1+\lambda\mu}, \qquad iY = a\frac{\lambda-\mu}{1+\lambda\mu}, \qquad Z = a\frac{1-\lambda\mu}{1+\lambda\mu},$$

real points of the surface being given by conjugate complex values of λ and μ.

Discuss the resulting representation of the real sphere on the Argand diagram for the complex variable λ.

24. Show, by use of stereographic projection of a sphere, that two inversions of the plane with respect to circles c_1 and c_2 commute with each other if and only if c_1 and c_2 are orthogonal.

25. A self-transformation τ of the plane is generated as the product of the inversions defined by two given circles c_1 and c_2. Show that τ can be generated, in infinitely many ways, as the product of the inversions defined by two other circles c_1' and c_2' of the coaxal system determined by c_1 and c_2. State precisely how the pairs of circles (c_1, c_2) and (c_1', c_2') are related within the coaxal system.

CHAPTER XII

THE TWISTED CUBIC CURVE AND CUBIC SURFACES

§ 1. The Twisted Cubic

We come, in this section, to the second of the manifolds in space which are analogous to the conic in the plane, namely the twisted cubic curve. In the latter part of the discussion we shall make use occasionally of results which are not established until Chapter XIV or Chapter XV. This is quite harmless, as these chapters do not depend upon the present one, and the reader may refer to the later results as they are required.

Basic properties of the twisted cubic

DEFINITION. A *twisted cubic* in S_3 is a curve which is represented, in terms of some allowable representation $\mathscr{R}$, by parametric equations of the form

$$x_0 : x_1 : x_2 : x_3 = f_0(\theta) : f_1(\theta) : f_2(\theta) : f_3(\theta),$$

where the functions $f_i(\theta)$ are linearly independent cubic polynomials in the parameter θ.

Since the polynomials $f_i(\theta)$ are four linearly independent linear combinations of θ^3, θ^2, θ, 1, say

$$f_i(\theta) \equiv \sum_{k=0}^{3} a_{ik}\,\theta^{3-k} \quad (i = 0, 1, 2, 3),$$

their leading coefficients a_{i0} are not all zero; i.e. at least one of the polynomials is actually of the third degree. Inverting the equations just given, we have

$$|a_{rs}|\,\theta^{3-k} = \sum_{i=0}^{3} A_{ik}\, f_i(\theta),$$

where $|A_{rs}| = |a_{rs}|^3 \neq 0$; and θ^3, θ^2, θ, 1 are accordingly proportional to four linearly independent linear combinations of x_0, x_1, x_2, x_3:

$$\theta^{3-k} = \rho \sum_{i=0}^{3} A_{ik}\, x_i.$$

If, therefore, we introduce a new representation $\mathscr{R}'$ of S_3 by the transformation

$$x'_k = \sum_{i=0}^{3} A_{ik}\, x_i \quad (k = 0, 1, 2, 3),$$

the twisted cubic has the canonical representation

$$x_0':x_1':x_2':x_3' = \theta^3:\theta^2:\theta:1.$$

As may be inferred from the form of its equations, the twisted cubic is an algebraic space curve of the third order, which meets a general plane of S_3 in three points. It is, of course, a rational curve.

The above algebra shows that by a suitable non-singular linear transformation of the coordinates, which we have interpreted as a change of representation from $\mathscr{R}$ to $\mathscr{R}'$, the equations of any given twisted cubic may be reduced to

$$x_0':x_1':x_2':x_3' = \theta^3:\theta^2:\theta:1.$$

Exactly the same algebra, interpreted now in terms of transformation of points, also shows that by means of a suitable (non-singular) collineation of space we can transform any given twisted cubic into the standard twisted cubic whose equations referred to $\mathscr{R}$ are

$$x_0:x_1:x_2:x_3 = \theta^3:\theta^2:\theta:1.$$

Since the set of all collineations is a group, it follows at once that *any twisted cubic can be transformed into any other twisted cubic by means of a suitably chosen space collineation.* This result may be compared with the analogous result for proper conics.

Before going any farther, we shall establish a connexion between twisted cubics and nets of quadrics which will be of considerable use later on.

THEOREM 1. *Through any twisted cubic there pass ∞^2 quadrics, forming a net; and the twisted cubic is the common intersection of all the quadrics of this net.*

Proof. Consider the twisted cubic c whose equations are

$$x_0:x_1:x_2:x_3 = \theta^3:\theta^2:\theta:1.$$

Since the coordinates of any point of c satisfy the equations

$$\frac{x_0}{x_1} = \frac{x_1}{x_2} = \frac{x_2}{x_3},$$

c lies on each of the quadrics

$$Q_1 \equiv x_1^2 - x_0 x_2 = 0,$$

$$Q_2 \equiv x_1 x_2 - x_0 x_3 = 0, \qquad Q_3 \equiv x_2^2 - x_1 x_3 = 0,$$

and therefore on every quadric of the net

$$Q \equiv \lambda_1 Q_1 + \lambda_2 Q_2 + \lambda_3 Q_3 = 0.$$

Now suppose, conversely, that

$$S \equiv \sum_i \sum_k a_{ik} x_i x_k = 0$$

represents an arbitrary quadric through c. Then

$$\sum_i \sum_k a_{ik} \theta^{3-i}\theta^{3-k} \equiv 0;$$

and, taking account of the symmetry conditions $a_{ki} = a_{ik}$, we may replace this identity by the set of equations

$$\begin{aligned} a_{00} &= 0, & 2a_{13}+a_{22} &= 0, \\ 2a_{01} &= 0, & 2a_{23} &= 0, \\ 2a_{02}+a_{11} &= 0, & a_{33} &= 0. \\ 2a_{03}+2a_{12} &= 0, \end{aligned}$$

The equation $S = 0$ may accordingly be written

$$a_{11} Q_1 + 2a_{12} Q_2 + a_{22} Q_3 = 0,$$

and the quadric therefore belongs to the net already defined.

Finally, the residual intersections of Q_1, Q_2, Q_3 in pairs are the lines $x_2 = 0 = x_3$, $x_1 = 0 = x_2$, $x_0 = 0 = x_1$; and the quadrics of the net therefore have no common point which does not lie on c. This completes the proof of the theorem.

The canonical parametric representation of the twisted cubic is similar to that of the conic, and the two representations lead to more or less similar consequences. First we give a few useful algebraic results.

(i) Let the coordinate vector $\mathbf{u}$ represent the plane which joins the points of c whose parameters are θ_1, θ_2, θ_3. Then θ_1, θ_2, θ_3 are the roots of the cubic equation

$$u_0 \theta^3 + u_1 \theta^2 + u_2 \theta + u_3 = 0,$$

and consequently

$$\frac{u_0}{1} = \frac{-u_1}{\theta_1+\theta_2+\theta_3} = \frac{u_2}{\theta_2\theta_3+\theta_3\theta_1+\theta_1\theta_2} = \frac{-u_3}{\theta_1\theta_2\theta_3}.$$

The equation of the plane $(\theta_1, \theta_2, \theta_3)$ is therefore

$$x_0 - (\theta_1+\theta_2+\theta_3)x_1 + (\theta_2\theta_3+\theta_3\theta_1+\theta_1\theta_2)x_2 - \theta_1\theta_2\theta_3 x_3 = 0.$$

(ii) The plane which meets a twisted curve three times at a point P is called the *osculating plane* at P. The equation of the osculating plane to the twisted cubic at (θ_1) is therefore

$$x_0 - 3\theta_1 x_1 + 3\theta_1^2 x_2 - \theta_1^3 x_3 = 0.$$

Since this equation is cubic in θ_1, there are three osculating planes of c which pass through a general point of space. We say that c is of *class* 3.

(iii) The chord (θ_1, θ_2) lies in the plane $(\theta_1, \theta_2, \theta)$ for all values of θ, i.e. it is the axis of the pencil of planes

$$x_0-(\theta_1+\theta_2)x_1+\theta_1\theta_2x_2+\theta\{x_1-(\theta_1+\theta_2)x_2+\theta_1\theta_2x_3\} = 0.$$

The chord (θ_1, θ_2) therefore has equations

$$x_0-(\theta_1+\theta_2)x_1+\theta_1\theta_2x_2 = 0 = x_1-(\theta_1+\theta_2)x_2+\theta_1\theta_2x_3.$$

It follows from this that through any point P of space, not a point of c, there passes a unique chord of c. For suppose the coordinates of P are (y_0, y_1, y_2, y_3). Then the equations

$$y_0-py_1+qy_2 = 0 = y_1-py_2+qy_3$$

determine p and q uniquely, and the roots of the quadratic equation

$$\theta^2-p\theta+q = 0$$

determine the end-points of a unique chord through P.

(iv) The tangent line to c at (θ_1) is the chord (θ_1, θ_1), and it is accordingly represented by the equations

$$x_0-2\theta_1x_1+\theta_1^2x_2 = 0 = x_1-2\theta_1x_2+\theta_1^2x_3.$$

If we eliminate θ_1 between these equations, we obtain the equation of the ruled surface of tangents:

$$(x_0x_3-x_1x_2)^2-4(x_1^2-x_0x_2)(x_2^2-x_1x_3) = 0,$$

i.e. $$Q_2^2-4Q_1Q_3 = 0.$$

This is a quartic surface, and there are therefore four tangents to c which meet a general line of space. We say that c is of *rank* 4.

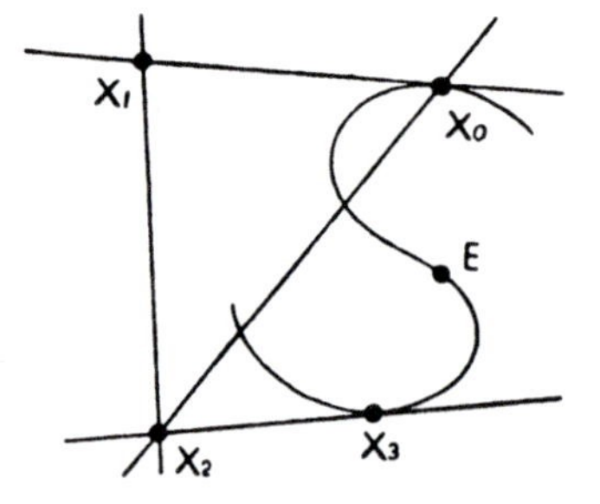

Summing up the above conclusions, we have:

THEOREM 2. *The twisted cubic is of order* 3, *class* 3, *and rank* 4, *and the number of chords of the cubic which pass through a general point of space is* 1.

We can now see how the tetrahedron of reference is related to the twisted cubic c when the representation of c is in canonical form. Let P be the variable point $(\theta^3, \theta^2, \theta, 1)$. When θ has the values ∞, 0, 1, P takes the positions X_0, X_3, E respectively. Thus c passes through two vertices of the tetrahedron of reference and the unit point. The osculating plane at X_0, where $\theta = \infty$, is $x_3 = 0$, and the tangent at this point is $x_2 = 0 = x_3$. Thus X_1 lies on the tangent at X_0, and $X_0 X_1 X_2$ is the osculating plane at X_0. Similarly X_2 lies on the tangent at X_3, and $X_1 X_2 X_3$ is the osculating plane at X_3. Hence *a canonical representation of the twisted cubic is uniquely defined by taking* X_0, X_3 *to be any two distinct points of the curve,* X_1, X_2 *to be the points in which the tangents at* X_0, X_3 *are respectively met by the osculating planes at* X_3, X_0, *and* E *to be any point of the curve other than* X_0 *and* X_3.

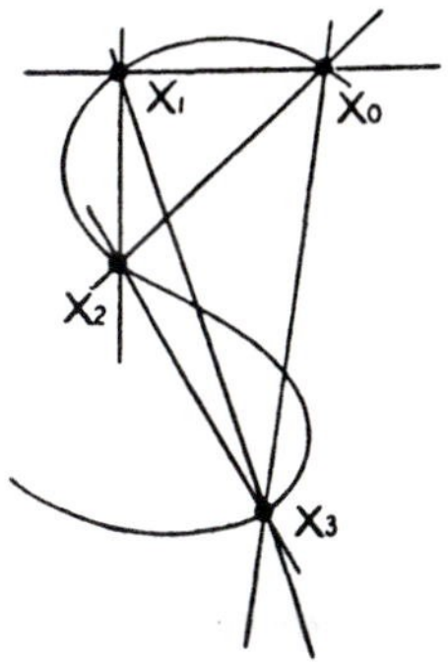

X₁ X₀ X₂ X₃

A second special representation of the twisted cubic that is often useful is obtained by taking an arbitrary inscribed tetrahedron as tetrahedron of reference. Suppose the equations of the twisted cubic c are then

$$x_0:x_1:x_2:x_3 = f_0(\theta):f_1(\theta):f_2(\theta):f_3(\theta),$$

and the parameter of X_i is θ_i ($i = 0, 1, 2, 3$). Then

$$f_0(\theta_1) = f_0(\theta_2) = f_0(\theta_3) = 0,$$

and hence $$f_0(\theta) \equiv c_0(\theta-\theta_1)(\theta-\theta_2)(\theta-\theta_3).$$

The equations of c may accordingly be written

$$x_0:x_1:x_2:x_3 = c_0(\theta-\theta_0)^{-1}:c_1(\theta-\theta_1)^{-1}:c_2(\theta-\theta_2)^{-1}:c_3(\theta-\theta_3)^{-1}.$$

Projection of the twisted cubic on to a plane

Suppose a twisted cubic c is projected from a vertex V on to a plane π. It must project into an algebraic plane curve c', and in order to find out the nature of this curve we need to determine its order. There are two cases to be distinguished, according to whether V does or does not lie on c.

Case 1: when V does not lie on c.

The order of c', the number of points in which it is met by a general line l in its plane, is equal to the number of points in which c is met by the plane Vl, namely 3. Thus c' is a plane cubic curve. Furthermore, there is a unique chord $A_1 A_2$ of c which passes through V, and this gives rise to a double point A' of c'. If A_1 and A_2 are distinct A' is a node of c', and the nodal tangents are the projections of the tangents at A_1 and A_2 to c. If A_1 and A_2 coincide, A' is a cusp of c', and the cuspidal tangent is the line in which π is met by the osculating plane to c at A_1. Finally, the class of c' is equal to the number of tangents of c which meet the line joining V to a general point of π; it is therefore equal to the rank of c, i.e. 4. In the special case when A_1 and A_2 coincide, VA_1 is a tangent which meets every line through V, and in this case the class of c' is reduced to 3.

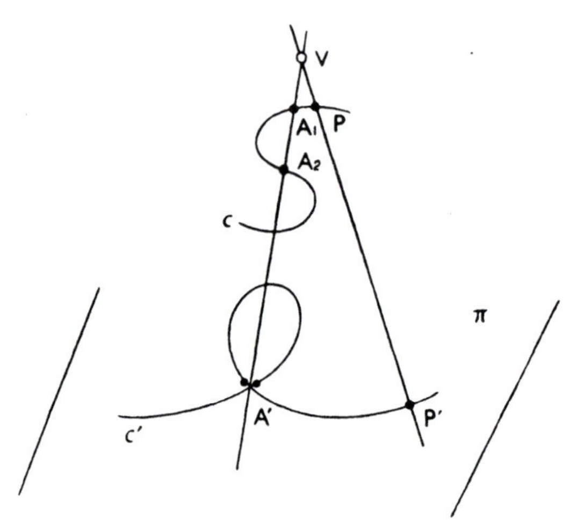

Case 2: when V lies on c.

In this case the plane Vl meets c in V itself and two free points. The curve c' is therefore an algebraic curve of order 2, i.e. a conic. We may compare with this projection the projection of a conic (in S_2) from a point of itself on to a line.

When V lies on c, the lines which project the points of c from V all lie on a quadric cone. This cone is a degenerate quadric of the net $\lambda_1 Q_1+\lambda_2 Q_2+\lambda_3 Q_3 = 0$, and there exists such a cone with any arbitrarily chosen point of c as its vertex.

Fundamental projective properties of the twisted cubic

We come now to the fundamental theorems which provide a basis for the whole projective geometry of the twisted cubic, and these are closely analogous to the corresponding theorems for the conic.

THEOREM 3. *If a variable point of a twisted cubic c is joined to two fixed chords of c, the pencils of planes so defined are homographically related.*

Proof. Let l be the fixed chord (θ_1, θ_2) and m the fixed chord (ϕ_1, ϕ_2), and let P be the variable point (θ). Then the equations of the planes lP and mP are

$$x_0-(\theta_1+\theta_2)x_1+\theta_1\theta_2 x_2+\theta\{x_1-(\theta_1+\theta_2)x_2+\theta_1\theta_2 x_3\} = 0$$

and

$$x_0-(\phi_1+\phi_2)x_1+\phi_1\phi_2x_2+\theta\{x_1-(\phi_1+\phi_2)x_2+\phi_1\phi_2x_3\} = 0,$$

and therefore $$l(P) \barwedge m(P).$$

COROLLARY. *If P_1, P_2, P_3, P_4 are four fixed points of c and l is a chord of c, the value of the cross ratio $l\{P_1, P_2; P_3, P_4\}$ is independent of the choice of l.*

Remark. The cross ratio $l\{P_1, P_2; P_3, P_4\}$ may be taken as the cross ratio $\{P_1, P_2; P_3, P_4\}$ *on the cubic*.

EXERCISE. Deduce Theorem 3 from Theorem 1 of Chapter VI. [If l, m have a common end-point, project c from this point on to a plane; if not, join an extremity of l to an extremity of m by a third chord n.]

THEOREM 4. *The projective geometry of S_3 induces a subordinate one-dimensional projective geometry on any twisted cubic c, and the canonical parameter θ is an allowable parameter in this geometry.*

This is a theorem of a type that is by now very familiar, and the details of the proof will be left to the reader. The importance of the theorem resides, of course, in the fact that it makes possible the consideration of homographic and other algebraic correspondences on the twisted cubic; and since some of these correspondences are associated with simple geometrical constructions, we can use them in investigating the projective properties of the curve. The theorems which now follow will establish the main connexions between algebraic correspondences and geometrical constructions.

Correspondences on the twisted cubic

THEOREM 5. *If τ is an involution on a twisted cubic c, the chords which join corresponding pairs (P, P') of τ form one system of generators of a quadric Q_τ through c. Every generator of the opposite system is a unisecant of c and is met by all the joins PP'. Any such generator is sufficient to determine the involution τ (of which it is said to be a directrix). The involution has a directrix through each point of c.*

Proof. Let PP' be a chord of c, with end-points (θ) and (θ'). Then the coordinates of every point of PP' satisfy the equations

$$x_0-(\theta+\theta')x_1+\theta\theta' x_2 = 0,$$

and $$x_1-(\theta+\theta')x_2+\theta\theta' x_3 = 0,$$

and these equations give

$$\frac{1}{x_1x_3-x_2^2} = \frac{\theta+\theta'}{x_0x_3-x_1x_2} = \frac{\theta\theta'}{x_0x_2-x_1^2},$$

i.e. $$\theta\theta' : \theta+\theta' : 1 = Q_1 : Q_2 : Q_3.$$

The point (x_i) therefore lies on a line joining a common pair of the involution given by

$$a\theta\theta'+b(\theta+\theta')+d = 0$$

if and only if $$Q_\tau \equiv aQ_1+bQ_2+dQ_3 = 0,$$

and there is thus a one–one correspondence between involutions τ on c and quadrics Q_τ through c.

The joins PP' all belong to the same system of generators of Q_τ, since if two of them belonged to opposite systems they would be coplanar, and their plane would cut c in four points.

If g' is a generator of Q_τ of the opposite system, it meets every generator of the first system, i.e. every join PP'. If g is such a join, g and g' are coplanar, and their plane meets c in three points, which lie on Q_τ. Two of these points are P, P', lying on g, and the third point must therefore lie on g'. Thus g' is a unisecant of c. The unisecant directrix g' determines τ uniquely, because there is a unique chord of c which passes through any general point of it.

Finally, one generator of Q_τ of the second system passes through every point of c, and τ therefore has a unisecant directrix through every point of the twisted cubic.

COROLLARY. *Two quadrics which pass through a twisted cubic intersect residually in a chord of the twisted cubic.*

For the involutions which they determine have a unique common pair.

Remarks

(i) The united points of the involution τ arise from those generators of Q_τ of the first system which touch c.

(ii) If the united points of τ are taken as X_0 and X_3, the equation of the involution is $\theta' = -\theta$, and then $Q_\tau \equiv Q_2$.

(iii) Among the quadrics which pass through c there are, as we have already seen, ∞^1 cones, each of which projects c from a point of itself. It may be verified that these correspond to degenerate (i.e. singular) involutions on c.

Non-symmetric homographic correspondences on the twisted cubic are not of much interest, and we shall prove only one theorem concerning them.

THEOREM 6. *Every homography ϖ_0 on c is subordinate to a unique self-collineation of S_3 which leaves c invariant.*

Proof. Suppose the frame of reference is chosen in such a way that the equation of ϖ_0 is in canonical form $\theta' = k\theta$, and let $x_i' = \sum_{k=0}^{3} a_{ik} x_k$ $(i = 0, 1, 2, 3)$ be a collineation ϖ which transforms c into itself and induces the homography ϖ_0 on it. Then

$$\rho k^{3-i}\theta^{3-i} \equiv \sum_{j=0}^{3} a_{ij}\,\theta^{3-j} \quad (i = 0, 1, 2, 3)$$

i.e.
$$\rho \sum_j \delta_{ij}\, k^{3-i}\theta^{3-j} \equiv \sum_j a_{ij}\,\theta^{3-j};$$

and hence
$$a_{ij} = \rho\delta_{ij}\, k^{3-i} \quad (i,j = 0, 1, 2, 3).$$

Thus the ratios of the a_{ij} are uniquely determined, and ϖ is given by

$$\frac{x_0'}{k^3 x_0} = \frac{x_1'}{k^2 x_1} = \frac{x_2'}{k x_2} = \frac{x_3'}{x_3}.$$

In general, this collineation has the four united points X_0, X_1, X_2, X_3. In the particular case in which ϖ_0 is an involution, however, $k = -1$; and ϖ is then a harmonic biaxial collineation† with $X_0 X_2$ and $X_1 X_3$ as axes.

EXERCISE. Examine the case in which the united points of ϖ_0 coincide at X_0, and show that ϖ then has only one united point, namely X_0.

Theorem 5 gives a geometrical procedure for generating the general involution or symmetrical (1, 1) algebraic correspondence on the twisted cubic, and the next four theorems will be concerned with procedures for generating symmetrical (2, 2) and (3, 3) correspondences.

THEOREM 7. *The symmetrical* (2, 2) *algebraic correspondences on a given twisted cubic c may be paired off against the linear complexes‡ in S_3 in such a way that the corresponding pairs in any given correspondence are cut by the chords of c which belong to the associated linear complex.*

† Chapter XIV, p. 351. ‡ Chapter XV, p. 372.

Proof. Let P, P' be two points of c, with parameters θ, θ'. The Grassmann coordinates of the line PP' are then given by

$$\frac{p_{01}}{\theta^2\theta'^2} = \frac{p_{02}}{\theta\theta'(\theta+\theta')} = \frac{p_{03}}{\theta^2+\theta\theta'+\theta'^2} = \frac{p_{23}}{1} = \frac{p_{31}}{-(\theta+\theta')} = \frac{p_{12}}{\theta\theta'},$$

and it follows that the most general linear relation between the p_{ij} is equivalent to the most general symmetric relation between θ and θ' that is quadratic in each parameter.

THEOREM 8. *If a symmetrical* (2, 2) *correspondence on* c *has one cyclic triad* (P_1, P_2, P_3), *such that each of the points gives rise in the correspondence to the other two, then every point of* c *belongs to a unique cyclic triad.*

Proof. The associated linear complex contains the three lines P_2P_3, P_3P_1, P_1P_2, which intersect in pairs but are not concurrent, and it is therefore special. The lines PP' joining pairs of the correspondence are then the chords of c which meet a fixed line l of space. If P is any point of c, the plane Pl meets c in two further points P', P'', and P belongs to the cyclic triad (P, P', P'').

A cyclic correspondence of the type just mentioned arranges the points of c into ∞^1 mutually exclusive triads, cut on c by the planes of a pencil. If a general plane of this pencil is $\mathbf{u}^T\mathbf{x}+\lambda\mathbf{v}^T\mathbf{x} = 0$, the parameters of the points of the corresponding triad are the roots of the cubic equation

$$(u_0\theta^3+u_1\theta^2+u_2\theta+u_3)+\lambda(v_0\theta^3+v_1\theta^2+v_2\theta+v_3) = 0,$$

i.e.

$$f(\theta)+\lambda g(\theta) = 0,$$

where $f(\theta)$ and $g(\theta)$ are two cubic polynomials. (Compare Theorem 15 of Chapter III.) When the points of a rational curve are grouped in triads in this manner we say that we have a *ternary involution* on the curve.

EXERCISES

(i) By projecting c from a point of itself into a conic k, deduce Theorem 21 of Chapter VI from Theorem 8.

(ii) Prove that the chords of c which belong to a given linear complex (possibly special) generate a quartic surface which passes doubly through c.

THEOREM 9. *The symmetrical* (3, 3) *algebraic correspondences* ω *on a given twisted cubic* c *may be paired off against the quadrics* ψ_ω *in* S_3 *in such a way that the three points of* c *which correspond in* ω *to any given point* P *of* c *are the points in which* c *is met by the polar plane of* P *with respect to* ψ_ω.

Proof. Take a quadric $\sum_i \sum_k a_{ik} x_i x_k = 0$ $(a_{ki} = a_{ik})$ and let the points (θ) and (θ') of c be conjugate with respect to this quadric. Then

$$\sum_i \sum_k a_{ik}(\theta^{3-i}\theta'^{3-k}+\theta^{3-k}\theta'^{3-i}) = 0,$$

i.e.

$$a_{00}\,\theta^3\theta'^3+a_{11}\,\theta^2\theta'^2+a_{22}\,\theta\theta'+a_{33}+a_{12}\,\theta\theta'(\theta+\theta')+a_{20}\,\theta\theta'(\theta^2+\theta'^2)+$$
$$+a_{01}\,\theta^2\theta'^2(\theta+\theta')+a_{03}(\theta^3+\theta'^3)+a_{13}(\theta^2+\theta'^2)+a_{23}(\theta+\theta') = 0,$$

and this is the most general symmetric equation that is cubic in θ and also in θ'. The united points of ω are, of course, the six points in which c is met by ψ_ω.

THEOREM 10. *If the symmetrical* (3, 3) *correspondence* ω *on* c *has one proper cyclic tetrad* (P_1, P_2, P_3, P_4), *then* ω *is cyclic, and every point of* c *belongs to a unique cyclic tetrad. The tetrads so determined form a quaternary involution on* c, *given, for varying* λ, *by an equation of the form*

$$f(\theta)+\lambda g(\theta) = 0,$$

where $f(\theta)$ *and* $g(\theta)$ *are quartic polynomials.*

Proof. Since the four points P_i all lie on c, they are not all in one plane, and we may therefore take them as vertices of the tetrahedron of reference. The equation of ψ_ω may then be taken to be

$$x_0^2+x_1^2+x_2^2+x_3^2 = 0.$$

If, further, the parameter of P_i on c is θ_i, the equations of c are of the form

$$x_i = c_i(\theta-\theta_i)^{-1} \quad (i = 0, 1, 2, 3).$$

Two points P, P', with parameters θ, θ', correspond in ω if and only if they are conjugate for ψ_ω, i.e. if

$$\sum_{i=0}^{3} \frac{c_i}{\theta-\theta_i}\,\frac{c_i}{\theta'-\theta_i} = 0.$$

Disregarding the trivial factor $\theta-\theta'$, we may write this relation as

$$\sum_{i=0}^{3} \frac{c_i^2}{\theta-\theta_i} - \sum_{i=0}^{3} \frac{c_i^2}{\theta'-\theta_i} = 0,$$

or

$$F(\theta)-F(\theta') = 0, \text{ say.}$$

If (θ_*, θ'_*) and (θ_*, θ''_*) are two solutions of this equation, then (θ'_*, θ''_*) is also a solution; and it follows that the (3, 3) correspondence defined by $F(\theta)-F(\theta') = 0$ is cyclic, in the sense that if P gives rise to P', P'', P''', each of the four points P, P', P'', P''' gives rise to the other three.

If the parameters of such a tetrad of points are θ, θ', θ'', θ''', then $F(\theta) = F(\theta') = F(\theta'') = F(\theta''')$; and the cyclic tetrads are given, for varying λ, by the quartic equation $F(\theta) = \lambda$. If this equation is cleared of fractions, it assumes the form $f(\theta)+\lambda g(\theta) = 0$, where $f(\theta)$ and $g(\theta)$ are polynomials in θ, $f(\theta)$ being cubic and $g(\theta)$ properly quartic.

THEOREM 11. *If a twisted cubic c has one inscribed tetrahedron that is self-polar for a quadric* ψ, *then it has an infinity of inscribed tetrahedra that are self-polar for* ψ.

This theorem follows immediately from Theorem 10, and it gives us a space analogue of the relation of apolarity between conics. When a twisted cubic c and a quadric ψ are related as in Theorem 11, we say that c is *outpolar* to ψ. When this is so, the faces of all the inscribed tetrahedra which are self-polar for ψ form a cubic envelope (the reciprocal of c with respect to ψ) and this envelope is said to be *inpolar* to ψ.

Alternative definitions of the twisted cubic

Just as the conic may be defined in several different ways (for instance, as an algebraic curve of the second order, or as the locus of the point of intersection of two homographically related pencils of lines) so there are various properties of the twisted cubic which may be taken as defining properties. For some purposes it is convenient to use one of these alternative definitions in place of the algebraic definition with which we began, and we shall now discuss the more important ones, establishing their equivalence with our original definition. We are led in this way to generalize the definition of a twisted cubic slightly. The curve, as we have defined it, is analogous to the plane curve $(\theta^2, \theta, 1)$, i.e. the proper conic, and we shall now refer to it as the *proper* twisted cubic, to distinguish it from certain composite space curves which are also of order 3. Among the possible kinds of degenerate twisted cubic are (i) a conic together with a unisecant line, and (ii) two skew lines and a transversal line which meets them both.

THEOREM 12. *Every proper twisted cubic may be generated as the locus of the point of intersection of corresponding planes of three homographically related pencils; and conversely, the locus of the point of intersection of corresponding planes of three related pencils is, in general, a proper twisted cubic.*

Proof. (i) If c is a proper twisted cubic, l_1, l_2, l_3 are any three fixed chords, and P is a variable point of the curve, then, by Theorem 3, $l_1(P) \barwedge l_2(P) \barwedge l_3(P)$. The twisted cubic is thus generated by three homographic pencils as required.

(ii) Conversely, let three homographic pencils of planes be given. By suitable choice of the base planes, the equations of the planes of a general mutually corresponding triad may be written in the form $\pi_i + \lambda\pi'_i = 0$ $(i = 1, 2, 3)$; and if we solve these three equations for the ratios $x_0 : x_1 : x_2 : x_3$ we obtain a solution of the form

$$x_0 : x_1 : x_2 : x_3 = f_0(\lambda) : f_1(\lambda) : f_2(\lambda) : f_3(\lambda),$$

where the $f_i(\lambda)$ are cubic polynomials in λ.

Except in the special case in which these polynomials are linearly dependent, the locus of the point of intersection of a general corresponding triad is therefore a proper twisted cubic.

Remarks

(i) The axes of the generating pencils are chords of the twisted cubic, for the homography cut on l_1, say, by corresponding pairs of planes belonging to the pencils with axes l_2 and l_3 has two united points.

(ii) The equations which define the cubic may be written in the form

$$\frac{\pi_1}{\pi'_1} = \frac{\pi_2}{\pi'_2} = \frac{\pi_3}{\pi'_3};$$

and since each of these ratios is equal to

$$\frac{\alpha_1\pi_1 + \alpha_2\pi_2 + \alpha_3\pi_3}{\alpha_1\pi'_1 + \alpha_2\pi'_2 + \alpha_3\pi'_3}$$

for every choice of α_1, α_2, α_3, we are able at once to write down equations for the full system of ∞^2 chords of the curve.

THEOREM 13. *Every proper twisted cubic may be obtained as the residual intersection of two proper quadrics which have a generator in common; and conversely, every such residual intersection is a twisted cubic (possibly composite).*

Proof. (i) The first half has already been proved, for it was shown above that the proper twisted cubic c is the residual intersection of the quadrics Q_1 and Q_2, which have the common generator

$$x_0 = 0 = x_1.$$

(ii) Let ψ_1 and ψ_2 be two proper quadrics with a common generator g. Take a fixed generator g_1 of ψ_1 and a fixed generator g_2 of ψ_2, both skew to g. Then any plane π_1 through g_1 meets ψ_1 in a second generator g'_1, and g'_1 lies in a uniquely defined plane π through g. This plane meets ψ_2 in g and a second generator g'_2, and g'_2 lies in a uniquely defined plane π_2 through g_2. By Theorem 10 of Chapter XI there is a homographic relation $(\pi_1) \barwedge (\pi) \barwedge (\pi_2)$ between the three pencils of planes whose axes are g_1, g, g_2, and the common point of any corresponding triad of planes is clearly a point of both quadrics. If, conversely, P is any point which lies on each of the quadrics, but not on g, there is a generator g'_1 of ψ_1 and also a generator g'_2 of ψ_2, both of which pass through P, and these generators determine a triad of corresponding planes π_1, π, π_2 which meet in P.

Remarks

(i) Two quadrics intersect in a quartic curve, but even when they are both proper this curve can break up in various ways (see Chapter XIII). When the quadrics have a common generator the quartic curve breaks up into this line and a residual curve of order 3, and we find it convenient to regard every such residual intersection of two proper quadrics as a twisted cubic. If the cubic is proper, then, by what has already been proved above, the common generator is one of its chords. Various types of degenerate twisted cubic are possible, the most important being those referred to on p. 306. The reader should look carefully into the proof of Theorem 13 in order to see how these special cases arise.

(ii) If g is a generator of a proper quadric ψ, every quadric through g cuts ψ residually in a twisted cubic; and in this way we obtain the two families of twisted cubics on ψ, already referred to on p. 281. The twisted cubics of the one family are $(1, 2)$ curves, having every u-generator as a unisecant and every v-generator as a chord, while the twisted cubics of the other family are $(2, 1)$ curves. This result is connected in an obvious way with Theorem 5, on the generation of involutions on a proper twisted cubic.

Theorem 12 gives a generation of the twisted cubic by means of three homographic pencils of planes. A second projective generation of the curve of a somewhat different kind is also possible, namely the generation by collinear stars. Two collinear stars (cf. p. 254) have ∞^2 pairs of corresponding rays, and although in general two corresponding rays are skew, there are ∞^1 special pairs which intersect. The ∞^1 points of intersection obtained in this way are the points of a twisted cubic.

THEOREM 14 (*The star generation*). *Every proper twisted cubic may be obtained as the incidence curve of intersecting pairs of corresponding rays of two collinear stars; and conversely, the incidence curve of intersecting pairs of corresponding rays of two collinear stars is, in general, a twisted cubic.*

Proof. Let c be a twisted cubic, and A, A' any two points on it; and consider the correspondence between the planes π, π' which join A and A' to a variable chord P_1P_2 of c. Let A, A' be taken as the reference points X_0, X_3 in a canonical parametric representation $(\theta^3, \theta^2, \theta, 1)$ of c. If θ_1, θ_2 are the parameters of P_1, P_2, the equations of π, π' are

$$x_1 - px_2 + qx_3 = 0 \quad \text{and} \quad x_0 - px_1 + qx_2 = 0,$$

where $p = \theta_1 + \theta_2$ and $q = \theta_1\theta_2$; and clearly the relation between π and π' is a collineation between the stars (A) and (A'). In this collineation, corresponding rays are given by

$$\frac{x_1}{\lambda} = \frac{x_2}{\mu} = \frac{x_3}{1} \quad \text{and} \quad \frac{x_0}{\lambda} = \frac{x_1}{\mu} = \frac{x_2}{1},$$

since π passes through the first of these rays if and only if π' passes through the second. A necessary and sufficient condition for two such corresponding rays to meet is $\mu^2 = \lambda$, and the rays then join A and A' to the point of c whose parameter is μ. Thus c is the incidence locus of intersecting pairs of corresponding rays of the collinear stars (A) and (A').

Suppose, conversely, that we are given two collinear stars (A) and (A'). Let $\pi_i = 0$ $(i = 1, 2, 3)$ be the equations of three base planes of the first star, and let $\pi'_i = 0$ $(i = 1, 2, 3)$ be the equations of the corresponding planes of the second star. By replacing π'_1, π'_2, π'_3 by suitable fixed multiples of themselves, we can arrange (cf. p. 317) that a general pair of corresponding rays is given by

$$\frac{\pi_1}{l} = \frac{\pi_2}{m} = \frac{\pi_3}{n} \quad \text{and} \quad \frac{\pi'_1}{l} = \frac{\pi'_2}{m} = \frac{\pi'_3}{n}.$$

If these rays meet, their point of intersection satisfies the equations

$$\frac{\pi_1}{\pi'_1} = \frac{\pi_2}{\pi'_2} = \frac{\pi_3}{\pi'_3},$$

and these are, in general, the equations of a twisted cubic c through A and A'. Since any point of c, other than A or A', defines a set of ratios $l:m:n$, it follows that c is the incidence locus of intersecting pairs of corresponding rays of the collinear stars.

Remark. It has appeared incidentally in the above proof that when a twisted cubic c is generated by collinear stars its full system of ∞^2 chords is generated simultaneously as the system of lines of intersection of corresponding planes of the stars. This result will prove useful later (p. 316).

The dual of the twisted cubic

As we have already mentioned on p. 268, the space-dual of a curve is a developable, or envelope of ∞^1 planes. If the curve is a plane curve, the dual developable is correspondingly special, being in fact a cone, and this was the case that interested us in Chapter XI. Now that we are concerned with the dual of the twisted cubic, it is convenient to describe briefly the nature of developables in general.

A developable is said to be of *class m* if m of its planes pass through a general point of space. Any two planes of the developable meet in a line, which is called an *axis* of the developable, and in the limiting case in which the two planes coincide, the line is called a *generating line* or *focal line* of the developable. Three planes of the developable meet in a point, and when they come to coincide, the point is called a *focal point*. Thus the axes, focal lines, and focal points of a developable are dual to the chords, tangent lines, and osculating planes of a twisted curve.

The osculating planes of a twisted curve constitute a developable; and dually, the focal points of a developable constitute a curve, the *cuspidal edge* of the developable. It may be proved that every twisted curve is the cuspidal edge of the developable formed by its osculating planes.

The developable that is dual to the twisted cubic c is the cubic developable δ. It may be represented parametrically by cubic polynomials; and by choosing the tetrahedron of reference suitably we can reduce the representation to the canonical form

$$u_0:u_1:u_2:u_3 = \theta^3:\theta^2:\theta:1.$$

The developable δ may be generated projectively as the envelope of a plane which joins corresponding points of three homographically related ranges.

The important feature of the (proper) twisted cubic is, however, not that it has a dual but that, like the proper conic in the plane, it is self-dual; *the osculating planes of a proper twisted cubic form a cubic developable.* This is an immediate consequence of the fact that the osculating plane of c at the point $(\theta^3, \theta^2, \theta, 1)$ is the plane $(1, -3\theta, 3\theta^2, -\theta^3)$.

Polarity with respect to a twisted cubic

It was shown in Chapter V that every proper conic s defines a polarity in the plane, i.e. a non-singular linear point–line transformation

$$\mathbf{u} = \mathbf{A}\mathbf{x} \quad (|\mathbf{A}| \neq 0),$$

with a symmetric matrix. The line corresponding in this polarity to any point P is the line which joins the points of contact of the two tangents of s which pass through P. We have already seen that these results may be extended in a natural way from the plane to three-dimensional space by replacing the conic s by a proper quadric ψ, and we may wonder whether any interesting results are obtainable by using the other space-analogue of the conic, that is to say the twisted cubic. This curve does in fact define a point–plane transformation, but instead of being an ordinary polarity the transformation is of the special kind known as a null polarity. A null polarity (see p. 361) is a point–plane transformation given by an equation

$$\mathbf{x} \rightarrow \mathbf{u} = \mathbf{A}\mathbf{x} \quad (|\mathbf{A}| \neq 0),$$

where the matrix $\mathbf{A}$ is skew-symmetric.

From this equation, we have

$$\begin{aligned} \mathbf{u}^T\mathbf{x} &\equiv \mathbf{x}^T\mathbf{A}^T\mathbf{x} \\ &\equiv a_{00}x_0^2+\ldots+(a_{01}+a_{10})x_0x_1+\ldots \\ &\equiv 0 \end{aligned}$$

since $$a_{ki} = -a_{ik} \quad (i,k = 0,1,2,3).$$

Thus the null polarity has the property that the polar plane of a point always passes through the point.

Let us now see how every (proper) twisted cubic defines a null polarity.

THEOREM 15. *If c is a given twisted cubic, the correspondence between a general point P of S_3 and the plane π which joins the points of contact of the three osculating planes of c which pass through P is a null polarity ν. The polar plane of every point of c in ν is the osculating plane at the point; and all the tangent lines of c are self-polar with respect to ν.*

Proof. Let P be the point $\mathbf{y}$. Then, if the osculating plane at (θ) passes through P,

$$y_0-3\theta y_1+3\theta^2 y_2-\theta^3 y_3 = 0.$$

The parameters θ_1, θ_2, θ_3 of the points of contact of the three osculating planes of c that pass through P are the roots of this cubic equation in θ, and hence

$$\theta_1+\theta_2+\theta_3 = \frac{3y_2}{y_3},$$

$$\theta_2\theta_3+\theta_3\theta_1+\theta_1\theta_2 = \frac{3y_1}{y_3},$$

$$\theta_1\theta_2\theta_3 = \frac{y_0}{y_3}.$$

The equation of the plane π is then

$$x_0-(\theta_1+\theta_2+\theta_3)x_1+(\theta_2\theta_3+\theta_3\theta_1+\theta_1\theta_2)x_2-\theta_1\theta_2\theta_3 x_3 = 0,$$

i.e.
$$y_3 x_0-3y_2 x_1+3y_1 x_2-y_0 x_3 = 0.$$

Thus if P is the point (y_0, y_1, y_2, y_3), π is the plane

$$(y_3, -3y_2, 3y_1, -y_0);$$

and P and π therefore correspond in the null polarity ν whose matrix is

$$\begin{pmatrix} 0 & 0 & 0 & 1 \\ 0 & 0 & -3 & 0 \\ 0 & 3 & 0 & 0 \\ -1 & 0 & 0 & 0 \end{pmatrix}$$

The polar plane of $(\theta^3, \theta^2, \theta, 1)$ is $(1, -3\theta, 3\theta^2, -\theta^3)$, i.e. the osculating plane at (θ).

The lines of S_3 which are self-polar with respect to ν form a linear complex $\mathscr{L}$ (cf. p. 372), the set of all lines l such that l lies in the polar plane of any one of its points. Since every tangent line of c lies in the osculating plane at its point of contact, it belongs to the complex $\mathscr{L}$ of self-polar lines.

In addition to setting up a null polarity ν, which is a linear $(1, 1)$ correspondence between the points and planes of S_3, the twisted cubic c also sets up a correspondence between points and points of S_3. This correspondence is $(1, 1)$ but not linear. Let P be a general point of space, not lying on c. Then there is a unique chord

AB of c which passes through P, and a unique point P' of this chord which is harmonically conjugate to P with respect to A and B. We call two points P, P', related in this manner, a pair of *harmonic points* with respect to c.

P' is, of course, the point which is conjugate to P with respect to all the quadrics of the net defined by c, and it may be constructed as the common point of the polar planes of P with respect to three of these quadrics, say Q_1, Q_2, Q_3.

THEOREM 16. *If P describes a line l, its harmonic point P' with respect to c describes a twisted cubic c'.*

Proof. As P describes l its polar planes with respect to Q_1, Q_2, Q_3 generate three homographically related pencils, and their point of intersection P' therefore describes a twisted cubic c'.

Remarks

(i) The twisted cubic c' meets the original twisted cubic c in four points, the points of contact of the four tangents of c which meet l.

(ii) Theorem 16 reflects the non-linearity of the relation between points which are harmonic with respect to c.

The twisted cubic in affine space

Although all proper twisted cubics are *projectively* equivalent (see p. 296), it is not possible to transform every twisted cubic into every other by an *affine* transformation. Just as, in the real affine plane, we are able to classify conics as hyperbolas, parabolas, and ellipses, so in much the same way we can devise an affine classification of real twisted cubics. The distinctions which we make are valid *a fortiori* in euclidean space.

A twisted cubic c meets the plane at infinity ι in three points H, K, L, and either all three are real or one is real and two are conjugate complex. It is also possible for two or all three of the points to coincide—when c touches or osculates the plane at infinity.

Consider the case in which L is real while H and K are complex. Then c has one real asymptote l and one real asymptotic plane λ, namely the tangent and osculating plane at L, and l lies in λ. Since c already meets λ three times at L, it cannot meet it in any finite point, and this means that it lies wholly on one side of its asymptotic plane. It consists of a single branch, which approaches the asymptote l at each end. Any general plane through l meets the curve in a unique finite point; and, if the plane of the paper is taken as λ, the appearance of the curve is as follows.

If H, K, L are all real, the twisted cubic has three asymptotes, each of which lies in an asymptotic plane. In this case the curve consists of three branches which link the asymptotes together (cf. the two branches of a hyperbola).

EXERCISE. Give a more detailed discussion of the above types of twisted cubic, showing by means of sketches how the curve projects from any point of itself into a conic. Discuss also the types of twisted cubic for which H, K, L are not all distinct.

A very special kind of twisted cubic that is met with in euclidean space is the *rectangular twisted cubic*, whose asymptotes are all real and mutually orthogonal. This curve is analogous to the rectangular hyperbola in the plane.

Suppose a twisted cubic c in affine space is met by a system of parallel planes. Each of the planes cuts the curve in three points, and the triangles so formed have many remarkable properties. The planes form a pencil whose axis is a line in the plane at infinity, and by applying Theorem 16 to this line we see at once that *the mid-points of the sides of the triangles all lie on a second twisted cubic c', and c' meets c at the points of contact of the four tangents to c which are parallel to the planes.* The parallel planes also cut the new twisted cubic c' in triangles, and the mid-points of the sides of these triangles lie on a third twisted cubic c''. Proceeding in this way, we define a sequence of twisted cubics $c, c', c'',...$; and since the triangles in each plane form a nest with a common centroid, the cubics have as their limit a triple line which passes through all the centroids. Thus *the centroids of the triangles cut on c by the system of parallel planes all lie on a line.*†

Some enumerative problems

THEOREM 17. *There is a unique twisted cubic that passes through six general points of space.*

Proof. Let four of the points be taken as vertices of the tetrahedron of reference. Then the equations of a twisted cubic c

† For an algebraic proof of this result see Ex. 14 on p. 325; and see also Ex. 15 for other properties of the triangles.

through them assume the form

$$x_i = c_i(\theta-\theta_i)^{-1} \quad (i = 0, 1, 2, 3).$$

Now suppose we apply the reciprocal transformation of space into itself, i.e. the (1, 1) transformation given by

$$x_i = \frac{1}{x'_i} \quad (i = 0, 1, 2, 3).$$

Then the twisted cubic c is transformed into the line

$$x'_i = \frac{\theta-\theta_i}{c_i} \quad (i = 0, 1, 2, 3).$$

In this way we obtain a (1, 1) correspondence between the twisted cubics through X_0, X_1, X_2, X_3 and the lines of space. But there is a unique line through two points, and therefore a unique twisted cubic passes through X_0, X_1, X_2, X_3 and two general points.

Theorem 17 may also be proved quite easily from the other definitions of the twisted cubic, for example from the definition by three homographic pencils of planes or from the star generation. Indeed we have a whole series of enumerative problems which can be solved by using the star generation of the twisted cubic.

The twisted cubic is a space curve with twelve degrees of freedom. The general twisted cubic may be represented, in terms of an arbitrary coordinate system $\mathscr{R}$, by equations of the form

$$\rho x_i = \sum_{k=0}^{3} a_{ik}\,\theta^{3-k} \quad (i = 0, 1, 2, 3).$$

There are sixteen coefficients a_{ik}, but since ρ is arbitrary, and θ admits of ∞^3 transformations

$$\theta' = \frac{\alpha\theta+\beta}{\gamma\theta+\delta} \quad (\alpha\delta-\beta\gamma \neq 0),$$

which leave the curve unaltered, only twelve of the coefficients contribute to the freedom of the curve.

If the curve is required to pass through a fixed point of space, the a_{ik} have to satisfy three independent linear conditions. These conditions, however, involve θ; and when θ is eliminated two conditions on the a_{ik} remain. Thus the condition of passing through a fixed point is a double (non-linear) condition on the twisted cubic.

Having a fixed line as unisecant is a simple condition, since the twisted cubic has merely to contain some one of a set of ∞^1 points; but having a fixed line as chord is again a double condition. Suppose we represent the composite condition of passing through α fixed general points and having β fixed general lines as chords, where $\alpha+\beta = 6$, by the symbol $P^\alpha C^\beta$. Then, corresponding to each possible choice of α and β, we have an enumerative problem—to find the number $[P^\alpha C^\beta]$ of twisted cubics which satisfy the condition $P^\alpha C^\beta$. Theorem 17 has already given the result $[P^6] = 1$.

If the value of α is not less than two, we can take two of the given points as vertices of generating stars and then make use of the star generation of the twisted cubic. Instead of working directly with the stars it is perhaps easier to define collineations between them by the plane collineations ϖ which they cut on a fixed plane π. Since an assigned point P gives a pair of corresponding rays of the two stars, and an assigned chord C gives a pair of corresponding planes (p. 310), the two conditions yield respectively a pair of corresponding points and a pair of corresponding lines in π. Let us now consider the separate cases which can arise.

Case 1: P^6

The required collineation ϖ in π has four assigned pairs of corresponding points, and is uniquely determined. Thus $[P^6] = 1$, as we have already seen.

Case 2: P^5C

In this case, ϖ has three assigned pairs of corresponding points, which determine three pairs of corresponding lines, and one assigned pair of corresponding lines. Once again, therefore, ϖ is uniquely determined; and $[P^5C] = 1$.

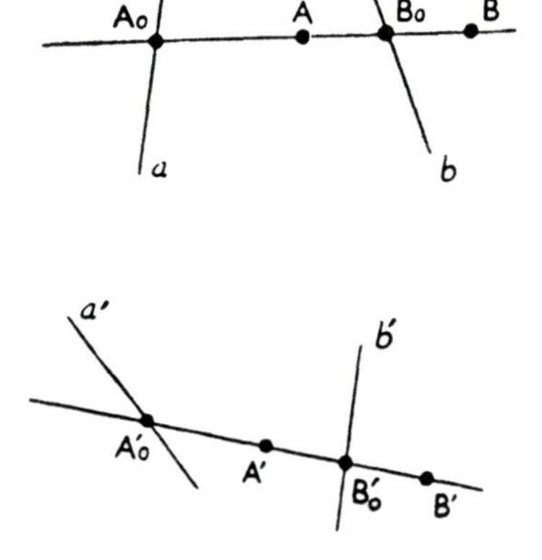

Case 3: P^4C^2

Here ϖ has two assigned pairs of corresponding points, say (A, A'), (B, B'), and two assigned pairs of corresponding lines, say (a, a'), (b, b'). The problem is poristic. For let AB meet a, b, in A_0, B_0, and let $A'B'$ meet a', b' in A_0', B_0'. Then, clearly, no solution is possible unless

$$\{A_0, B_0; A, B\} = \{A_0', B_0'; A', B'\}.$$

If this condition is satisfied, the four conditions are no longer independent, and there exists an infinity of solutions. Thus $[P^4C^2] = 0$ or ∞.

Case 4: P^3C^3 and *Case* 5: P^2C^4

The conditions to be satisfied by ϖ are dual to those in cases 2 and 1 respectively, and therefore $[P^3C^3] = [P^2C^4] = 1$.

To sum up: the problems P^6, P^5C, P^3C^3, P^2C^4 each have a unique solution, while P^4C^2 is poristic.

§ 2. Cubic Surfaces

In addition to the quadric surface and the twisted cubic curve we have now to introduce a third locus, the *projectively generated cubic surface*, which must be accepted as yet another analogue for space of the projectively generated conic. This surface, which we shall denote for brevity by F, is defined as follows:

DEFINITION: F is the surface generated by the point of intersection of corresponding planes of three collinearly related stars.

Consider first of all a pair of stars with vertices A and A'. If π_1, π_2, π_3 are three linearly independent planes through A and π_1', π_2', π_3' are three independent planes through A', a general pair of planes of the two stars is represented by the pair of equations $\lambda\pi_1+\mu\pi_2+\nu\pi_3 = 0$ and $\lambda'\pi_1'+\mu'\pi_2'+\nu'\pi_3' = 0$; and for these two planes to correspond in a collineation, λ', μ', ν' must be connected with λ, μ, ν by a fixed non-singular linear transformation. If we now replace λ', μ', ν' in the equation $\lambda'\pi_1'+\mu'\pi_2'+\nu'\pi_3' = 0$ by their expressions in terms of λ, μ, ν we obtain an equation of the form $\lambda\bar{\pi}_1'+\mu\bar{\pi}_2'+\nu\bar{\pi}_3' = 0$, where $\bar{\pi}_1'$, $\bar{\pi}_2'$, $\bar{\pi}_3'$ are three fixed linearly independent linear combinations of π_1', π_2', π_3'. We may accordingly take the planes $\bar{\pi}_1' = 0$, $\bar{\pi}_2' = 0$, $\bar{\pi}_3' = 0$ as new base planes of the second star; and when this is done, the equations of a general pair of corresponding planes of the two stars may be written as

$$\lambda\pi_1+\mu\pi_2+\nu\pi_3 = 0 \quad \text{and} \quad \lambda\pi_1'+\mu\pi_2'+\nu\pi_3' = 0.$$

Reverting now to the generation of the cubic surface F, we may suppose that a general triad of corresponding planes of the three related stars is given by

$$\begin{aligned} \lambda\pi_1+\mu\pi_2+\nu\pi_3 &= 0, \\ \lambda\pi_1'+\mu\pi_2'+\nu\pi_3' &= 0, \\ \lambda\pi_1''+\mu\pi_2''+\nu\pi_3'' &= 0. \end{aligned} \qquad (1)$$

As λ, μ, ν vary, the point of intersection of these three planes describes the locus whose equation is

$$D \equiv \begin{vmatrix} \pi_1 & \pi_2 & \pi_3 \\ \pi_1' & \pi_2' & \pi_3' \\ \pi_1'' & \pi_2'' & \pi_3'' \end{vmatrix} = 0, \tag{2}$$

and the equation of F is therefore $D = 0$.

A cubic form in x_0, x_1, x_2, x_3 which is expressed, like D, as a determinant of linear forms is said to be a *determinantal cubic form*. It can be shown, though we do not prove it here, that any general cubic form in x_0, x_1, x_2, x_3 can be expressed, in infinitely many ways, as a determinantal cubic form; and this implies that any general cubic surface, given by the vanishing of such a form, can be generated projectively in the manner defined above.

Consider, then, the surface represented by an equation of the form (2). Since this equation is equivalent to the set of parametric equations (1), the surface can be generated projectively. But unless the three linear forms which make up any given row of D are linearly independent, the projective relation between the planes given by (1) will not have the general character that we have had in mind so far. We shall nevertheless look upon equations (1) as furnishing a projective generation of the surface as long as the three planes given by a general set of values of λ, μ, ν meet in a unique point.

Any one projective generation of F leads to an equation (2); and from this equation we can pass to infinitely many other projective generations of the same surface by replacing the rows and columns of D by linear combinations of themselves and also by interchanging the rows and the columns. This means that, subject to certain restrictions, the vertices A, A', A'' of a projective generation can be chosen arbitrarily on F.

The plane representation of the cubic surface F

The projectively generated cubic surface is a rational surface—i.e. it can be mapped birationally on a plane—and the plane representation of the surface is of great value in the investigation of its properties. We shall discuss this representation very briefly; and what we say here may be compared with what has already been said about Cremona transformations of the plane in Chapter IX and about the projection of a quadric on to a plane in Chapter XI. The quadric is a rational surface, and a birational mapping of this surface on a plane is very easily set up by means of the geometrical procedure of projection from a vertex on the surface. Since the cubic surface is of order 3, projection from an arbitrary point of the surface would not yield a (1, 1) mapping, and in this case it is natural to define a plane representation algebraically.

Let us consider once again the surface F defined by equations (1). Since $\pi_1,\ldots,\pi''_3$ are linear forms in x_0, x_1, x_2, x_3, equations (1) may be rewritten as

$$\sum_{k=0}^{3} \lambda_{ik} x_k = 0 \quad (i = 1, 2, 3), \tag{3}$$

where the λ_{ik} are all linear forms in λ, μ, ν. If the rank of the 3×4 matrix (λ_{rs}) is 3, these equations may be solved for the ratios of x_0, x_1, x_2, x_3 in terms of λ, μ, ν, and we have a solution of the form

$$\rho x_i = \phi_i(\lambda, \mu, \nu) \quad (i = 0,\ldots, 3). \tag{4}$$

The $\phi_i(\lambda, \mu, \nu)$ are cubic forms in λ, μ, ν, being in fact the 3×3 determinants (with appropriate signs) formed from the matrix (λ_{rs}).

If the values of λ, μ, ν are such that not all the ϕ_i are zero, the triad of parameters (λ, μ, ν) gives rise to a unique point of F; but if all the ϕ_i are zero, and (λ_{rs}) is of rank 2, we have a simply infinite system of corresponding points, making up a line which lies on F. It will appear shortly that, for general choice of the coefficients in equations (1), there are six special triads (λ, μ, ν), each of which gives rise to a line on F. Thus F *contains, in general, six lines* $a_1,\ldots, a_6$, *arising from triads of parameters* $(\lambda_i, \mu_i, \nu_i)$ $(i = 1,\ldots, 6)$ *for which the three related planes of the generating stars meet in a line instead of a point.*

If, now, $P\,(x_0, x_1, x_2, x_3)$ is an assigned point of F, the corresponding values of the ratios $\lambda:\mu:\nu$ are given by the three linear equations (1). Since the point P lies on F, the rank of this set of equations is at most 2. If it is exactly 2, the ratios are uniquely determined, but if it is 1 there will be an infinity of solutions. In the latter case the requirement that the general plane shall pass through the point P imposes the same condition on λ, μ, ν in each star, and this means that the three stars have a triad of concurrent corresponding rays $(AP, A'P, A''P)$. This, however, implies the existence of a special relation between the stars, and we can accordingly say that in the case of three general related stars a point P of F always corresponds to a unique parameter triad (λ, μ, ν).

We now obtain the plane representation of F, for which we are looking, by taking (λ, μ, ν) as allowable coordinates of a variable point of a plane π. We have thus the following theorem:

THEOREM 18. *The points P of a general projectively generated cubic surface F can be mapped on the points P' of a plane π in such a way that the correspondence between P and P' is* $(1, 1)$ *in general, and is expressible algebraically in terms of polynomials.*

For general choice of the related stars which generate F there are six fundamental points $A_1,..., A_6$ in π, to which correspond exceptionally six lines $a_1,..., a_6$ on F; but every point of F corresponds to a unique point of π. The exceptional lines $a_1,..., a_6$ on F are of necessity skew to one another, for if two of them, say a_1 and a_2, were to meet, the point $a_1 a_2$ on F would correspond to both A_1 and A_2 in π.

We have still to prove that the number of fundamental points in π is six, and this is easily done by considering the representation in π of the plane sections of F. Since every point P of F is mapped by a point P' of π, a curve C drawn on F must have as its image a curve C' in π. If C is the plane cubic cut on F by the plane

$$\sum_{i=0}^{3} u_i x_i = 0,$$

the coordinates (λ, μ, ν) of every point of C' must satisfy the equation

$$\Phi \equiv \sum_{i=0}^{3} u_i \phi_i(\lambda, \mu, \nu) = 0.$$

Thus C' is a plane cubic. If, now, we take two plane sections C_1, C_2 of F they meet in three points, namely the points of intersection of the common line of the two planes with the cubic surface F. The corresponding curves C'_1, C'_2 then also meet in three free points (i.e. points distinct from base points, common to all curves $\Phi = 0$ in π); and since two plane cubics have nine points in common, the system of Φ-curves must have six fundamental points $A_1,..., A_6$. We may say, therefore, that *in the representation of points of F by points of π, the ∞^3 plane sections of F are represented by the ∞^3 cubic curves through the six fundamental points $A_1,..., A_6$ of the representation.*

Two Φ-curves C'_1, C'_2 meet, as we have seen, in the points $A_1,..., A_6$ and also in three free points P'_1, P'_2, P'_3, corresponding to the three points P_1, P_2, P_3 in which the line of intersection of the planes of C_1 and C_2 meets F. Now any plane in space which passes through two of the collinear points P_1, P_2, P_3 must also pass through the third, and therefore any Φ-curve through two of the points P'_1, P'_2, P'_3 passes also through the third. We say that P'_1, P'_2, P'_3 form with $A_1,\dots, A_6$ a set of nine associated points, the base points of a pencil of plane cubics. Thus *the three points of intersection of F with a line are represented by a triad of points forming with $A_1,\dots, A_6$ a set of nine associated points in π.*

From the parametric equations (4), by taking A_1 to be one of the reference points in π the reader will readily verify the following result (cf. p. 279): *The ∞^1 points P' infinitely near to a base point A_i (i.e. the directions through A_i) correspond to the individual points P of the corresponding line a_i on F.* Thus the curves on F that are represented in π by two given curves through A_i meet a_i in the same point if and only if the tangents (or branch tangents) at A_i coincide.

Curves on the surface F

We now indicate very briefly how the plane representation of F can be made to reveal the whole structure of systems of curves on F, just as in the previous chapter the plane representation of the quadric surface yielded corresponding information about curves on that surface.

If C is any curve on F, which we suppose to be represented by a curve C' in π, the order N of C (i.e. the number of points in which it is met by a general plane or, what comes to the same thing, by a general plane section of F) is equal to the number of free intersections (not at $A_1,\dots, A_6$) of C' with a general Φ-curve. Thus if C' is a curve $C'^m(A_1^{k_1},\dots, A_6^{k_6})$ of order m, with multiplicities $k_1,\dots, k_6$ at $A_1,\dots, A_6$, then $N = 3m-k_1-\dots-k_6$. If we wish, in particular, to find all the lines on F, other than those of the set $a_1,\dots, a_6$ which are represented in π by the neighbourhoods of $A_1,\dots, A_6$, we must put $N = 1$ in this relation. For the curves in π which represent lines on F we then have $3m-k_1-\dots-k_6 = 1$. This Diophantine equation yields the following solutions of the problem: fifteen lines c_{ij} of F, represented in π by the lines $A_i A_j$; and six lines b_i, represented by conics through five of the six points $A_1,\dots, A_6$. We thus have the well-known result:

THEOREM 19. *The general projectively generated cubic surface F contains in all twenty-seven lines. Of these, six are represented by the neighbourhoods of the fundamental points $A_1,\dots, A_6$, fifteen by the lines joining pairs of these points, and six by conics through sets of five of the six points.*

The reader may now investigate for himself, by means of the plane representation of F, the incidence relations of these twenty-seven lines. A full discussion of this subject is to be found in Baker, *Principles of Geometry*, volume iii, chapter iv.

We mention, in conclusion, a few more results concerning curves on F that may be obtained in a similar way. Any conic on the surface lies in a plane through one of the twenty-seven lines. If we take $N = 3$, we find that (the plane sections being excluded from consideration) F possesses no fewer than 72 exactly similar doubly infinite families of twisted cubics, represented in π by various systems of curves of degrees 1, 2, 3, 4, and 5. A typical one of these systems in π consists of all the lines of π; and it may be noted that any two of the corresponding twisted cubics meet in a unique point.

Any quadric section of F is represented in π by a curve

$$C^6(A_1^2,\ldots, A_6^2);$$

and, more generally, any section of F by a surface of order n is represented by a curve $C^{3n}(A_1^n,\ldots, A_6^n)$.

Other cubic surfaces

F was defined at the beginning of this section as a surface that is generated by three collinear stars, but we can now see that, quite independently of the idea of projective generation, any set of parametric equations of the form (4), in which the $\phi_i(\lambda, \mu, \nu)$ are four linearly independent cubic forms which vanish at six assigned points $A_1,\ldots, A_6$ of π define, in general, a cubic surface F; and by imposing special conditions on the points A_i we can obtain various special types of cubic surface. In particular, F can be made in this way to acquire 1, 2, 3, or 4 nodes.

Thus, for example, if we take the points A_i to lie at the vertices of a complete quadrilateral, then F is a four-nodal cubic surface, whose explicit equation may readily be obtained from the parametric representation in the form

$$x_1 x_2 x_3 + x_2 x_3 x_0 + x_3 x_0 x_1 + x_0 x_1 x_2 = 0,$$

the reference points being the nodes. In this case all the points of any side of the quadrilateral represent (exceptionally) the same node of F.

It should be added, however, that there is another totally different type of cubic surface, namely the *cubic ruled surface* or *cubic scroll*, which is more naturally approached in other ways. The simplest definition of the general cubic scroll is as the surface generated by a line which joins corresponding points of two ranges, in (1, 2) algebraic correspondence, whose axes are skew lines. Its equation may be reduced to the simple form $x_0 x_2^2 - x_1 x_3^2 = 0$; and it admits of a general parametric representation

$$x_0 : x_1 : x_2 : x_3 = S_0(\lambda, \mu, \nu) : S_1(\lambda, \mu, \nu) : S_2(\lambda, \mu, \nu) : S_3(\lambda, \mu, \nu),$$

where the equations $S_i(\lambda, \mu, \nu) = 0$ are those of four linearly independent conics through a common point in π (cf. Exercise 22 below). One of the two skew lines, namely $X_0 X_1$, is a double line on the surface.

We shall encounter, in the next chapter especially, important examples of the natural intrusion of projectively generated cubic surfaces into the general structure of projective geometry, side by side with the quadric and the twisted cubic.

For a fuller discussion of cubic surfaces and their properties the reader should turn to the following more advanced works—Reye, *Geometrie der Lage*; Baker, *Principles of Geometry*, volume iii; Todd, *Projective and Analytical Geometry*; Semple and Roth, *Introduction to Algebraic Geometry*.

EXERCISES ON CHAPTER XII

1. A twisted cubic curve has parametric equations

$$x : y : z : t = \theta : \theta^2 : \theta^3 + 1 : \theta^3 - 1.$$

Find the equation of the plane through the points whose parameters are θ_1, θ_2, θ_3, and deduce that the osculating plane at the point (θ) has equation

$$6\theta^2 x - 6\theta y + (1 - \theta^3)z + (1 + \theta^3)t = 0.$$

Show that the line $x = 0 = y$ is a chord and the line $z = 0 = t$ is the line of intersection of two osculating planes, and prove that the curve projects from the reference point T into the plane cubic $x^3 + y^3 = xyz$ in the reference plane XYZ.

Find all the quadrics which pass through the curve, and select from among them the two cones whose vertices are on ZT.

2. Find parametric equations for the unique twisted cubic which passes through the four reference points, the point (a, b, c, d), and the point (a', b', c', d').

Find the condition for the pairs of reference points (X, Y) and (Z, T) to separate each other harmonically on the curve.

3. $AB'CA'BC'$ is a skew hexagon inscribed in a twisted cubic c, and ψ_1, ψ_2, ψ_3 are the three quadrics through c which contain the pairs of opposite edges $(BC', B'C)$, $(CA', C'A)$, $(AB', A'B)$ respectively. Prove that ψ_1, ψ_2, ψ_3 have a common generator, and that this line is a chord of c.

4. Prove that the four tangents of a twisted cubic c which meet a general line p also meet the *polar line* p' of p, i.e. the axis of the pencil of polar planes of points of p in the null polarity defined by c.

If the twisted cubic is $c\,(\theta^3, \theta^2, \theta, 1)$, and p meets the tangents XY and ZT in the points $(1, \lambda, 0, 0)$ and $(0, 0, \mu, 1)$ respectively, prove that the parameters of the two remaining points of c at which the tangents meet p are the roots of the equation $2\lambda\theta^2-(3\lambda\mu+1)\theta+2\mu = 0$.

Deduce that a line is self-polar for c if and only if the points of contact of the four tangents which meet it form an equianharmonic tetrad on c.

5. Prove that the points (x, y, z, t) and (x', y', z', t') are *conjugate* with respect to the cubic $c\,(\theta^3, \theta^2, \theta, 1)$—i.e. that each lies in the polar plane of the other—if and only if $xt'-x't = 3(yz'-y'z)$. If two points satisfy this condition, show that the line which joins them is *self-polar* for c, any two of its points being conjugate to each other.

Show that the only self-polar chords of the curve are the tangents, and the only self-polar unisecants are those which lie in osculating planes.

6. Two twisted cubics c, c' are given parametrically by the matrix equations $\mathbf{x} = \mathbf{A}\boldsymbol{\theta}$, $\mathbf{x} = \mathbf{B}\boldsymbol{\theta}$, where $\mathbf{A}$, $\mathbf{B}$ are non-singular 4×4 matrices and $\boldsymbol{\theta}$ is the 4×1 matrix whose elements are $(\theta^3, \theta^2, \theta, 1)$. If $\mathbf{S}$ is the matrix

$$\begin{pmatrix} 0 & 0 & 0 & 1 \\ 0 & 0 & -3 & 0 \\ 0 & 3 & 0 & 0 \\ -1 & 0 & 0 & 0 \end{pmatrix}$$

and $\mathbf{C}=\mathbf{B}^{-1}\mathbf{A}$, prove that c and c' define the same null polarity if and only if

$$\mathbf{C}^T\mathbf{S}\mathbf{C} = \rho\mathbf{S},$$

where ρ is a scalar.

7. On a twisted cubic c, let the pairs (P, P') of an involution be cut by one system of generators of a quadric ψ, and let Q, Q' be the points in which the other generators of ψ through P and P' meet a fixed general plane π. Show that (i) the lines QQ' all pass through a fixed point R, and (ii) when π turns about a given line l, the locus of R is a line which meets the two lines PP' which meet l.

8. Find the equation of the quadric through the tangents to $c\,(\theta^3, \theta^2, \theta, 1)$ at the points whose parameters are 0, 1, ∞.

Hence, or by using the results of Exercise 4, prove that four tangents to a twisted cubic have only one transversal if their points of contact form an equianharmonic tetrad on the curve.

9. Find the equation of the quadric ψ through the twisted cubic c

$$(\theta^3, \theta^2, \theta, 1)$$

which contains the tangents to c at the reference points X and T.

Show that (i) the osculating planes of c at X and T are the tangent planes to ψ at these points; (ii) if π is the osculating plane of c at a variable point P, the locus of the pole Q of π with respect to ψ is another twisted cubic c' through X and T; (iii) the tangent plane to ψ at P is the osculating plane of c' at Q.

10. Show that the plane-equation of any quadric which touches all the osculating planes of the twisted cubic c $(\theta^3, \theta^2, \theta, 1)$ is of the form

$$\lambda(v^2-3wu)+2\mu(vw-9up)+\nu(w^2-3vp) = 0.$$

Show that the quadrics of this system which also touch a general plane π do so at the points of a line g, and that any two of them meet in g and a twisted cubic k.

Find the equations of g and k when π is the plane $x = t$ and the two quadrics touch π at points in the planes $y+z = 0$ and $y-z = 0$ respectively.

11. A twisted cubic c and two of its tangents a, b are given, and a transversal p of a and b varies in such a way that the two remaining tangents to c which meet it are coincident. If p does not meet c, prove that it lies on a certain fixed quadric through a and b.

12. If p, q are chords and A, B, C, D are points of a twisted cubic c, prove that $p(A,B,C,D) \barwedge q(A,B,C,D)$. Explain how the result is to be interpreted when p or q passes through one of the four points A, B, C, D, and also when p or q is a tangent to c.

A skew quadrilateral is formed by a chord UV of c, the tangents to c at U and V, and the line of intersection of the osculating planes at these points. If l, m are the diagonals of the quadrilateral, prove that every chord of c which meets l also meets m, and that its points of intersection with these lines are separated harmonically by its end-points on c.

13. Show that any plane cubic curve with a double point can be regarded as the projection of a twisted cubic, and deduce that it has three points of inflexion, which are collinear.

Prove that the node and the line of collinearity of the three points of inflexion are harmonic pole and polar with respect to the triangle formed by the three inflexional tangents.

[*Hint.* Represent the plane cubic parametrically by writing its equation in the form $xyz-u_3(x,y) = 0$, where $u_3(x,y)$ is a cubic polynomial, and putting $y = \theta x$.]

14. A twisted cubic c in affine space passes through the origin of coordinates O and has asymptotes parallel to the axes OX, OY, OZ. Show that it has parametric equations of the form $X = a\theta/(\theta-\alpha)$, $Y = b\theta/(\theta-\beta)$, $Z = c\theta/(\theta-\gamma)$, and find the equations of its asymptotes.

Prove that the plane $uX+vY+wZ = p$ meets c and its asymptotes in two triads of points with the same centroid (X_0, Y_0, Z_0), given by

$$3uX_0 = p+(au\beta+bv\alpha)/(\beta-\alpha)+(au\gamma+cw\alpha)/(\gamma-\alpha)$$

and two similar equations; and deduce that the locus of centroids of triads of points cut on c by a system of parallel planes is a straight line.

15. A system of parallel planes cuts a system of triangles on a given twisted cubic. Prove that the circumcentres of the triangles lie on a line and the orthocentres lie on another line. Deduce that the locus of centroids is also a line.

16. Show that if there exists one triad of mutually perpendicular chords OP, OQ, OR of a twisted cubic c, then there exists an infinity of such triads through O. Prove that, when this is the case, all the planes PQR meet the normal plane of c at O in the same fixed line.

17. A twisted cubic c passes through the vertices of a tetrahedron $ABCD$ which is self-polar for a quadric ψ. Prove that every point of c is one vertex of a tetrahedron inscribed in c and self-polar for ψ.

If P, Q are any two points of a twisted cubic c which has three perpendicular asymptotes, and X, Y, Z are the points of intersection of c with a plane perpendicular to PQ, prove that each of the tetrahedra $PXYZ$ and $QXYZ$ has three pairs of opposite edges at right angles.

18. Show that a rectangular twisted cubic c (one with three mutually perpendicular asymptotes) contains an infinity of sets of six points such that the plane of any three points of a set is perpendicular to the plane of the other three. [*Hint.* Take a general quadric Q which meets ι in a conic outpolar for Ω, and consider the six points in which it meets c. A plane-pair which contains these six points is a quadric of the system

$$\lambda_1 Q_1+\lambda_2 Q_2+\lambda_3 Q_3+\lambda Q = 0,$$

and it therefore meets ι in a conic outpolar for Ω.]

19. Verify that the (non-singular) cubic surface whose equation is

$$x_0 x_1(x_0+x_1) = x_2 x_3(x_2+x_3)$$

possesses the parametric representation

$$x_0 : x_1 : x_2 : x_3 = x(z^2-xy) : y(x^2-yz) : x(y^2-zx) : y(z^2-xy),$$

and that the plane sections of the surface are represented in the (x,y,z) plane by cubic curves through the three vertices of the triangle of reference and the three points $(1,1,1)$, $(1,\omega,\omega^2)$, $(1,\omega^2,\omega)$, where ω is a complex cube root of unity.

20. By resolving the left-hand side of the equation $x^3+y^3+z^3+t^3 = 0$ into a sum of products of linear factors in three different ways, obtain the equations of the twenty-seven lines on the cubic surface which the equation represents.

21. Show that the four-nodal cubic surface $1/x_0+1/x_1+1/x_2+1/x_3 = 0$ can be transformed into a plane by the reciprocal transformation, and hence obtain its plane parametric representation.

Show that the surface contains nine lines, of which six are the joins of the nodes, while the remaining three form a triangle.

22. Show that the cubic scroll whose equation is $x_0 x_2^2 = x_1 x_3^2$ has the plane representation $x_0:x_1:x_2:x_3 = y^2:z^2:zx:xy$, and find the representation in the (x,y,z) plane of (a) the generating lines of the surface, and (b) the double line and the simple line on the surface which are met by all the generators.

CHAPTER XIII

LINEAR SYSTEMS OF QUADRICS

THE theory of linear systems of quadrics, as we might expect, is very similar to the corresponding theory for conics but more complicated. In this chapter we shall present the theory in outline only, introducing the main ideas but not going into detailed discussion of all possible special cases. We deal first with the two ∞^1 systems, the pencil of loci and the range of envelopes, and then we say something about the ∞^2 linear system, or net, of quadric loci.

§ 1. PENCILS OF QUADRICS

Let S, S' be two linearly independent quadratic forms in x_0, x_1, x_2, x_3. Then the equations $S = 0$, $S' = 0$ represent two quadric loci ψ, ψ', and for every value of λ the equation $S+\lambda S' = 0$ represents a quadric ψ_λ which passes through all the common points of ψ and ψ'. Such a system of ∞^1 quadrics is called a *pencil*, and it has the property that there is a unique quadric of the system through any given point which is not common to ψ and ψ'.

The points which are common to all quadrics of the pencil are simply the points which make up the curve of intersection of ψ and ψ', and this curve is referred to as the *base curve* of the pencil. Since ψ and ψ' each meet a given general plane in a conic, and two conics have four points in common, the base curve is met by a general plane in four points, i.e. it is a quartic curve C^4. We do not assert that this quartic curve is proper, and it will soon appear that it can break up, even in cases which are by no means trivial. The quadrics ψ and ψ' might, for instance, be two proper quadrics with four generators in common, and then C^4 would be a skew quadrilateral. We do, however, wish to exclude trivial cases—for example the one which arises when ψ and ψ' are plane-pairs with one plane in common.

A general plane π, then, cuts the base curve C^4 in four points, and it therefore cuts every quadric of the pencil in a conic through these four points. In other words, *a pencil of quadrics is cut by a general plane in a pencil of conics.*

For the quadric ψ_λ to be degenerate, λ must be a root of the quartic equation $|\mathbf{A}+\lambda\mathbf{A}'| = 0$. If the four roots are distinct, we say that the pencil is *general*, and for the present we shall confine our attention to such pencils. A general pencil, then, contains precisely four cones κ_i ($i = 1,\ldots, 4$) with vertices V_i, say. If $i \neq j$, the points V_i, V_j are conjugate for both κ_i and κ_j, i.e. for two quadrics of the pencil, and therefore for every quadric of the pencil. Thus the four points V_i are such that each is conjugate to all the others, and the polar plane of each, for every quadric of the pencil, is the plane of the other three. It follows that the four points are not coplanar; for, if they were, their plane would touch every quadric ψ_λ at the four points V_i. Thus *the points V_i are the vertices of a proper tetrahedron, self-polar for every quadric of the pencil.*

EXERCISE. Prove directly that the four points V_i are linearly independent. [*Hint.* The original quadrics ψ, ψ' may be taken to be proper quadrics, and the equation $(\mathbf{A}-\lambda\mathbf{A}')\mathbf{x} = 0$ is then equivalent to $\mathbf{A}'^{-1}\mathbf{A}\mathbf{x} = \lambda\mathbf{x}$. Now apply the algebraic theorem given in Chapter XIV, Exercise 16.]

By taking $V_1 V_2 V_3 V_4$ as tetrahedron of reference and choosing the unit point suitably, we can reduce the equation of the general pencil to the form

$$(a_0+\lambda)x_0^2+(a_1+\lambda)x_1^2+(a_2+\lambda)x_2^2+(a_3+\lambda)x_3^2 = 0.$$

The values of λ which correspond to the four cones are $-a_0$, $-a_1$, $-a_2$, $-a_3$, and the four numbers a_i are therefore all different. It is at once apparent from the above equation that the vertices of the four cones are the vertices of a common self-polar tetrahedron; and we see further that the common self-polar tetrahedron of a general pencil is unique.

The pencil of conics in which the plane $x_0 = 0$ is cut by the quadrics ψ_λ is given by

$$x_0 = 0 = (a_1+\lambda)x_1^2+(a_2+\lambda)x_2^2+(a_3+\lambda)x_3^2,$$

and the base points of this pencil are accordingly the four points

$$\{0, \sqrt{(a_2-a_3)}, \pm\sqrt{(a_3-a_1)}, \pm\sqrt{(a_1-a_2)}\}.$$

In this way we obtain the coordinates of the sixteen points in which the base curve C^4 cuts the faces of the tetrahedron of reference. Many other properties of C^4 are equally easy to derive, and we now give a few examples.

(i) At each of its points, C^4 has a well-defined tangent line, the axis of the pencil of tangent planes to the quadrics ψ_λ at this point. If the point is (y_0, y_1, y_2, y_3), the equations of the tangent line may be written

$$\sum_{i=0}^{3} a_i y_i x_i = 0 = \sum_{i=0}^{3} y_i x_i.$$

(ii) Any tangent plane to a quadric ψ_λ at a point of C^4 clearly touches C^4. If, now, ψ_λ is one of the four cones κ_i, any tangent plane to κ_i touches κ_i all along a generator. But the generator meets C^4 in two points (since the plane determined by any two generators has four points in common with C^4) and the corresponding tangent plane to κ_i must touch C^4 at each of these points. Thus *C^4 has four families of ∞^1 bitangent planes.* This may also be inferred from the fact that a tangent plane to κ_i cuts the quadrics ψ_λ in a pencil of conics containing a repeated line, i.e. a double-contact pencil of conics.

(iii) Since C^4 has the equations

$$\sum_{i=0}^{3} a_i x_i^2 = 0 = \sum_{i=0}^{3} x_i^2,$$

we see that if (y_0, y_1, y_2, y_3) lies on C^4 then the eight points

$$(y_0, \pm y_1, \pm y_2, \pm y_3)$$

all lie on C^4. This means that C^4 is invariant with respect to the group of eight collineations

$$\frac{x_0'}{x_0} = \frac{x_1'}{\pm x_1} = \frac{x_2'}{\pm x_2} = \frac{x_3'}{\pm x_3},$$

a group comprising the identical collineation ϵ, four harmonic homologies, and three harmonic biaxial collineations (cf. p. 350).

(iv) Suppose l is a chord of C^4, meeting the curve in A and B. If P is a general point of l, there is a unique value of λ for which ψ_λ passes through P, and there is therefore a unique quadric ψ_λ of the pencil which has l as a generator. Conversely, provided C^4 is a proper quartic curve, any generator of a quadric of the pencil is a chord of C^4. For the plane of a pair of generators of ψ_λ cuts C^4 in four points; and unless these points lie two on each generator, one generator cuts C^4 in three points and therefore lies wholly on every quadric of the pencil—which contradicts the hypothesis that the base curve C^4 does not break up. We see then that the chords of C^4 are the generators of the quadrics ψ_λ. It follows immediately that *C^4 has two chords through a general point of space,* namely the two generators of the unique quadric ψ_λ that passes through the point.

General properties of a pencil of quadrics

We now leave the base curve C^4, and pass on to the consideration of the pencil of quadrics itself. We have first of all four theorems which are obvious generalizations of the corresponding theorems of Chapter VII.

THEOREM 1 (*Desargues's Theorem*). *A general line of space is met by the quadrics of a pencil in pairs of points in involution.*

The proof is formally the same as that of Theorem 1 of Chapter VII. We have also the corollary that there are two quadrics of a pencil which touch a given line, their points of contact being the united points of the Desargues involution.

Since, further, the quadrics cut a general plane in the conics of a pencil, there are three of the quadrics which touch a given plane. Their points of contact are the vertices of the three line-pairs of the pencil of conics.

THEOREM 2. *The polar planes of a fixed point P with respect to the quadrics of a pencil all pass through a fixed line ϑ_P.*

This theorem is an immediate consequence of the linearity in the parameter λ of the equation $S+\lambda S' = 0$.

The lines ϑ_P corresponding to the different points P of space are called the *axes* of the pencil. Since there are ∞^4 lines of space but only ∞^3 points, not every line is an axis of a given pencil.†

THEOREM 3. *If l is a fixed line, the polar lines l'_λ of l with respect to the quadrics ψ_λ and the axes ϑ_P of the different points P of l form the two systems of generators of a quadric ϕ_l, and this quadric passes through the vertices of the cones of the pencil (ψ_λ).*

Proof. Take two points A, B arbitrarily on l. Then the equations of their polar planes with respect to ψ_λ may be written

$$\alpha_\lambda \equiv \alpha+\lambda\alpha' = 0, \qquad \beta_\lambda \equiv \beta+\lambda\beta' = 0;$$

and, as λ varies, the two planes describe homographic pencils with axes ϑ_A, ϑ_B. Their line of intersection l'_λ therefore describes a regulus, and the axes ϑ_A, ϑ_B belong to the complementary regulus. Since the polar line of l with respect to the cone κ_i passes through the vertex V_i of this cone, the quadric ϕ_l which contains the two reguli passes through the four points V_i.

EXERCISE. Show that an exceptional case occurs when l is itself an axis ϑ_{P_0}, and that in this case ϕ_l is a cone with vertex P_0.

THEOREM 4. *The poles of a fixed plane π with respect to the quadrics of a pencil all lie on a fixed twisted cubic, which passes through the vertices of the four cones.*

† The system of ∞^3 axes is a tetrahedral complex—see p. 374.

Proof. Take three points A, B, C arbitrarily in π. Then the pole of π for ψ_λ is the common point of the polar planes of A, B, C, and as λ varies these polar planes describe homographic pencils with axes $\vartheta_A, \vartheta_B, \vartheta_C$. The locus of poles is therefore a twisted cubic curve; and since the pole of π for κ_i is V_i, this cubic contains the four points V_i.

COROLLARY. The ∞^2 axes ϑ_P of the points P of π are the ∞^2 chords of the twisted cubic.

Special types of pencil

So far we have had in mind primarily the general pencil. When the roots of the quartic equation $|\mathbf{A}+\lambda\mathbf{A}'| = 0$ are not all distinct, the pencil is said to be special; and there are many special types of pencil, corresponding to different modes of coincidence of the roots. We do not propose to examine all these types,† and shall confine our attention to the ones which are of particular geometrical interest.

We shall also leave the reader to look into the modifications that have to be made in the preceding general theory when the pencil concerned is of one of the special types.

Type 1: *The simple-contact pencil*

Suppose ψ and ψ' touch at a single point A. Any general plane through A then cuts ψ and ψ' in conics which touch at A and have two further points of intersection. The plane therefore meets C^4 in A (twice) and in two other points; and this means that A is a double point of C^4.

If we take A as X_0 and three other points of C^4 as X_1, X_2, X_3, then ψ and ψ' both circumscribe the tetrahedron of reference. Since they have the same tangent plane at X_0, their equations may be written as

$$S \equiv a_{23}x_2x_3+a_{31}x_3x_1+a_{12}x_1x_2+x_0(p_1x_1+p_2x_2+p_3x_3) = 0$$

and

$$S' \equiv a'_{23}x_2x_3+a'_{31}x_3x_1+a'_{12}x_1x_2+x_0(p_1x_1+p_2x_2+p_3x_3) = 0.$$

Thus

$$S-S' \equiv (a_{23}-a'_{23})x_2x_3+(a_{31}-a'_{31})x_3x_1+(a_{12}-a'_{12})x_1x_2,$$

and the pencil contains a cone κ_1 with vertex X_0, i.e. A. If we now

† For an exhaustive classification see Todd, *Projective and Analytical Geometry*.

take this cone in place of the quadric ψ' and change the unit point suitably, we can put the equation of the pencil in the form

$$(a_{23}+\lambda)x_2x_3+(a_{31}+\lambda)x_3x_1+(a_{12}+\lambda)x_1x_2+ \\ +a_{01}x_0x_1+a_{02}x_0x_2+a_{03}x_0x_3 = 0.$$

The condition for degeneracy is now

$$\begin{vmatrix} 0 & a_{01} & a_{02} & a_{03} \\ a_{01} & 0 & a_{12}+\lambda & a_{31}+\lambda \\ a_{02} & a_{12}+\lambda & 0 & a_{23}+\lambda \\ a_{03} & a_{31}+\lambda & a_{23}+\lambda & 0 \end{vmatrix} = 0.$$

This is a quartic equation in λ with apparent degree 2, and it therefore has ∞ as a double root. In other words, the cone with vertex A counts twice in the set of four cones of the pencil. The pencil has no common self-polar tetrahedron.

If the quadrics ψ and ψ' which define a pencil (ψ_λ) have two distinct points of contact A, B, the base curve C^4 must have each of these points as a double point. It follows at once that C^4 breaks up in some way; for if P is any point of C^4, other than A and B, the plane ABP cuts the quartic curve in five points and therefore contains a whole component of it. If this component is a conic, the residual component is also a conic; while if it is a line, the residual component is a twisted cubic, proper or degenerate. We shall consider first of all the case in which C^4 consists of two proper conics, lying in different planes but meeting in two points A, B. Such pencils do in fact exist, for we can obtain one by taking an arbitrary quadric for ψ and a plane-pair for ψ'. Since the tangent lines at A and B to both conics touch every quadric of the pencil, all the quadrics have a common tangent plane at A and also at B, i.e. they have double contact. We have, then:

Type 2: The double-contact pencil whose base is a pair of conics

By taking A, B as X_1, X_2, the poles of AB for the two conics as X_0, X_3, and a suitable point E as unit point, we can put the equations of the two conics in the forms

$$x_0^2-x_1x_2 = 0 = x_3 \quad \text{and} \quad x_3^2-x_1x_2 = 0 = x_0$$

respectively. A general quadric through these two curves then has an equation which may be expressed indifferently in the two forms

$$x_0^2-x_1x_2+x_3(\alpha_0x_0+\alpha_1x_1+\alpha_2x_2+\alpha_3x_3) = 0$$

and

$$x_3^2-x_1x_2+x_0(\beta_0x_0+\beta_1x_1+\beta_2x_2+\beta_3x_3) = 0,$$

and the equation may therefore be written

$$x_0^2+x_3^2-x_1x_2+2\lambda x_0x_3 = 0.$$

The discriminant of the quadratic form is $\frac{1}{4}(\lambda^2-1)$, and the degenerate quadrics of the pencil are accordingly given by $\lambda = -1$, $\lambda = 1$, and $\lambda = \infty$ (twice). They are the two proper cones

$$(x_0-x_3)^2-x_1x_2 = 0 \quad \text{and} \quad (x_0+x_3)^2-x_1x_2 = 0,$$

and the plane-pair $x_0x_3 = 0$ (the pair of planes of the conics) which counts twice. The vertices of the cones are clearly the points $(1,0,0,1)$ and $(1,0,0,-1)$, and they lie on X_0X_3 and separate X_0, X_3 harmonically.

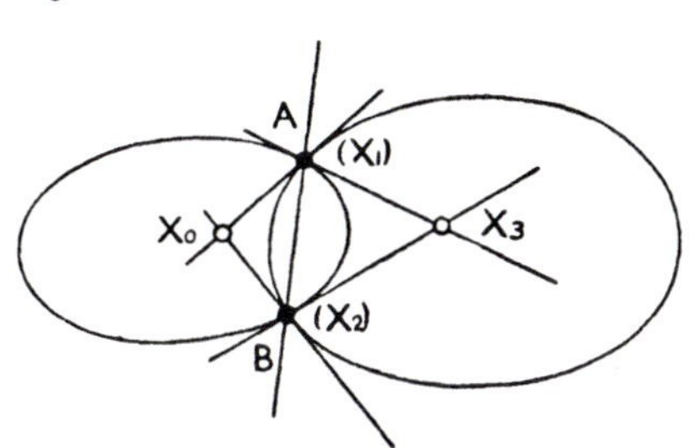

For any non-zero value of the arbitrary constant k, the equation of the general quadric of the pencil can be written in the form

$$(1+\lambda)(x_0+x_3)^2+(1-\lambda)(x_0-x_3)^2-\frac{1}{2k}\{(x_1+kx_2)^2-(x_1-kx_2)^2\} = 0.$$

Thus the tetrahedron formed by the four planes $x_0 \pm x_3 = 0$, $x_1 \pm kx_2 = 0$ is a common self-polar tetrahedron for all the quadrics of the pencil. The pencil possesses therefore ∞^1 such common self-polar tetrahedra, each of which has the vertices of the two cones as two of its vertices, while the remaining two are any pair of points of the chord of contact which separate A, B harmonically.

Now suppose we begin with an arbitrary pair of conics k_1, k_2, lying in different planes but having two points A, B in common. We can make a quadric contain both conics by making it pass through A and B and three further points of each conic—eight points in all. The quadrics through k_1 and k_2 thus form a pencil of type 2, and so we have the theorem:

THEOREM 5. *If two conics in space meet in two points, then there exist two other points from each of which they are in perspective.*

Type 3: *The double-contact pencil whose base is a twisted cubic and one of its chords*

As we saw on p. 332, two quadrics which touch at A and B may have as their curve of intersection the line AB together with a twisted cubic through A and B. That this case can actually arise has been shown in the previous chapter (p. 307) and it is clear that the equation of the corresponding pencil may be expressed in the form

$$Q_1+\lambda Q_3 \equiv x_1^2-x_0x_2+\lambda(x_2^2-x_1x_3) = 0.$$

The condition for degeneracy gives $\lambda = 0$ (twice) and $\lambda = \infty$ (twice). The pencil therefore has two cones—the quadrics Q_1 and Q_3—each of which is to be counted twice.

Type 4: *The pencil with a base quadrilateral*

If the twisted cubic breaks up into three lines, we obtain a still more special pencil whose base is a skew quadrilateral. Such a pencil is the system of all quadrics with two given common u-generators and two given common v-generators. Its equation may be written (cf. p. 275)

$$x_0x_3-\lambda x_1x_2 = 0;$$

and the degenerate members are the plane-pairs $x_0 = 0$, $x_3 = 0$ and $x_1 = 0$, $x_2 = 0$, each counted twice.

Type 5: *The ring-contact pencil*

Another way in which the curve C^4 can degenerate is by becoming a repeated conic k, lying in a plane π, say. In this case, if $S = 0$ is the equation of any one quadric of the pencil, the equation of a general member can be written

$$S+\lambda\pi^2 = 0.$$

The quadrics all touch at every point of k, and we say that they have *ring contact* along k. If P_0 is the pole of π for S, it is conjugate to every point of π for two quadrics of the pencil, namely S and π^2, and therefore for every quadric. If $P_1P_2P_3$ is any triangle self-polar for k, then $P_0P_1P_2P_3$ is clearly a common self-polar tetrahedron for the pencil, which accordingly possesses ∞^3 common self-polar tetrahedra. If such a tetrahedron is taken as tetrahedron of reference, the equation of the pencil may be written in the form

$$\lambda x_0^2+x_1^2+x_2^2+x_3^2 = 0,$$

and it follows that the degenerate quadrics are the repeated plane

π^2, counted three times, and the proper cone $x_1^2+x_2^2+x_3^2 = 0$. This cone is, of course, the enveloping cone of the system.

§ 2. Ranges of Quadrics

Two distinct quadric envelopes ψ and ψ', represented by equations $\Sigma = 0$ and $\Sigma' = 0$, determine an ∞^1 linear system of quadric envelopes, or *range* of quadrics, the system of all quadrics having a plane-equation of the form

$$\Sigma+\lambda\Sigma' = 0.$$

From a general point of space, enveloping cones can be drawn to ψ and ψ', and these cones have four common tangent planes. Since ψ and ψ' are both systems of ∞^2 tangent planes, they have ∞^1 planes in common, and these common planes form a developable δ^4 of class 4, the base developable of the range. The range of quadrics is, of course, the space-dual of the pencil, and the developable δ^4 is dual to the base quartic curve C^4 of the pencil.

A range of quadrics has four degenerate members, in general four distinct disk quadrics; but in particular cases either one or two of these disk quadrics may degenerate further, and it then counts multiply in the set of four degenerate quadrics.

We say that a range is general when the four degenerate quadrics are all distinct. The planes of the four disk quadrics then form a proper tetrahedron, which is self-polar for every quadric of the range.

The general properties of ranges of quadrics may be written down by dualizing the corresponding properties of pencils, and we shall confine ourselves to a brief summary of the results.

First of all, Desargues's Theorem states that the pairs of tangent planes from a given line to the quadrics of a range all belong to an involution pencil of planes. And with this we may couple the result that the enveloping cones drawn from a given point P to the quadrics of a range form a range of cones (in the star with vertex P).

The poles of a fixed plane π with respect to the quadrics of a range all lie on a line ϑ_π, called an *axis* of the range.

The polar lines of a fixed line l form a regulus; and the axes of the planes through l make up the complementary regulus.

Finally, the polar planes of a fixed point P all belong to a cubic developable, which contains the planes of the four disk quadrics of the range.

Special ranges

Each of the special types of pencil considered above gives rise, by duality, to a correspondingly special type of range. The pencil with a base quadrilateral (type 4) and the ring-contact pencil (type 5) are evidently self-dual systems, as may easily be verified algebraically. The double-contact range dual to the pencil of type 2 is perhaps worthy of mention. It consists of all quadrics which touch all the tangent planes of two given cones which have two tangent planes in common.

§ 3. Affine and Euclidean Specializations

A general plane is touched by three quadrics of any pencil and by one quadric of any range; and by taking the plane to be the plane at infinity ι we see at once that a pencil of quadrics in affine space in general contains three paraboloids (of which either one is real or all three are real), while a range of quadrics contains one (real) paraboloid. The centre-locus of a pencil is a twisted cubic curve, while the centre-locus of a range is a line.

In euclidean space we have one very special kind of pencil of quadrics, namely the coaxal system of spheres, obtained by rotating a coaxal system of circles about its line of centres. This is a double-contact pencil of type 2, the two common conics being the absolute conic Ω and the circle (real or virtual) traced out by the common points of the coaxal circles. This circle meets Ω, of course, in the two absolute points in its plane.

Confocal quadrics

The theory of confocal quadrics is very similar to the theory of confocal conics, given in Chapter VII. In the present section we shall state the main results for quadrics, emphasizing the points of difference between confocal quadrics and confocal conics, and omitting proofs wherever the reader should be able to construct them by analogy with what has gone before.

A confocal system of conics can be defined, it will be remembered, in either of two ways: as the system of all conics which have the same foci as a given fixed conic, or as a range of conics which has the absolute point-pair Ω as one member. In the case of quadrics we naturally choose a definition of the second kind, since in elementary geometry there is no obvious analogue for quadrics of the foci of a conic. A confocal system of quadrics, then, is a range

$$\Sigma + \lambda\Omega = 0,$$

where Ω represents the absolute disk quadric. It is easily seen that there are two distinct types of confocal system, one consisting entirely of central quadrics and the other of paraboloids. It will be sufficient here to consider confocal central quadrics.

Using non-homogeneous rectangular cartesian coordinates (X, Y, Z), let us take the ellipsoid ψ whose equation is

$$\frac{X^2}{a^2}+\frac{Y^2}{b^2}+\frac{Z^2}{c^2} = 1 \quad (a^2 > b^2 > c^2).$$

The plane-equation of ψ is then

$$a^2U^2+b^2V^2+c^2W^2 = 1,$$

and the plane-equation of a general quadric confocal with ψ is therefore

$$(a^2+\lambda)U^2+(b^2+\lambda)V^2+(c^2+\lambda)W^2 = 1.$$

The confocal system determined by ψ is thus given by the point-equation

$$\frac{X^2}{a^2+\lambda}+\frac{Y^2}{b^2+\lambda}+\frac{Z^2}{c^2+\lambda} = 1.$$

It may now be shown (cf. p. 167) that if P is a general point of space, three quadrics of the family pass through P, and the parameters of these quadrics are all real. The quadrics are of different types, one being an ellipsoid, one a hyperboloid of one sheet, and the third a hyperboloid of two sheets.

The values of λ which correspond to degenerate quadrics of the range are $-a^2$, $-b^2$, $-c^2$, ∞. The value ∞ gives Ω itself, while the other three values give the three *focal conics* f_1, f_2, f_3, one of which lies in each of the three principal planes of ψ. The point-equations of these focal conics are plainly

$$-\frac{Y^2}{a^2-b^2}-\frac{Z^2}{a^2-c^2} = 1, \qquad X = 0;$$

$$\frac{X^2}{a^2-b^2}-\frac{Z^2}{b^2-c^2} = 1, \qquad Y = 0;$$

$$\frac{X^2}{a^2-c^2}+\frac{Y^2}{b^2-c^2} = 1, \qquad Z = 0;$$

and the focal conics are respectively a virtual conic, a hyperbola, and an ellipse.

If l is a given line, there are just two confocals which touch l, and their tangent planes at the points of contact are the united planes of the involution of pairs of tangent planes drawn from l to the quadrics. Since these two planes are conjugate for Ω, they are perpendicular (cf. p. 167).

Now let P be a general point of space, and let π_1, π_2, π_3 be the tangent planes at P to the three confocals ψ_1, ψ_2, ψ_3 which pass through P. The line $\pi_2\pi_3$ is clearly a tangent line to ψ_2 and ψ_3, so that these are the two confocals which touch it. It now follows from what was said immediately above that π_2 and π_3 are perpendicular; and hence, by symmetry, π_1, π_2, π_3 form an orthogonal trihedral of planes. This gives us the important orthogonality property of confocal quadrics: *the three quadrics of a confocal system which pass through a point cut each other orthogonally*. The reader will easily verify that, further, the planes π_1, π_2, π_3 are the common principal planes of the range of enveloping cones drawn from P to the quadrics of the system.

If, finally, we take a general plane π, the poles of π for the various confocals all lie on the corresponding axis ϑ_π, and since ϑ_π contains the pole of π with respect to Ω it is normal to π. It is, of course, the normal, drawn at the point of contact, to the unique confocal which touches π. Conversely, if a line is normal to some confocal ψ_λ, it is an axis of the confocal system, namely the axis of the tangent plane to ψ_λ at its foot. Many geometrical properties of the confocal system can be obtained by considering the system of axes ϑ_π. Since this system of lines is a tetrahedral complex, further discussion of its properties is best deferred to Chapter XV, where other tetrahedral complexes will be dealt with at the same time.

Foci of a quadric

As we have already remarked, there is no definition of focus of a quadric as natural as that of focus of a conic; but it is nevertheless possible to generalize the familiar notion in various ways. Whatever method we use, the points that we are led to regard as foci of a quadric ψ are all the points of the three focal conics belonging to the confocal system determined by ψ. A quadric thus has an infinite number of foci. For a discussion of the properties of these points, the reader is referred to Sommerville, *Analytical Geometry of Three Dimensions* (Cambridge, 1934), chapter xii.

EXERCISES

(i) A *focus* of a quadric ψ may be defined as (*a*) any point which lies on one of the focal conics of ψ, or (*b*) a point such that the tangent cone from it to ψ is right circular (a cone of revolution), or (*c*) a point-sphere which has double contact with ψ. Prove that the three definitions are equivalent.

(ii) If F is a focus of a quadric ψ, show that there exist two planes π_1, π_2 and a constant k such that ψ is the locus of a point P for which

$$FP^2 = kPN_1 . PN_2,$$

where N_1, N_2 are the feet of the perpendiculars from P on to π_1 and π_2.

§ 4. LINEAR NETS OF QUADRIC LOCI

We now turn to the *net*, or doubly infinite linear system of quadric loci. If $S_1 = 0$, $S_2 = 0$, $S_3 = 0$ are the equations of three linearly independent quadric loci, the ∞^2 quadric loci with equations of the forms $\lambda S_1 + \mu S_2 + \nu S_3 = 0$ are said to form a linear net. Since the three equations $S_i = 0$ are all quadratic in the coordinates x, y, z, t they will in general have eight common solutions; and this means that the general net has eight base points $A_1,..., A_8$, common to every quadric of the system. When it is special, of course, a net may have a different kind of base. Thus the net of quadrics introduced on p. 296 has a twisted cubic curve as base; and there also exist nets with less than eight distinct base points. That nets do in fact exist with exactly eight base points is seen at once by considering the net determined by three general plane-pairs.

EXERCISE. Show that the following systems are nets:

(i) the set of all quadrics through a conic and two general points;
(ii) the set of all quadrics through a line and four general points.

Eight associated points

Consider a net whose base consists of eight distinct points $A_1,..., A_8$. Each of these points, taken by itself, imposes a linear condition on a quadric which is required to pass through it, but the eight linear conditions so obtained need not be independent. Indeed they cannot be; for quadrics subject to eight linearly independent linear conditions constitute a linear system with one degree of freedom only—i.e. a pencil, and not a net. Seven general points of space, on the other hand, do in general impose seven linearly independent conditions on quadrics required to contain them; and such quadrics therefore form a net whose base includes the seven given points. We thus have the theorem:

THEOREM 6. *If seven points of space impose linearly independent conditions on quadrics required to contain them then the set of all quadrics through them is a net. The quadrics of the net may all pass through an eighth point, which completes the base, or they may pass through a curve which, with those of the original seven points which do not lie on it, makes up the complete base of the net.*

In the general case any seven of the eight points $A_1,..., A_8$ determine the net of quadrics and, in consequence, the eighth point. The points are therefore symmetrically related, and we say that they form a set of *eight associated points*. We naturally wish to be able to decide whether eight given points of space do or do not form such a set, and the following theorem provides a convenient algebraic criterion.

THEOREM 7. *A necessary and sufficient condition for k points to impose less than k linearly independent conditions on quadrics required to contain them is that the squares of their plane-equations should be linearly dependent.*

Proof. Let the points have equations

$$P_i \equiv ux_i+vy_i+wz_i+pt_i = 0 \quad (i = 1,..., k).$$

The condition for the quadric

$$S \equiv (a,b,c,d,f,g,h,l,m,n \between x,y,z,t)^2 = 0$$

to contain P_i may be written $S_i = 0$; and the conditions imposed by the k points fail to be independent if and only if there exist constants $\lambda_1,..., \lambda_k$, not all zero, such that $\sum_{i=1}^{k} \lambda_i S_i \equiv 0$, identically in $a, b,..., n$. But this identity is equivalent to the ten equations $\sum_i \lambda_i x_i^2 = 0,..., \sum_i \lambda_i t_i z_i = 0$, and these are in turn equivalent to the single identity

$$\sum_{i=1}^{k} \lambda_i(ux_i+vy_i+wz_i+pt_i)^2 \equiv 0$$

in u, v, w, p. The theorem is thus completely proved.

COROLLARY 1. *If a set of eight points $P_1,..., P_8$, but no subset of seven of the eight, is such that the squares of the plane-equations of the points are linearly dependent, then $P_1,..., P_8$ are either associated or lie on a twisted cubic; and conversely.*

COROLLARY 2. *If the vertices of two tetrahedra form a set of eight associated points, there exists a quadric for which the two tetrahedra are self-polar* (*cf. Chapter* VIII, *Theorem* 8, *Corollary* 1).

THEOREM 8. *The twisted cubic through any six of a set of eight associated points has the line joining the remaining two as a chord.*

Proof. Let $A_1,..., A_8$ be associated, and let B be a general point of $A_1 A_2$. Then quadrics through $A_1,..., A_8$ form a net, and quadrics through $A_1,..., A_8$, B therefore form a pencil. But since all these latter quadrics have the line $A_1 A_2 B$ as a common generator, their residual intersection is a twisted cubic—the unique twisted cubic through $A_3,..., A_8$—and this cubic therefore has $A_1 A_2$ as a chord.

The mode of argument used here is very general, and it is worth giving a second illustration of its use. There is, in general, a unique quadric of the net through $A_1,..., A_8$ which contains both the lines $A_1 A_2$ and $A_3 A_4$; and the generators of the opposite system of this quadric which are drawn through $A_5,..., A_8$ meet the first two generators in related ranges. Hence:

THEOREM 9. *If eight points $A_1,..., A_8$ are associated, the transversals drawn to $A_1 A_2$ and $A_3 A_4$ (if these are skew) from A_5, A_6, A_7, A_8 are generators of a quadric. The four points A_5, A_6, A_7, A_8 subtend related pencils of planes with axes $A_1 A_2$ and $A_3 A_4$.*

General properties of a net of quadrics

We shall conclude this chapter by discussing briefly some of the general properties of nets of quadrics. We consider for simplicity a net (S) with eight distinct base points, given by

$$S \equiv \lambda S_1 + \mu S_2 + \nu S_3 = 0.$$

The necessary modifications for more special nets are often simple, however, and the reader will be able to supply them.

The parameters (λ, μ, ν) may be looked upon as homogeneous coordinates of S in the net, and by taking them as coordinates of a point in a plane π we can define a one–one mapping of the quadrics of (S) on the points of π.

Any two quadrics of (S) define a pencil of quadrics of the net, whose members are represented in π by the points of a line. Thus, for example, the quadrics of (S) which pass through a general fixed point P form the pencil which corresponds to the line

$$\lambda(S_1)_{PP} + \mu(S_2)_{PP} + \nu(S_3)_{PP} = 0$$

in π; and we see at once that there is a unique quadric of the net through two general points of space.

The net contains ∞^1 cones, namely all those quadrics S whose coordinates satisfy the quartic equation $|\lambda A_1+\mu A_2+\nu A_3| = 0$; and the cones of the net are accordingly represented in π by the points of a quartic curve D. The four points in which D is met by any line in π represent the four cones in the corresponding pencil of quadrics.

The locus of vertices of the ∞^1 cones is a curve, called the *Jacobian curve* J of the net. Its equations are obtained by eliminating λ, μ, ν between the four equations

$$\lambda\frac{\partial S_1}{\partial x}+\mu\frac{\partial S_2}{\partial x}+\nu\frac{\partial S_3}{\partial x} = 0,$$

$$\cdot \quad \cdot \quad \cdot \quad \cdot \quad \cdot \quad \cdot \quad \cdot$$

$$\cdot \quad \cdot \quad \cdot \quad \cdot \quad \cdot \quad \cdot \quad \cdot$$

$$\lambda\frac{\partial S_1}{\partial t}+\mu\frac{\partial S_2}{\partial t}+\nu\frac{\partial S_3}{\partial t} = 0.$$

The points of J are those points which lie on both the cubic surfaces

$$\frac{\partial(S_1, S_2, S_3)}{\partial(x, y, z)} = 0 \quad \text{and} \quad \frac{\partial(S_1, S_2, S_3)}{\partial(x, y, t)} = 0,$$

but not on all the three quadrics

$$\frac{\partial(S_i, S_j)}{\partial(x, y)} = 0.$$

Since the three quadrics have as their common intersection a twisted cubic, J is the sextic curve which, with this cubic, makes up the complete intersection of the two cubic surfaces. Hence:

THEOREM 10. *The Jacobian curve of a general net of quadrics—the locus of vertices of cones of the net—is a sextic curve J, the residual intersection of two cubic surfaces through a twisted cubic curve; and the points of J are in* $(1, 1)$ *correspondence with those of the plane quartic curve D in π which maps the cones of* (S).

The further development of the properties of the net (S) centres almost entirely, as might be expected, round the Jacobian curve J, and most of the important properties of the net can be expressed as properties of J. It is no part of our purpose here to describe these developments in detail. This has been done, and in a manner which has ever inspired the utmost admiration, by Reye in his

Geometrie der Lage (second edition, Leipzig, 1882: Part II, Lecture 27) and we shall content ourselves with referring to a few of his methods and results.

The key to Reye's development is the symmetrical correspondence, defined by the net (S), between points P, P^* which are conjugate for all quadrics of the net. To a general point P corresponds a unique point P^*, the common point of the polar planes of P with respect to S_1, S_2, S_3; but if P lies on J, so that it is a vertex of a cone of (S), the three polar planes in question meet in a line l. This line is said to be an exceptional line of the correspondence, and P^* can be any point of it. If we denote the correspondence by τ, we may say that τ is a $(1, 1)$ involutory self-transformation of space, with the points of J and their associated lines as exceptional elements.

It is easy to see that if P describes a general plane α, P^* describes a cubic surface ϕ, the locus of the point of intersection of corresponding planes of three collinear stars. Every such surface ϕ passes through J, for the plane from which it arises has a point in common with every exceptional line.

Again, if P describes a general line l, P^* describes a twisted cubic c, the locus of the point of intersection of corresponding planes of three homographic pencils; and this curve is evidently the residual base curve of the pencil of cubic surfaces ϕ through J which correspond to planes through l. If, however, l meets J in one, two, or three points, the curve corresponding to it is reduced to a conic, a line, or a single point. The point in the last case is a point of J, and the line is the corresponding exceptional line. The exceptional lines of the correspondence τ are thus identified as the trisecant lines of J. Three such lines, as may easily be shown, pass through a general point of J.

The quadrics of (S) meet a general plane α in a net (s) of conics. The Jacobian curve of (s), the locus of vertices of line-pairs, is plainly the locus of points in which α is touched by quadrics of (S), and it is therefore the cubic curve j in which α is met by its corresponding surface ϕ.

The essential properties of the correspondence may be summed up in the following general theorem:

THEOREM 11. *A net of quadrics* (S) *determines an involutory* $(1, 1)$ *correspondence* τ *of pairs of points* P, P^* *of space which are* *conjugate for every quadric of the set. In this correspondence, the points of J correspond exceptionally to lines, such lines being trisecant to J. To the planes of space correspond cubic surfaces through J, each of which meets its corresponding plane in the Jacobian curve of the net of conics in which the plane is met by (S). Finally, to a line of space corresponds either a twisted cubic (8-secant to J), or a conic (5-secant to J), or a line (a chord of J), or a point of J, according as the number of points in which the original line meets J is* 0, 1, 2, *or* 3.

In conclusion, we refer again to the representation of the quadrics of (S) by the points of a plane π. In this representation, as we have already noticed, the quadrics of any pencil (S') contained in the net (S) are represented by the points of a line in π, the four cones of (S') being represented by the points of intersection of the line with the quartic curve D. Clearly any line which touches D represents a pencil (S') for which two of the four cones coincide, i.e. in general a simple-contact pencil of quadrics of (S). Such a pencil is formed by all the quadrics of (S) which pass through a given point of J.

A double tangent of D, by the same argument, must represent a pencil (S') in which the four cones coincide in pairs, i.e. a double-contact pencil, whose base is either a pair of conics which meet in two points or a twisted cubic and one of its chords. In general, a net (S) does not contain any plane-pairs; and when this is the case it cannot contain double-contact pencils of the first type; but, by Theorem 8 above, every line which joins two of the base points $A_1,\ldots, A_8$ forms with the twisted cubic through the remaining six the base curve of a pencil (S') of the second type. In this way, then, we arrive at the following result:

THEOREM 12. *The quartic curve D has twenty-eight bitangents, which represent double-contact pencils in (S), each of which has as base curve the line joining two of the points $A_1,\ldots, A_8$ together with the twisted cubic through the remaining six of these points.*

EXERCISES ON CHAPTER XIII

1. Show that the equation

$$x^2+y^2+z^2-t^2+2\lambda(yz+zx+xy) = 0$$

represents a pencil of quadrics of type 2 (p. 332). Find the plane-pair, the two proper cones, and the two base conics of the system.

Find also the ∞^1 common self-polar tetrahedra.

2. Give a dual interpretation of the results obtained in Exercise 1 for the range of quadrics whose equation is

$$u^2+v^2+w^2-p^2+2\lambda(vw+wu+uv) = 0.$$

3. Show that the equation $x^2-y^2-z^2+2\lambda(xt-yz) = 0$ represents a pencil of quadrics with one fixed point of contact. Find the three cones of the pencil, and prove that the base curve admits of the parametric representation

$$x:y:z:t = (1+\theta^2)^2 : 1-\theta^4 : 2\theta(1+\theta^2) : 2\theta(1-\theta^2).$$

Verify that this curve passes twice through the reference point T, in the directions of the lines $x = 0 = y \pm iz$.

4. Find the faces of the common self-polar tetrahedron of the two quadric envelopes whose equations are

$$u^2+2vw = 0, \qquad u^2+2v^2+w^2-p^2 = 0.$$

Find also the four disk quadrics of the range defined by these two quadric envelopes.

5. If two proper quadrics touch along a common generator, show that they intersect residually in two lines (either skew or coincident) which meet the generator of contact.

If two quadric cones touch along a common generator, show that every quadric of the pencil defined by them is a cone. Show also that (i) if the vertices of the two given cones are distinct the residual intersection of the cones is, in general, a conic which meets the generator of contact, and every point of this generator is the vertex of a cone of the pencil, and (ii) if the vertices are coincident, all the cones of the pencil have the same vertex.

Discuss the pencils whose equations are

(i) $xt-yz+\lambda zt = 0$;

(ii) $y^2-zx+\lambda zt = 0$.

6. A variable plane π touches two fixed conics s_1, s_2, which lie in different planes. Show that, in general, it touches two other conics s_3, s_4, and that its four points of contact with s_1, s_2, s_3, s_4 are in line; but that if s_1 and s_2 meet in two points then π passes through one or other of two fixed points, lying in planes which are harmonically conjugate with respect to the planes of s_1 and s_2.

If s_1 and s_2 are the circles given by

$$X = 0 = Y^2+Z^2-a^2, \quad \text{and} \quad Y = 0 = Z^2+X^2-b^2,$$

find s_3 and s_4. Discuss the special case in which $a = b$.

7. There are four conics which are touched by all the common tangent planes of the two quadrics

$$X^2/a^2+Y^2/b^2 = 2Z, \qquad X^2+Y^2-Z^2 = 1.$$

Find their point-equations.

8. Two lines l, m are given, and also a quadric ψ for which l, m are polar lines. Variable planes through l and m respectively meet ψ in conics s and t. Show that the vertices of the cones through s and t lie on a fixed quadric ϕ which meets ψ in four lines; and show also that the relation between ϕ and ψ is symmetrical.

9. If a general quadric ψ is met by another quadric ψ' in a proper quartic curve C^4, show that C^4 is a (2, 2) curve on ψ.

A correspondence between variable generators g, g' of the two reguli on ψ is set up by making two generators correspond when they meet in a point of C^4. Show that the correspondence is (2, 2) and algebraic, and also that every (2, 2) algebraic correspondence between the two reguli arises in this way from a quartic curve cut on ψ by another quadric.

By projecting C^4 on to an arbitrary plane π from the vertex of one of the cones of the pencil determined by ψ and ψ', show that the cross ratio of the four critical generators g is equal to the cross ratio of the four critical generators g', when the two tetrads are ordered suitably.

10. If c is a plane cubic curve and A, B are two fixed points on it, show that the correspondence between rays through A and B respectively which meet on c is (2, 2) and algebraic. Deduce the theorem that four tangents can be drawn to a plane cubic from a general point P of the curve and their cross ratio is independent of the choice of P. [*Hint.* Use the result proved in Exercise 9.]

11. The quadrics of a given net have in common a set of eight associated points. Show that the centres of the quadrics of the net lie on a cubic surface through the mid-points of the twenty-eight joins of the eight points.

What special quadrics of the net have their centres at the other points in which the cubic surface meets the twenty-eight joins?

12. If $\Omega = 0$ is the plane-equation of the absolute conic, and $P = 0$, $Q = 0$ are those of two points, show that the equation $\Omega + \lambda PQ = 0$ represents a quadric of revolution with P, Q as principal foci, and $\Omega + \lambda P^2 = 0$ represents a sphere with P as centre.

The coordinate system being rectangular cartesian, find the general plane-equation or equations of

(i) a quadric of revolution whose principal foci are the points $(\pm a, 0, 0)$;

(ii) a quadric which passes through the conic whose plane-equation is

$$a^2u^2 + c^2w^2 = p^2;$$

(iii) a quadric of which this conic is a focal conic;

(iv) a circle of unit radius in the plane $X + Y + Z = 0$, with its centre at the origin;

(v) the asymptotic cone of the quadric $vw + wu + uv = p^2$;

(vi) a quadric inscribed in the cone whose point-equation is

$$YZ + ZX + XY = 0.$$

13. If the focal hyperbola of a confocal system of quadrics has equations

$$2XY + Z^2 + 2Z + 2 = 0 = X + Y,$$

show that the focal ellipse of the same system has equations

$$3X^2 + 2XY + 3Y^2 - 4 = 0 = Z + 1.$$

14. Show that a central quadric has six right circular enveloping cylinders, and that if the quadric is real at most two of the cylinders are real.

Find the equations of the two real right circular cylinders which envelop the quadric whose equation is

$$X^2/6+Y^2/2+Z^2/1 = 1,$$

and verify that they are of radius $\sqrt{2}$.

15. Show that the points of contact of a system of parallel tangent planes to the quadrics of a confocal system lie on a rectangular hyperbola.

16. If l is the locus of poles of a plane π for the quadrics of a confocal system, prove that the normals to quadrics of the system which lie in π envelop a parabola, and that they are the polar lines of l for quadrics of the system.

Deduce that the points of contact of tangent planes from l to the confocals lie on a nodal cubic curve in π (namely the pedal curve of the parabola from the point of intersection of l with π).

17. Show that the equation in rectangular cartesian coordinates of any quadric with the point (X_0, Y_0, Z_0) as a focus is of the form

$$(X-X_0)^2+(Y-Y_0)^2+(Z-Z_0)^2 = LM,$$

where $L = 0$ and $M = 0$ are the equations of two planes.

Discuss the position of the *directrix line* $L = 0 = M$ when the quadric is the ellipsoid $X^2/a^2+Y^2/b^2+Z^2/c^2 = 1$ and the focus is a point of the focal ellipse.

18. Show that the polar planes of a general point P for the quadrics of a confocal system with centre O generate a cubic developable (dual of a twisted cubic curve) and that the three planes of this developable through any point of OP are mutually perpendicular.

19. Show that the planes of parabolic section of a central quadric ψ which touch a quadric confocal with ψ all touch a sphere concentric with ψ.

20. The coordinates being rectangular cartesian, find the plane-equation of the pair of absolute points in the plane $aX+bY+cZ+d = 0$.

If this plane-equation is $\Xi = 0$ and $\Sigma = 0$ is the plane-equation of an ellipsoid ψ, and if a source of light is placed so that ψ throws a circular shadow on the given plane, show that every ellipsoid of the system

$$\Sigma+\lambda\Xi = 0$$

also throws a circular shadow on this plane.

CHAPTER XIV

LINEAR TRANSFORMATIONS OF SPACE

§ 1. SPACE COLLINEATIONS

IF S_3 and S'_3 are two projective spaces, in which allowable representations $\mathscr{R}$ and $\mathscr{R}'$ have been chosen, an equation $\mathbf{x}' = \mathbf{Ax}$ defines a linear transformation of the points of S_3 into the points of S'_3, and such a transformation is called a *space collineation*. If the matrix $\mathbf{A}$ is singular, the points of S_3 transform into points of a certain proper submanifold of S'_3, which may be a plane, a line, or a single point, according to the rank of $\mathbf{A}$. This case is special and, as usual, we shall use the term 'collineation', unless otherwise stated, to mean 'non-singular collineation'. We shall also confine ourselves to self-collineations of a single space S_3, supposing always that S'_3 coincides with S_3 and that the same representation is used as both $\mathscr{R}$ and $\mathscr{R}'$. It will not be necessary to discuss the general properties of space collineations in great detail, since many of the results follow directly from the linearity of the transformation and may be proved by arguments similar to those already used in Chapter IX.

THEOREM 1. *Every space collineation ϖ is a self-dual transformation, which transforms points into points, lines into lines, and planes into planes. It is represented by equations*

$$\mathbf{x}' = \mathbf{Ax} \quad \textit{and} \quad \mathbf{u}' = \mathbf{A}^{-1T}\mathbf{u} \quad (|\mathbf{A}| \neq 0),$$

each of which determines the other.

THEOREM 2. *The set of all self-collineations of S_3 is a group, isomorphic with $PGL(3)$.*

THEOREM 3. *There is a unique collineation which transforms five given points, no four of which are coplanar, into five given points, no four of which are coplanar; and dually, there is a unique collineation which transforms five given planes, no four of which are concurrent, into five given planes, no four of which are concurrent.*

As in the case of plane collineations, we can show that a given collineation transforms any range of points or pencil of planes into a homographically related range or pencil, and that it transforms any projectively generated manifold—for example a quadric or a twisted cubic—into another manifold of the same kind.

EXERCISE. If ϖ transforms a plane π into a plane π', show that it sets up a plane collineation between π and π'.

The united points of the collineation ϖ whose equation is $\mathbf{x}' = \mathbf{A}\mathbf{x}$ are those points whose coordinate vectors are characteristic vectors of the matrix $\mathbf{A}$. In general there are four isolated united points, but in particular cases there may be a line or even a plane of united points. The nature of the united points of ϖ depends upon the algebraic properties of $\mathbf{A}$. The various possibilities may be classified by elementary argument, using the rank-multiplicity relation $\rho[\mathbf{A}-\lambda_i \mathbf{I}] \geqslant 4-\mu(\lambda_i)$ as in Chapter IX; or alternatively the classification may be derived from the general theory of elementary divisors.† There are considerably more possibilities than in the case of plane collineations, and it would be tedious to go into all the details of the classification. We shall accordingly select a few particularly important cases which are of geometrical interest. The other types of collineation may all be regarded as specializations of those which we discuss.

I. The general collineation

In the general case the characteristic equation $|\mathbf{A}-\lambda\mathbf{I}| = 0$ has four distinct roots $\lambda_0, \lambda_1, \lambda_2, \lambda_3$. Each of these roots, being simple, gives rise to an isolated united point; and since the four points cannot be coplanar‡ we can take them as vertices of the tetrahedron of reference. The equations of the collineation then assume the canonical form

$$\begin{aligned} x_0' &= \lambda_0 x_0, \\ x_1' &= \qquad \lambda_1 x_1, \\ x_2' &= \qquad\qquad \lambda_2 x_2, \\ x_3' &= \qquad\qquad\qquad \lambda_3 x_3. \end{aligned}$$

EXERCISE. Show that the ratios $\lambda_0 : \lambda_1 : \lambda_2 : \lambda_3$ may be interpreted as moduli of certain homographies, and also that the point which corresponds to any assigned point P may be constructed by drawing suitable planes.

II. The collineation with a line of united points and two isolated united points

If the characteristic roots of $\mathbf{A}$ are $\lambda_0, \lambda_0, \lambda_2, \lambda_3$, with

$$\rho[\mathbf{A}-\lambda_0 \mathbf{I}] = 2,$$

the collineation has isolated united points corresponding to λ_2

† See Todd, *Projective and Analytical Geometry*, chapter v.

‡ Cf. p. 211. See also Exercise 16 at the end of this chapter.

and λ_3 and a line of united points corresponding to λ_0. The line joining the isolated points is necessarily skew to the line of united points [*Exercise.* Verify this] and we may take as X_0, X_1, X_2, X_3 two points of the line of united points and the two isolated united points. The equations of ϖ then become

$$\frac{x_0'}{\lambda_0 x_0} = \frac{x_1'}{\lambda_0 x_1} = \frac{x_2'}{\lambda_2 x_2} = \frac{x_3'}{\lambda_3 x_3}.$$

The same collineation may also be represented by

$$\frac{x_0'}{x_0} = \frac{x_1'}{x_1} = \frac{x_2'}{ax_2} = \frac{x_3'}{bx_3},$$

where $a = \lambda_2/\lambda_0$ and $b = \lambda_3/\lambda_0$, and consequently $a \neq b$.

An example of this general type of collineation is provided, in euclidean space, by a rotation, of angle $\alpha \neq n\pi$, about a fixed axis l. The axis l is a line of united points, and two isolated united points are the points in which the absolute conic Ω is met by a plane orthogonal to l. This follows from the fact that the collineation induced in any such plane is a rotation about a fixed point. This space collineation is not projectively the most general one of type II, however, since the characteristic roots λ_2, λ_0, λ_3 are in the ratio $e^{i\alpha}:1:e^{-i\alpha}$

III. The biaxial collineation

If the characteristic roots of $\mathbf{A}$ are λ_0, λ_0, λ_2, λ_2, with

$$\rho[\mathbf{A}-\lambda_0\mathbf{I}] = \rho[\mathbf{A}-\lambda_2\mathbf{I}] = 2,$$

the collineation has two lines of united points, and these lines are necessarily skew. If we take them as X_0X_1 and X_2X_3, the equations of ϖ may be written as

$$\frac{x_0'}{\lambda_0 x_0} = \frac{x_1'}{\lambda_0 x_1} = \frac{x_2'}{\lambda_2 x_2} = \frac{x_3'}{\lambda_2 x_3},$$

or as

$$\frac{x_0'}{x_0} = \frac{x_1'}{x_1} = \frac{x_2'}{kx_2} = \frac{x_3'}{kx_3},$$

where $k = \lambda_2/\lambda_0 \neq 1$.

This type of collineation is known as the *biaxial collineation*, or sometimes the biaxial homography or skew perspective. It is characterized completely by the two lines of united points—m and n, say—called its *axes*, and the number k, called its *modulus*. The point P' corresponding to any given point P may be found by drawing the unique transversal from P to m and n, meeting them in M and N, and then taking as P' the unique point of MN for which $\{M, N; P', P\} = k$. For if

$$P \equiv (y_0, y_1, y_2, y_3) \quad \text{then} \quad P' \equiv (y_0, y_1, ky_2, ky_3),$$

and hence

$$M \equiv (y_0, y_1, 0, 0) \quad \text{and} \quad N \equiv (0, 0, y_2, y_3).$$

Thus $$\{M, N; P', P\} = \{0, \infty; k, 1\} = k.$$

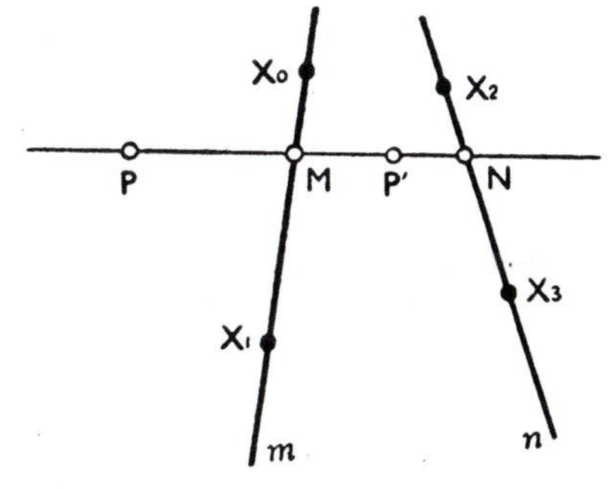

It should be noted that the modulus k is associated with the *ordered* pair of axes, and that if the axes are taken in the reverse order the modulus is changed to $1/k$.

In the particular case when $k = -1$ the biaxial collineation is said to be *harmonic*. It is then an involutory transformation, and is sometimes referred to as a *skew involution*.

A simple example of the biaxial collineation in euclidean space may be obtained as follows. Take an arbitrary line as m, and let it meet the plane at infinity in a point T. Now let n be the polar line of T with respect to Ω, and take an arbitrary real number as the modulus k. Then the corresponding pairs (P, P') are such that PP' is perpendicular to m and $MP' = kMP$. This particular transformation may be described as a radial expansion about the axis m; and for the special value $k = -1$ it becomes reflection in m.

IV. The space homology

If the characteristic roots of $\mathbf{A}$ are $\lambda_0, \lambda_0, \lambda_0, \lambda_3$, with

$$\rho[\mathbf{A} - \lambda_0 \mathbf{I}] = 1,$$

the collineation has a plane of united points and an isolated united point. If we take the isolated united point as X_3 and three points in the plane of united points as X_0, X_1, X_2, the equations of the collineation may be written in either of the forms

$$\frac{x_0'}{\lambda_0 x_0} = \frac{x_1'}{\lambda_0 x_1} = \frac{x_2'}{\lambda_0 x_2} = \frac{x_3'}{\lambda_3 x_3},$$

or $$\frac{x_0'}{x_0} = \frac{x_1'}{x_1} = \frac{x_2'}{x_2} = \frac{x_3'}{kx_3},$$

where $$k = \lambda_3/\lambda_0 \neq 1.$$

This type of space collineation is known as a *space homology* or space perspective. It is characterized completely by its vertex A, its axial plane α, and its modulus k. If P, P' are any two corresponding points, then PP' passes through A; and if PP' meets α in M, $\{M, A; P', P\} = k$. If $k = -1$, ϖ is said to be a *harmonic homology*, and it is an involutory transformation.

In euclidean space, reflection in a given plane is a harmonic homology with the point at infinity in the direction of the normal as vertex, radial expansion from a point is a homology with the plane at infinity as axial plane, and reflection in a point is a harmonic homology with the plane at infinity as axial plane.

V. The special space homology

If the characteristic roots of $\mathbf{A}$ are λ_0, λ_0, λ_0, λ_0, with

$$\rho[\mathbf{A}-\lambda_0 \mathbf{I}] = 1,$$

the collineation has a plane of united points and no united point outside this plane. It is a special space homology and, like the special plane homology, it may be thought of as a homology in which the vertex falls accidentally on the axial plane.

If X_0, X_1, X_2 are taken in the axial plane, the equations of the special homology take the form

$$\begin{aligned} x_0' &= x_0 \qquad\qquad\quad +ax_3, \\ x_1' &= \qquad\quad x_1 \qquad +bx_3, \\ x_2' &= \qquad\qquad\quad x_2+cx_3, \\ x_3' &= \qquad\qquad\qquad\quad x_3, \end{aligned}$$

and from these equations it is obvious that all joins of pairs of corresponding points pass through the vertex $(a, b, c, 0)$.

A typical example of such a transformation, in euclidean space, is the translation.

We conclude this discussion of space collineations by proving two theorems on involutory collineations which give further insight into the possible transformations of this kind.

THEOREM 4. *If a collineation has three involutory pairs of distinct corresponding points, lying on distinct lines that are not all in one plane, then it is an involutory collineation; and every point of space (if not a united point) then belongs to an involutory pair.*

Proof. If (P, P') is an involutory pair of a collineation ϖ, PP' is a self-corresponding line; and since the homography ϖ_0 induced on this line by ϖ has an involutory pair, it is an involution. Then every point of PP' is a united point for ϖ_0^2, and therefore for ϖ^2.

The collineation ϖ^2 now has three lines of united points, which do not all lie in one plane. But a line of united points that does not lie in a plane of united points corresponds to a double characteristic root of the matrix of ϖ^2, while a plane of united points corresponds to a triple root. It follows that the characteristic equation of the matrix is an identity, and ϖ^2 is accordingly the identical collineation ϵ, i.e. ϖ is an involutory transformation.

THEOREM 5. *The only involutory space collineations are the harmonic biaxial collineation and the harmonic space homology.*

Proof.† Let ϖ be a space collineation such that $\varpi^2 = \epsilon$. If P_1 is any point which is not a united point of ϖ, ϖ induces an involution on the line joining P_1 to ϖP_1, and the united points M_0, M_1 of this involution are united points of ϖ. If P_2 is now any point which does not lie on $M_0 M_1$ and which is not a united point of ϖ, there are two united points of ϖ on the line joining P_2 to ϖP_2. If the line is skew to $M_0 M_1$, we call these united points M_2 and M_3. If it is not skew to $M_0 M_1$ it contains at least one united point M_2 which does not lie on $M_0 M_1$; and in this case we determine a further united point M_3, not in the plane $M_0 M_1 M_2$, by applying the previous argument to a point P_3 which is outside this plane.

In any case, therefore, we are able to choose four united points M_0, M_1, M_2, M_3, which are the vertices of a proper tetrahedron; and if this tetrahedron is taken as tetrahedron of reference the matrix of ϖ assumes the canonical form

$$\mathbf{A} = \begin{pmatrix} \lambda_0 & 0 & 0 & 0 \\ 0 & \lambda_1 & 0 & 0 \\ 0 & 0 & \lambda_2 & 0 \\ 0 & 0 & 0 & \lambda_3 \end{pmatrix}.$$

Since $\mathbf{A}^2 = \lambda \mathbf{I}$, for some λ, we have $\lambda_0^2 = \lambda_1^2 = \lambda_2^2 = \lambda_3^2$. There are therefore only three essentially different forms which $\mathbf{A}$ can assume:

$$\mu \begin{pmatrix} 1 & 0 & 0 & 0 \\ 0 & 1 & 0 & 0 \\ 0 & 0 & -1 & 0 \\ 0 & 0 & 0 & -1 \end{pmatrix}, \quad \mu \begin{pmatrix} 1 & 0 & 0 & 0 \\ 0 & 1 & 0 & 0 \\ 0 & 0 & 1 & 0 \\ 0 & 0 & 0 & -1 \end{pmatrix},$$

† See also Exercise 15 at the end of the chapter.

and μI. These forms correspond respectively to the harmonic biaxial collineation, the harmonic space homology, and the identical collineation, which is, of course, inadmissible.

§ 2. Collineations which leave a Quadric Invariant

The ∞^{15} non-singular self-collineations of S_3 make up a group isomorphic with the projective group $PGL(3)$, and this group has many subgroups which it is worth while to single out for closer study. We might select, for instance, a subgroup which is of special significance in connexion with one of the geometries of the projective hierarchy or one which is specially related to some particular geometrical configuration. The most obvious cases which present themselves are the groups of collineations which leave invariant (*a*) a plane, (*b*) a conic, (*c*) a quadric, or (*d*) a twisted cubic. The first two of these give, as we have already seen, the affine and euclidean groups, and we shall show in the present section that the third gives the group of congruence transformations of a non-euclidean geometry. The fourth has no such application, but is of some intrinsic interest. We shall only discuss (*c*) here in any detail, but a reference to (*d*) will be found in the exercises at the end of the chapter.

Let us consider, then, a proper quadric ω, whose equation may be taken in the form

$$x_0 x_3 - x_1 x_2 = 0, \tag{1}$$

and let us denote by $G(\omega)$ the group of collineations ϖ of S_3 which transform ω into itself. Since a collineation cannot transform intersecting lines into skew lines or vice versa, any collineation ϖ of $G(\omega)$ either transforms each regulus on ω into itself or else transforms the two reguli into each other. In this way we arrive at two distinct kinds of transformation, which we call direct transformations and opposite transformations respectively. The direct transformations form a subgroup $D(\omega)$ of index 2 in $G(\omega)$. The opposite transformations do not form a group, but constitute the second coset of $D(\omega)$ in $G(\omega)$. If, in fact, η_0 is any fixed opposite transformation, and δ runs through $D(\omega)$, then $\eta_0 \delta$ runs through the full set of opposite transformations. From the point of view of the theory of groups, the distinction between direct and opposite transformations of ω into itself is strictly analogous to that between even and odd permutations.

The group $D(\omega)$ of direct transformations

Consider now any direct transformation δ. This transformation permutes among themselves (homographically, of course) the u-generators of ω and also the v-generators. It follows that if P is the general point $(\theta\phi, \theta, \phi, 1)$ of ω its transform P' has coordinates $(\theta'\phi', \theta', \phi', 1)$, where

$$\theta' = \frac{\alpha_1\theta+\beta_1}{\gamma_1\theta+\delta_1}, \qquad \phi' = \frac{\alpha_2\phi+\beta_2}{\gamma_2\phi+\delta_2}, \tag{2}$$

$\alpha_1,\ldots, \delta_2$ being constants such that

$$\alpha_1\delta_1-\beta_1\gamma_1 \neq 0 \quad \text{and} \quad \alpha_2\delta_2-\beta_2\gamma_2 \neq 0.$$

The argument already used in the proof of Theorem 8 of Chapter IX may now be used in order to show that δ is represented by the matrix equation

$$\begin{pmatrix} x_0' & x_1' \\ x_2' & x_3' \end{pmatrix} = \begin{pmatrix} \alpha_1 & \beta_1 \\ \gamma_1 & \delta_1 \end{pmatrix}\begin{pmatrix} x_0 & x_1 \\ x_2 & x_3 \end{pmatrix}\begin{pmatrix} \alpha_2 & \gamma_2 \\ \beta_2 & \delta_2 \end{pmatrix}. \tag{3}$$

The collineation δ is in fact determined uniquely by the homographies (2). For if δ_1 and δ_2 both induced the same homographies in the two reguli, $\delta_1^{-1}\delta_2$ would leave every point of ω invariant, and the only collineation which does this is the identical collineation ϵ.

Equation (3) may conveniently be written as

$$\mathbf{X}' = \mathbf{A}_1\mathbf{X}\mathbf{A}_2^T,$$

where $\mathbf{A}_1$ and $\mathbf{A}_2$ are the matrices of the two induced homographies.

If $\mathbf{A}_1 = \mathbf{I}$ we have a collineation $\mathbf{X}' = \mathbf{X}\mathbf{A}_2^T$, which permutes only the generators given by ϕ = constant, the v-generators say. Such a collineation will be called a *right translation*. In the same way we have also the left translations $\mathbf{X}' = \mathbf{A}_1\mathbf{X}$, which permute the u-generators while leaving the v-generators invariant. Every direct transformation δ can now be resolved uniquely into a product of a right translation $\mathbf{X} \to \mathbf{X}\mathbf{A}_2^T$ and a left translation $\mathbf{X} \to \mathbf{A}_1\mathbf{X}$; and we may sum up the results that have so far been obtained in the following comprehensive theorem.

THEOREM 6. *The space collineations which leave a proper quadric ω invariant form an ∞^6 group $G(\omega)$. This consists of ∞^6 direct transformations δ, forming a subgroup $D(\omega)$ of index 2, and ∞^6 opposite transformations η. If η_0 is any fixed opposite transformation, the set of all the opposite transformations is the coset $\eta_0 D(\omega)$ of $D(\omega)$ in $G(\omega)$.*

The right translations, which permute the v-generators among themselves while leaving the u-generators invariant, form an ∞^3 *subgroup* $D_r(\omega)$ *of* $D(\omega)$*; and the left translations form an* ∞^3 *subgroup* $D_l(\omega)$. *The only common member of* $D_l(\omega)$ *and* $D_r(\omega)$ *is* ϵ, *and* $D(\omega)$ *is the direct product of these two groups:* $D(\omega) = D_l(\omega)\times D_r(\omega)$.

The simplest kind of opposite transformation is a harmonic homology whose vertex and axial plane are pole and polar for ω. Such a transformation is called a *reflection* relative to ω, and a simple example of a such a reflection is furnished by the equation $\mathbf{X}' = \mathbf{X}^T$.

Right and left translations

Consider now the general right translation ρ, given by $\mathbf{X}' = \mathbf{X}\mathbf{A}^T$. This transforms every point (θ, ϕ) into a point (θ, ϕ') of the same u-generator, and for the two values of ϕ given by

$$\phi = \frac{\alpha\phi+\beta}{\gamma\phi+\delta}$$

we have $\phi' = \phi$. We see therefore that there are two v-generators, in general distinct, which are lines of united points for ρ; i.e. ρ is a biaxial collineation. If we take the two special generators to be those corresponding to $\phi = 0$ and $\phi = \infty$, the equation of the homography induced among the v-generators assumes the canonical form $\phi' = k\phi$ and the equation of ρ then reduces to

$$\begin{pmatrix} x_0' & x_1' \\ x_2' & x_3' \end{pmatrix} = \begin{pmatrix} x_0 & x_1 \\ x_2 & x_3 \end{pmatrix}\begin{pmatrix} k & 0 \\ 0 & 1 \end{pmatrix},$$

i.e.

$$\frac{x_0'}{kx_0} = \frac{x_1'}{x_1} = \frac{x_2'}{kx_2} = \frac{x_3'}{x_3}.$$

We recognize these equations as the equations of a biaxial collineation, and we see further that the modulus k is the ratio of the characteristic roots of the matrix $\mathbf{A}$. So we have the theorem:

THEOREM 7. *A right translation is in general a biaxial collineation whose axes are v-generators, and a left translation is in general a biaxial collineation whose axes are u-generators.*

Non-euclidean geometry

We now propose to indicate as briefly as possible something of the significance of the above results in the three-dimensional non-euclidean geometry obtained by taking ω as absolute quadric in real projective space. This kind of geometry, developed largely by Minkowski and Clifford, is in part a straightforward generalization of plane non-euclidean geometry; but it has important features which have no two-dimensional analogues, nor indeed any close analogues in space of any dimensionality.

We begin by defining some of the basic concepts.

(i) *Distance*. We say that the distance between two points P, Q is equal to the distance between P', Q' if $\{P, Q; M, N\}$ is equal to $\{P', Q'; M', N'\}$ or $\{P', Q'; N', M'\}$, where M, N and M', N' are the pairs of points in which ω is met by PQ and $P'Q'$.

(ii) *Angle*. Equality of angle between two pairs of planes is defined in the manner dual to (i).

(iii) *Orthogonality*. A line p is orthogonal to a plane π if it passes through the pole of π with respect to ω. Also two lines p, q are orthogonal if they are conjugate for ω (each meeting the polar line of the other).

From this it follows that two lines p_1, p_2 have in general two common perpendicular transversals, namely the lines which meet p_1, p_2 and their polar lines p_1', p_2'.

(iv) *Parallelism (in the sense of Clifford)*. Two lines are right parallel if they meet the same pair of v-generators of ω, and they are left parallel if they meet the same pair of u-generators.

A system of right parallels (or left parallels) is the system of lines meeting two fixed generators v_1, v_2 (or u_1, u_2).

Through any general point there can be drawn a unique line right parallel to a given line, and a unique line left parallel to it.

The characteristic properties of Clifford parallels (stated for definiteness in terms of right parallels) are given in the following theorem.

THEOREM 8. *Two right parallel lines p, q have an infinity of common perpendicular transversals, which form a regulus of left parallels. The distances between p and q along all these transversals are equal.*

Proof. Let p, q meet the same two v-generators v_1, v_2, and let the pairs of u-generators met by p, q be u_1, u_2 and u_1^*, u_2^* respectively. Also let P_{ij} and Q_{ij} denote the points of intersection of v_i with u_j and u_j^* respectively. Then p, q and their polar lines p', q' are the lines $P_{11}P_{22}$, $Q_{11}Q_{22}$ and $P_{12}P_{21}$, $Q_{12}Q_{21}$. Since the four u-generators cut related ranges on v_1 and v_2,

$$(P_{11}, Q_{11}, P_{12}, Q_{12}) \barwedge (P_{21}, Q_{21}, P_{22}, Q_{22}) \barwedge (P_{22}, Q_{22}, P_{21}, Q_{21}).$$

Thus p, q, p', q' cut related ranges on v_1, v_2; and therefore the four lines are generators of one system on a quadric ψ, v_1, v_2 being generators of the opposite system. Since the generators v_1 and v_2 are common to ω and ψ, the complete intersection of these quadrics consists of v_1, v_2 and two u-generators $u^{(1)}$, $u^{(2)}$ of ω. On ψ, therefore, the generators of the regulus which contains v_1 and v_2 all meet the

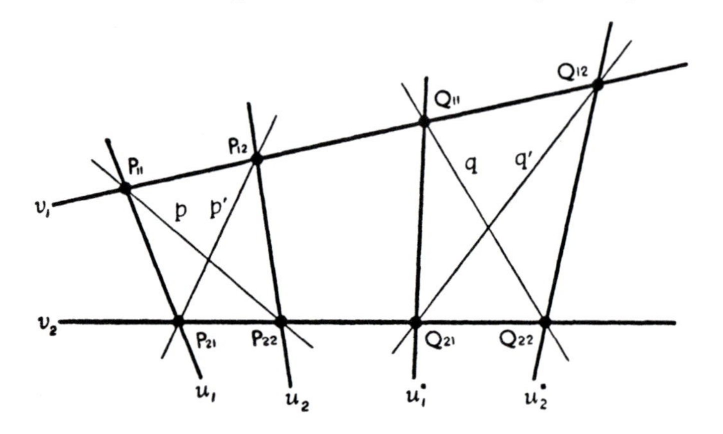

six generators p, q, p', q', $u^{(1)}$, $u^{(2)}$. In other words, this regulus consists of ∞^1 left parallel common perpendicular transversals of p and q; and since each one of them cuts a range of the same cross ratio on p, q, $u^{(1)}$, $u^{(2)}$, the same distance is intercepted on them all by p and q. This completes the proof of the theorem.

A quadric surface, such as ψ, which meets the absolute quadric ω in a skew quadrilateral, is called a *Clifford surface*.

Congruence transformations

The non-euclidean geometry defined by ω as absolute quadric may now be connected with the group $G(\omega)$ of self-collineations of space which leave ω invariant. Every transformation of the group preserves all properties of figures that are expressible in terms of non-euclidean distance or angle or the relation of non-euclidean orthogonality; and it either preserves the two kinds of parallelism or else interchanges them, according as it is direct or opposite. In view of these facts, we naturally call the transformations of $G(\omega)$ the *congruence transformations* of the non-euclidean geometry. The group $G(\omega)$ is related to this geometry in much the same way as the orthogonal group is related to euclidean geometry.

Consider now the meaning, in terms of non-euclidean geometry, of the two special types of direct transformation that we have called right and left translations. A right translation ρ is a biaxial congruence transformation with two generators v_1, v_2 of ω as axes, and its united lines (all those lines which meet v_1 and v_2) therefore form a complete system of right parallels. We may say shortly, then, that the join PP' of every pair of corresponding points is in a fixed right parallel direction. By assigning an order to the pair of generators (v_1, v_2) we can fix a sense along all the right parallels which meet them; and the displacement from P to P' in this sense is determined by the cross ratio $\{P, P'; M_1, M_2\}$, where M_1, M_2 are the points in which PP' is met by v_1, v_2. Since this cross ratio is constant for all corresponding pairs (P, P') we have the theorem:

THEOREM 9. *In any right translation, each point P of space is moved in a fixed right parallel direction, in the same sense, through a fixed non-euclidean distance.*

There is, of course, a corresponding property of left translations.

Since the most general direct congruence transformation δ is compounded of a left translation and a right translation, it has just four united points, namely the points of intersection of the pairs of generators associated with the translations. If the translations are chosen suitably, however, δ may be made to have any general line a as a line of united points, and in this case it is called a non-euclidean *rotation* about a. Or again, δ may be made to have both a and its polar line a' as lines of united points, and in this case it is called a *double rotation*.

For a detailed analysis of direct and opposite congruence transformations we refer the reader to the excellent exposition given by Coxeter in his *Non-euclidean Geometry* (Toronto, 1942).

EXERCISES

1. Show that, with the notation used on p. 355, the direct congruence transformation $\mathbf{X}' = \mathbf{A}_1\mathbf{X}\mathbf{A}_2^T$ is a rotation if and only if either (i) $\mathbf{A}_1$ and $\mathbf{A}_2$ both have coincident characteristic roots, or (ii) $\mathbf{A}_1$ and $\mathbf{A}_2$ both have distinct characteristic roots and the ratio between the roots is the same for both matrices; i.e. if and only if $\mathbf{A}_1$ is similar to a scalar multiple of $\mathbf{A}_2$.

2. Prove that the equation of any non-euclidean reflection may be written $\mathbf{X}' = \mathbf{A}\mathbf{X}^T\mathbf{A}^{-1T}$, and find the centre and axial plane of this reflection.

3. Prove that any non-euclidean rotation can be expressed, in infinitely many ways, as a product of two non-euclidean reflections.

§ 3. SPACE CORRELATIONS

In space, as in the plane, we often have occasion to consider dualizing linear transformations

$$\mathbf{x} \rightarrow \mathbf{u}' = \mathbf{A}\mathbf{x},$$

which transform points into planes, and we shall call such transformations *space correlations*. We shall confine our attention to non-singular self-correlations of space.

If κ is such a correlation, every point P is transformed by it into a plane $\pi' = \kappa P$; and conversely, every point P arises from a unique plane π' by the inverse transformation κ^{-1}.

If P describes a plane π, represented by the coordinate vector $\mathbf{u}$, then, by an argument similar to that used on p. 225, we see that κP envelops the point P' represented by $\mathbf{x}' = \mathbf{A}^{-1T}\mathbf{u}$. We write $P' = \kappa\pi$, saying that P' is the transform of π by the given correlation κ, and this enables us to regard κ as a dualizing transformation of the whole totality of points and planes of space into itself.

If P varies on a line l, describing a range of points (P), then $\pi' = \kappa P$ describes a pencil of planes homographically related to (P). If the axis of this pencil is l', we may say that κ transforms l into l'. It may be seen algebraically that the pencil of planes with axis l is transformed by κ into a range of points whose axis is also the line $l' = \kappa l$. For let $\mathbf{y}$, $\mathbf{z}$ represent two points of l and let $\mathbf{v}$, $\mathbf{w}$ represent two planes through l. Then, applying κ,

$$\mathbf{y}+\theta\mathbf{z} \rightarrow \mathbf{A}\mathbf{y}+\theta\mathbf{A}\mathbf{z}$$

and

$$\mathbf{v}+\phi\mathbf{w} \rightarrow \mathbf{A}^{-1T}\mathbf{v}+\phi\mathbf{A}^{-1T}\mathbf{w}.$$

Thus the axis of the range described by the point $\mathbf{y}+\theta\mathbf{z}$ becomes the line of intersection of the planes $\mathbf{Ay}$, $\mathbf{Az}$, while the axis of the pencil described by the plane $\mathbf{v}+\phi\mathbf{w}$ becomes the line joining the points $\mathbf{A}^{-1T}\mathbf{v}$, $\mathbf{A}^{-1T}\mathbf{w}$. But $(\mathbf{Ay})^T\mathbf{A}^{-1T}\mathbf{v} \equiv (\mathbf{v}^T\mathbf{y})^T$, etc.; and therefore if the planes $\mathbf{v}$, $\mathbf{w}$ are incident with the points $\mathbf{y}$, $\mathbf{z}$ the points $\mathbf{A}^{-1T}\mathbf{v}$, $\mathbf{A}^{-1T}\mathbf{w}$ are also incident with the planes $\mathbf{Ay}$, $\mathbf{Az}$. This proves the statement made above.

In general a correlation κ and its inverse κ^{-1} are distinct transformations, for they transform the point $\mathbf{x}$ into the planes $\mathbf{Ax}$ and $\mathbf{A}^T\mathbf{x}$ respectively, but in special cases the two transformations may be the same. A necessary and sufficient condition for this to be so is $\mathbf{A}^T = \lambda\mathbf{A}$, where λ is a scalar. If this condition holds, then $\mathbf{A} = \mathbf{A}^{TT} = \lambda^2\mathbf{A}$, and hence $\lambda^2 = 1$. There are then only two possibilities, $\lambda = 1$ and $\lambda = -1$, and the matrix $\mathbf{A}$ must be either symmetric or skew-symmetric.

If $\mathbf{A}$ is symmetric, the correlation is a polarity, namely the polarity defined by the proper quadric ψ given by $\mathbf{x}^T\mathbf{A}\mathbf{x} = 0$.

The polarity is also known sometimes as *reciprocation* with respect to ψ. This case is fully analogous to that of the polarity of the plane referred to on p. 227.

If $\mathbf{A}$ is skew-symmetric, we have a new kind of correlation, called the *null polarity*. This has no analogue in the plane, since a skew-symmetric 3×3 matrix is necessarily singular.

Consider the null polarity ν defined by

$$\mathbf{u}' = \mathbf{A}\mathbf{x}, \qquad \mathbf{A}^T = -\mathbf{A}.$$

Since $\mathbf{u}'^T\mathbf{x} = (\mathbf{A}\mathbf{x})^T\mathbf{x} = \mathbf{x}^T\mathbf{A}\mathbf{x} \equiv 0$, by the skew-symmetry of $\mathbf{A}$, the null polarity has the remarkable property that *the polar plane of every point passes through that point.* (Cf. the discussion, on p. 311, of the null polarity defined by a twisted cubic.)

THEOREM 10. *If, in the null polarity ν, the polar plane of P passes through Q, then the polar plane of Q passes through P.*

Proof. If $\mathbf{x}$, $\mathbf{y}$ are coordinate vectors of P, Q, we have

$$(\mathbf{A}\mathbf{x})^T\mathbf{y} = 0,$$

i.e.

$$\mathbf{x}^T\mathbf{A}^T\mathbf{y} = 0.$$

Transposing,

$$\mathbf{y}^T\mathbf{A}\mathbf{x} = 0,$$

i.e.

$$(\mathbf{A}^T\mathbf{y})^T\mathbf{x} = 0,$$

i.e.

$$(\mathbf{A}\mathbf{y})^T\mathbf{x} = 0,$$

since $\mathbf{A}^T = -\mathbf{A}$.

If P describes a line l, then its polar plane always passes through a second line l', and two lines l, l' which are related in this manner are said to be *polar lines* with respect to ν. Those lines l which coincide with their polar lines are said to be *self-polar*. If l' is distinct from l the relation between the two lines is symmetrical, and it may be expressed very simply by saying that the polar plane of any point of either line is the plane joining this point to the other line. [*Exercise*. Prove that when l' is distinct from l it is skew to l.]

THEOREM 11. *Every line which meets a pair of (distinct) lines l, l' which are polar in a null polarity ν is self-polar in ν; and conversely, every self-polar line which meets l also meets l'.*

Proof. If p meets l, l' in P, P', the polar line of p is the line of intersection of the polar planes of P and P'. But these are the planes Pl' and $P'l$, and they meet in p. Further, if p is any self-polar line which meets l, then its polar line (i.e. p itself) meets the polar line l' of l.

THEOREM 12. *The self-polar lines which pass through a given point A are all the lines through A which lie in the polar plane of A. Dually, the self-polar lines which lie in a plane α are all the lines in α which pass through the pole of α.*

Proof. If l is a self-polar line through A, the polar plane of A must be a plane of the pencil whose axis is l. If, conversely, l is any line which both passes through A and lies in the polar plane α' of A, then l is self-polar with respect to ν. For if P is any point of l, the polar plane of A passes through P and the polar plane of P therefore passes through A; and since this polar plane necessarily contains P it is a plane of the pencil with axis l. The polar line of l is thus l itself, i.e. l is a self-polar line, and this proves the theorem.

We see, then, that the self-polar lines of a null polarity ν form a triply infinite system. This system of lines is of the type known as a linear line complex, and we shall discuss the properties of such complexes in the next chapter.

The null polarity was studied, first of all, not in geometry but in statics, where it arises naturally in connexion with systems of forces in three dimensions. Such a system of forces, acting on a rigid body, can be replaced by a single force and a couple, the line of action of the force passing through an arbitrarily assigned point. Let the force be $\mathbf{F}$, acting at O, and let the couple be $\mathbf{G}_0$. Then if A is any point of space, with position vector $\overrightarrow{OA} = \mathbf{a}$, the system is also equivalent to a force $\mathbf{F}$ at A and a couple

$$\mathbf{G}_A = \mathbf{G}_0+\mathbf{F}\wedge\mathbf{a}.$$

If, now, l is any specified line through A, with parametric representation $\mathbf{r} = \mathbf{a}+\mathbf{l}t$, $\mathbf{l}$ being a unit vector, the moment of the system about the line l is given by $\mathbf{G}_A.\mathbf{l}$. If this moment vanishes, the line is called a *null line* for the system of forces, and the condition for this is

$$(\mathbf{G}_0+\mathbf{F}\wedge\mathbf{a}).\mathbf{l} = 0,$$

i.e. $$\mathbf{G}_0.(\mathbf{r}-\mathbf{a})+(\mathbf{F}\wedge\mathbf{a}).(\mathbf{r}-\mathbf{a}) = 0,$$

i.e. $$\mathbf{G}_0.\mathbf{r}-\mathbf{G}_0.\mathbf{a}+(\mathbf{F}\wedge\mathbf{a}).\mathbf{r} = 0,$$

where $\mathbf{r}$ represents a general point of l. In cartesian coordinates, therefore, every point of a null line through the point A (ξ,η,ζ)

satisfies the condition

$$Lx+My+Nz-L\xi-M\eta-N\zeta+(Y\zeta-Z\eta)x+$$
$$+(Z\xi-X\zeta)y+(X\eta-Y\xi)z = 0,$$

where $\mathbf{F} = (X, Y, Z)$ and $\mathbf{G}_0 = (L, M, N)$.

The null lines through A are accordingly the lines through A which lie in the plane

$$(-Z\eta+Y\zeta+L)x+(Z\xi-X\zeta+M)y+(-Y\xi+X\eta+N)z-$$
$$-L\xi-M\eta-N\zeta = 0.$$

This plane is called the *null plane* of A, and it is derived from A by the null polarity whose matrix is

$$\begin{pmatrix} 0 & -Z & Y & L \\ Z & 0 & -X & M \\ -Y & X & 0 & N \\ -L & -M & -N & 0 \end{pmatrix}.$$

EXERCISES ON CHAPTER XIV

1. Discuss in detail the united points, united planes, and geometrical character of each of the following collineations:

(i) $x' : y' : z' : t' = 2x-y : 2y-x : z : -z-t$;

(ii) $x' : y' : z' : t' = y : z : x : t$;

(iii) $x' : y' : z' : t' = t : -y : -z : x$;

(iv) $x' : y' : z' : t' = x+al : y+bl : z+cl : t+dl$, where

$$l \equiv ux+vy+wz+pt.$$

2. Discuss in detail the space collineation whose matrix is

$$\begin{pmatrix} k & 1 & -1 & 1 \\ 1 & k & -1 & 1 \\ 1 & -1 & k & 1 \\ 1 & -1 & 1 & k \end{pmatrix}.$$

3. Characterize projectively the following types of self-collineation of euclidean space:

(i) radial expansion from a point;

(ii) radial expansion from a line (perpendicular to the line);

(iii) translation;

(iv) reflection in a point;

(v) reflection in a line;

(vi) reflection in a plane;

(vii) rotation about a line.

4. Find the equations of the biaxial collineation with modulus k and with the lines $x = 0 = y-z$ and $t = 0 = y+z$ as axes.

5. Show that, for any non-zero value of λ, the collineation

$$x' : y' : z' : t' = \lambda^3 x : \lambda^2 y : \lambda z : t$$

leaves the twisted cubic $c(\theta^3, \theta^2, \theta, 1)$ invariant.

Discuss in detail the special cases in which (i) $\lambda = -1$, (ii) $\lambda = \omega$, where ω is a complex cube root of unity. Show in each case the special way in which the collineation is related to the curve.

Find also, in each of the two cases, what quadrics through the curve are left invariant by the collineation.

6. If ϖ is a general collineation of the type which transforms every generator of a given quadric ψ into a generator of the opposite system, show that there exists on ψ a skew quadrilateral $XYZT$ such that ϖ transforms each of the lines XY, XZ into the other and each of TY, TZ into the other.

If $XYZT$ is taken as tetrahedron of reference and ψ has the equation $xt = yz$, show that ϖ has equations

$$x' : y' : z' : t' = ax : bz : cy : dt,$$

where $ad = bc$.

7. The poles of a variable plane π with respect to two quadrics ψ and ψ' are P and P'. Show that P and P' correspond in a space collineation ϖ, and that ϖ is a homology if and only if ψ and ψ' touch along a plane section.

8. Show that any space collineation with a plane of united points is necessarily a homology.

Two space homologies ϖ, ϖ' have vertices A, A' and axial planes α, α' respectively. Show that, when the lines AA' and $\alpha\alpha'$ are skew, $\varpi\varpi' = \varpi'\varpi$ if and only if A and A' lie respectively in α' and α.

If this condition is satisfied, prove that $\varpi\varpi'$ is not involutory unless ϖ and ϖ' are harmonic homologies.

9. If two biaxial collineations ϖ_1, ϖ_2 have one common axis b, while their other axes a_1, a_2 intersect in O, prove that $\varpi_1 \varpi_2$ is in general a biaxial collineation with b as one axis and a line of the pencil defined by a_1, a_2 as the other.

In the special case when each of ϖ_1, ϖ_2 is involutory, show that $\varpi_1 \varpi_2$ is a space homology whose axial plane is the plane which joins O to b and whose vertex (which lies in the axial plane) is the intersection of b with the plane $a_1 a_2$.

[*Hint.* Consider the euclidean case in which Ob is the plane at infinity and O and b are pole and polar for Ω.]

10. A general space collineation ϖ has the reference points X, Y, Z, T as united points, and a pair of corresponding points P, P' varies in such a way that PP' always passes through the unit point E. Show that P and P' describe twisted cubics c and c' through X, Y, Z, T, E.

Show that the unique quadric through c and c' has an equation of the form

$$(a-d)(b-c)(yz+xt)+(b-d)(c-a)(zx+yt)+(c-d)(a-b)(xy+zt) = 0.$$

11. A space correlation has equations

$$u' : v' : w' : p' = at : bz : cy : dx.$$

Find its two coincidence quadrics, and verify that they intersect in a skew quadrilateral.

Investigate the space collineation which is the square of the given correlation.

12. Show that the relation between points $\mathbf{x}$ and $\mathbf{y}$ which are conjugate with respect to a null polarity is of the form

$$\sum_{i,j=0,\dots,3} \lambda_{ij}(x_i y_j - x_j y_i) = 0,$$

and that any equation of this type defines a null polarity (which may be singular).

Deduce that the line joining any pair of distinct conjugate points is self-polar in the polarity.

13. If a pair of polar lines in the null polarity are taken as edges $X_0 X_3$ and $X_1 X_2$ of the tetrahedron of reference, show that the equation given in the previous exercise takes the form

$$x_0 y_3 - x_3 y_0 = k(x_1 y_2 - x_2 y_1).$$

14. If σ and τ are two space correlations, show that there exist in general four points whose corresponding planes in σ and τ are identical.

Show that on a given quadric ψ there exist in general four points at each of which the tangent plane to ψ is the polar plane of the point in a given null polarity, and that the four points are the vertices of a skew quadrilateral of generators of ψ which are self-polar with respect to the null polarity.

15. If the characteristic roots of the $n \times n$ matrix $\mathbf{A}$ are $\lambda_1,\dots,\lambda_n$ (not necessarily distinct) then the characteristic roots of the matrix

$$a_0 \mathbf{A}^m + \dots + a_{m-1} \mathbf{A} + a_m \mathbf{I}$$

are the numbers $a_0 \lambda_i^m + \dots + a_{m-1}\lambda_i + a_m$ $(i = 1,\dots,n)$. Prove this theorem, and by applying it to the special polynomial $\mathbf{A}^2$ obtain an alternative proof of Theorem 5.

16. Prove algebraically that the four united points of the general space collineation are the vertices of a tetrahedron by establishing the following general theorem. If a non-singular $n \times n$ matrix $\mathbf{A}$ has distinct characteristic roots $\lambda_1,\dots,\lambda_n$, and $\mathbf{x}^{(i)}$ is a non-zero vector for which

$$\mathbf{A}\mathbf{x}^{(i)} = \lambda_i \mathbf{x}^{(i)} \quad (i = 1,\dots,n),$$

then the n vectors $\mathbf{x}^{(i)}$ are linearly independent.

[*Hint*. Suppose that $\sum_{i=1}^{n} \alpha_i \mathbf{x}^{(i)} = \mathbf{0}$, where not all the α_i are zero. Multiply this equation successively by $\mathbf{A}, \mathbf{A}^2,\dots, \mathbf{A}^{n-1}$, and obtain a contradiction by using the fact that n linear relations between n vectors are necessarily linearly dependent.]

CHAPTER XV

LINE GEOMETRY

§ 1. Line-coordinates

We have already seen in Chapter X how the ∞^4 lines of S_3 can be represented, in terms of any allowable representation $\mathscr{R}$, by sets of six homogeneous coordinates $(p_{23}, p_{31}, p_{12}, p_{01}, p_{02}, p_{03})$, connected by the fixed quadratic relation

$$\Omega_{pp} \equiv p_{01}p_{23}+p_{02}p_{31}+p_{03}p_{12} = 0.$$

We have also discussed the dual coordinates π_{ij}, which are such that

$$\frac{\pi_{23}}{p_{01}} = \frac{\pi_{31}}{p_{02}} = \frac{\pi_{12}}{p_{03}} = \frac{\pi_{01}}{p_{23}} = \frac{\pi_{02}}{p_{31}} = \frac{\pi_{03}}{p_{12}};$$

and we have shown that the condition for two lines p, q to intersect is the polarized form of the identical relation, i.e.

$$2\Omega_{pq} \equiv p_{01}q_{23}+p_{02}q_{31}+p_{03}q_{12}+p_{23}q_{01}+p_{31}q_{02}+p_{12}q_{03} = 0.$$

Our main purpose in the present chapter is to introduce the reader to the systematic study of those subsystems of the full system of ∞^4 lines of S_3 which are of geometrical interest; but before doing so we need to look more closely at the way in which lines are represented by the p_{ij}.

The first essential in the use of line-coordinates is to be able to find the points which lie on a line with given coordinates and also the planes which pass through it; and for this it is sufficient to specify, once for all, the points P_0, P_1, P_2, P_3 in which the given line p meets the four reference planes and the planes π_0, π_1, π_2, π_3 which join it to the four reference points.

It is sometimes convenient to work not with the six quantities $p_{23}, p_{31}, p_{12}, p_{01}, p_{02}, p_{03}$ only, but with the full set of sixteen quantities p_{ij}, connected by the antisymmetry relations $p_{ji} = -p_{ij}$ $(i,j = 0,..., 3)$, and this is so at present. If $\mathbf{P}$ denotes the skew-symmetric 4×4 matrix (p_{rs}) we have the following result, the verification of which is left to the reader:

The coordinate vectors of P_0, P_1, P_2, P_3 are the rows of the matrix $\mathbf{P}$, and dually, the coordinate vectors of π_0, π_1, π_2, π_3 are the rows of the matrix $\mathbf{\Pi} = (\pi_{rs})$.

If $\mathbf{p} = (p_{23}, p_{31}, p_{12}, p_{01}, p_{02}, p_{03})$ is taken as coordinate vector of the line p, the six basis vectors $(1, 0, 0, 0, 0, 0)$, etc., belong to the six reference lines of $\mathscr{R}$, i.e. the edges of the tetrahedron of reference. More precisely, that basis vector for which only the coordinate $p_{\alpha\beta}$ is non-zero represents the edge $X_\alpha X_\beta$.

If the line p lies in the reference plane $x_\alpha = 0$, then three of its coordinates $p_{\alpha\beta}$, $p_{\alpha\gamma}$, $p_{\alpha\delta}$ vanish; and the equation of the line in the subordinate geometry in the plane $x_\alpha = 0$ is

$$p_{\gamma\delta}x_\beta + p_{\delta\beta}x_\gamma + p_{\beta\gamma}x_\delta = 0.$$

(The conventions governing the use of suffixes in equations such as this will be obvious to the reader.)

Dually, if p passes through the reference point X_α, given by $x_\beta = x_\gamma = x_\delta = 0$, then $p_{\gamma\delta} = p_{\delta\beta} = p_{\beta\gamma} = 0$, and the equations of the line are

$$x_\beta : x_\gamma : x_\delta = p_{\alpha\beta} : p_{\alpha\gamma} : p_{\alpha\delta}.$$

We note finally that p meets the edge $X_\alpha X_\beta$, whose equations are $x_\gamma = x_\delta = 0$, if and only if $p_{\gamma\delta} = 0$.

EXERCISES

(i) Discuss the families of lines for which

$$(a)\ p_{01} = p_{12} = 0, \quad \text{and} \quad (b)\ p_{01} = p_{12} = p_{23}.$$

(ii) Find the conditions for a line to pass through the unit point $(1, 1, 1, 1)$.

Line systems

The lines of space, as has already been remarked, form a quadruply infinite system, and this means that we have to consider subsystems of three different kinds:

(*a*) line complexes, or triply infinite line systems;
(*b*) line congruences, or doubly infinite line systems;
(*c*) ruled surfaces, or simply infinite line systems.

The ruled surface is, of course, the surface generated by a variable line which has one degree of freedom.

We have already met with examples of all these kinds of system. Thus the lines which meet a fixed line and those which touch a fixed quadric form complexes, the chords of a twisted cubic form a congruence, and the plane pencil of lines and the regulus are familiar examples of simply infinite line systems.

Any equation

$$f(p_{23}, p_{31}, p_{12}, p_{01}, p_{02}, p_{03}) = 0,$$

in which f is not divisible by Ω_{pp}, imposes a condition on a variable line p and takes away one of its four degrees of freedom. A single equation thus defines a complex, two independent equations a congruence, and three independent equations a ruled surface.

The simplest line systems are naturally those which are given by one or more linear equations of the form

$$(\mathbf{a}, \mathbf{p}) \equiv a_{23}p_{23}+a_{31}p_{31}+a_{12}p_{12}+a_{01}p_{01}+a_{02}p_{02}+a_{03}p_{03} = 0.$$

Such an equation may also be written as

$$\sum_{i=0}^{3}\sum_{k=0}^{3} a_{ik}p_{ik} = 0$$

if we define the remaining a_{ik} by means of the relations $a_{ki} = -a_{ik}$ $(i, k = 0,..., 3)$.

It is important to realize, however, that even these simple systems are more awkward to handle than the linear subspaces of the space S_3 of points. They are in fact essentially non-linear, since the identical quadratic relation $\Omega_{pp} = 0$ has always to be taken in conjunction with the linear equations which serve to define the system. This peculiarity of the geometry of lines deserves to be examined a little more closely.

A linear subspace of the point space S_3 may be represented indifferently as the set of all points whose coordinate vectors $\mathbf{x}$ satisfy r linearly independent linear equations $(\mathbf{u}^{(i)}, \mathbf{x}) = 0$ or as the set of all points whose coordinate vectors are linearly dependent on $4-r$ independent vectors $\mathbf{x}^{(j)}$. In the geometry of lines, on the other hand, r linear equations $(\mathbf{a}^{(i)}, \mathbf{p}) = 0$ determine a line system which is made up of all the lines whose coordinate vectors $\mathbf{p}$ satisfy the $r+1$ equations

$$(\mathbf{a}^{(i)}, \mathbf{p}) = 0 \quad (i = 1,..., r); \qquad \Omega_{pp} = 0.$$

Such a system cannot, in general, be represented parametrically by an equation of the form

$$\mathbf{p} = \sum_{j=1}^{s} \lambda_j \mathbf{p}^{(j)};$$

and indeed if $\mathbf{p}^{(1)},..., \mathbf{p}^{(s)}$ are coordinate vectors of lines of space the vector $\lambda_1 \mathbf{p}^{(1)}+...+\lambda_s \mathbf{p}^{(s)}$ need not represent a line at all. Only in special circumstances will this vector satisfy the identical relation $\Omega_{pp} = 0$ for arbitrary values of the λ_i.

§ 2. Linear Line Spaces

Suppose p, q are given lines, represented by coordinate vectors $\mathbf{p}$, $\mathbf{q}$ respectively. Then, for $\mathbf{p}+\lambda\mathbf{q}$ to represent a line, we must have

$$\sum (p_{01}+\lambda q_{01})(p_{23}+\lambda q_{23}) = 0,$$

where the summation extends over three terms; and this condition may be written

$$\Omega_{pp}+2\lambda\Omega_{pq}+\lambda^2\Omega_{qq} = 0.$$

Since **p** and **q** represent lines, $\Omega_{pp} = \Omega_{qq} = 0$; and the condition is therefore satisfied for all λ if and only if $\Omega_{pq} = 0$, i.e. if and only if p and q intersect.

When this is the case, $\mathbf{p}+\lambda\mathbf{q}$ clearly represents a general line of the flat pencil determined by p, q. For if r is any line which meets both p and q then

$$\Omega_{p+\lambda q,r} \equiv \Omega_{p,r}+\lambda\Omega_{q,r} = 0,$$

and r also meets the line represented by $\mathbf{p}+\lambda\mathbf{q}$. It follows that this line $\mathbf{p}+\lambda\mathbf{q}$ lies in the plane of p and q and passes through their common point.

A system of lines which is such that the coordinate vectors **p** which represent lines of the system form a subspace of the vector space $V_6(K)$—i.e. the set of all vectors linearly dependent on a finite set of vectors in $V_6(K)$—will be called a *linear line space.* The simplest non-trivial linear line space is accordingly the pencil, represented by $\mathbf{p} = \lambda_1\mathbf{p}^{(1)}+\lambda_2\mathbf{p}^{(2)}$; and we shall now show that only two other linear line spaces are possible, namely the *point-star*, or system of ∞^2 lines through a point, and the *ruled plane*, or system of ∞^2 lines in a plane.

Suppose, in fact, that $\mathbf{p}^{(1)},\ldots,\mathbf{p}^{(m)}$ are coordinate vectors of m lines $p^{(1)},\ldots,p^{(m)}$, and that every linear combination $\sum_{j=1}^{m}\lambda_j\mathbf{p}^{(j)}$ also represents a line. Then, in particular, every linear combination of two of the vectors, $\mathbf{p}^{(i)}$, $\mathbf{p}^{(j)}$ say, must represent a line; and this means, as we have already seen, that the lines $p^{(i)}$, $p^{(j)}$ intersect. Thus the given lines $p^{(1)},\ldots,p^{(m)}$ all intersect in pairs. Now $p^{(1)}$, $p^{(2)}$ meet in a point and also lie in a common plane, and since $p^{(3)}$ meets them both it must either pass through the point or lie in the plane (but it cannot do both if the vectors $\mathbf{p}^{(1)}$, $\mathbf{p}^{(2)}$, $\mathbf{p}^{(3)}$ are linearly independent). It is easy to see that the equation

$$\mathbf{p} = \lambda_1\mathbf{p}^{(1)}+\lambda_2\mathbf{p}^{(2)}+\lambda_3\mathbf{p}^{(3)}$$

gives a parametric representation in the first case of the star and in the second of the ruled plane containing the three lines $p^{(1)}$, $p^{(2)}$, $p^{(3)}$. But in neither case can there be a fourth line $p^{(4)}$, not belonging to the star or the ruled plane, which meets all of $p^{(1)}$, $p^{(2)}$, $p^{(3)}$; and this means that no more ample linear line spaces can exist. This enables us to enunciate the theorem:

THEOREM 1. *The only linear line spaces in S_3 are the line pencil, the point-star, and the ruled plane.*

It will be observed that the star and the ruled plane are dual to each other in space, while the pencil is self-dual. Every pencil can of course be obtained as the intersection of a unique star and a unique ruled plane—i.e. the set of all lines common to the two systems.

The regulus

When m lines $p^{(j)}$ do not all intersect in pairs they certainly do not determine a linear line space; but we may still consider the system of lines p which are linearly dependent on the given lines. In other words, we may form the vector

$$\mathbf{p} = \sum_{j=1}^{m} \lambda_j \, \mathbf{p}^{(j)}$$

and then consider the system of all lines arising from sets of parameters $(\lambda_1,..., \lambda_m)$ for which the condition Ω_{pp} is satisfied.

If q, r are two given lines and $\mathbf{p} = \lambda\mathbf{q}+\mu\mathbf{r}$ then, as we have seen,

$$\Omega_{pp} \equiv \lambda^2\Omega_{qq}+2\lambda\mu\Omega_{qr}+\mu^2\Omega_{rr};$$

and this means that if q and r are skew no other line can be linearly dependent on them.

If q, r, s are three given lines and $\mathbf{p} = \lambda\mathbf{q}+\mu\mathbf{r}+\nu\mathbf{s}$, then

$$\Omega_{pp} \equiv (\Omega_{qq}, \Omega_{rr}, \Omega_{ss}, \Omega_{rs}, \Omega_{sq}, \Omega_{qr} \between \lambda, \mu, \nu)^2;$$

and since $\Omega_{qq} = \Omega_{rr} = \Omega_{ss} = 0$ this reduces to

$$\Omega_{pp} \equiv 2(\Omega_{rs}\,\mu\nu+\Omega_{sq}\,\nu\lambda+\Omega_{qr}\,\lambda\mu).$$

Even when the lines q, r, s are all skew there are therefore ∞^1 lines p which are linearly dependent on them, and these lines generate a ruled surface through q, r, s. But, by the linearity in $\mathbf{p}$ of the condition $\Omega_{pt} = 0$, every line t which meets q, r, s also meets any line given by $\mathbf{p} = \lambda\mathbf{q}+\mu\mathbf{r}+\nu\mathbf{s}$. The lines p which generate the ruled surface therefore meet all transversals of q, r, s; and this gives us the following theorem.

THEOREM 2. *The lines which are linearly dependent on three given skew lines q, r, s constitute the unique regulus which contains q, r, s.*

If λ, μ, ν are taken as coordinates of a variable point in a plane π, the lines of the regulus are represented in π by the points of a conic k whose equation is

$$\Omega_{rs}\mu\nu+\Omega_{sq}\nu\lambda+\Omega_{qr}\lambda\mu = 0. \tag{1}$$

Now a line given by $\lambda\mathbf{q}+\mu\mathbf{r}+\nu\mathbf{s}$ meets a given line t only if

$$\Omega_{qt}\lambda+\Omega_{rt}\mu+\Omega_{st}\nu = 0, \tag{2}$$

and the two lines of the regulus which meet t are accordingly represented in π by the points of intersection of the conic (1) with the line (2). These two lines and the two common transversals of q, r, s, t meet t in the same points The transversals coincide, therefore, if the line (2).touches k, i.e. if

$$(\Omega_{rs}^2, \Omega_{sq}^2, \Omega_{qr}^2, -\Omega_{sq}\Omega_{qr}, -\Omega_{qr}\Omega_{rs}, -\Omega_{rs}\Omega_{sq} \between \Omega_{qt}, \Omega_{rt}, \Omega_{st})^2 = 0,$$

i.e. if
$$(\Omega_{rs}\Omega_{qt})^{\frac{1}{2}} \pm (\Omega_{sq}\Omega_{rt})^{\frac{1}{2}} \pm (\Omega_{qr}\Omega_{st})^{\frac{1}{2}} = 0.$$

We could now go on to consider the systems of lines linearly dependent on four or five general lines; but since such systems are more simply defined by means of linear equations $(\mathbf{a}, \mathbf{p}) = 0$ satisfied by the coordinate vector $\mathbf{p}$ of a general line of the system, we leave them over until later.

Exercise. Discuss the systems of lines linearly dependent on q, r, s when (a) one pair of these lines intersect, and (b) two pairs intersect.

§ 3. Line Complexes

If $F(p_{23},\ldots,p_{03})$ is a form of degree n in the six line-coordinates, and Ω_{pp} is not a factor of this form, there are ∞^3 lines whose coordinates satisfy the equation

$$F(p_{23},\ldots,p_{03}) = 0.$$

Such a system of lines is called a *line complex* of order n. If K denotes the complex, the lines of K which pass through a general point P will form a cone—the *complex cone* C_P associated with P; and the lines of K which lie in a general plane π will form an envelope—the *complex envelope* E_π associated with π. Since the complex is a self-dual entity, whose equation can equally well be written

$$F(\pi_{01},\ldots,\pi_{12}) = 0,$$

the order of C_P must be the same as the class of E_π.

Suppose P has coordinates $(\xi_0, \xi_1, \xi_2, \xi_3)$. Then the line joining P to the general point (x_0, x_1, x_2, x_3) belongs to the complex if

$$F(\xi_2 x_3-\xi_3 x_2,\ldots, \xi_0 x_3-\xi_3 x_0) = 0.$$

Since the left-hand side of this equation is homogeneous of degree n in x_0, x_1, x_2, x_3, C_P is a cone of order n. Similarly, or by duality, E_π is an envelope of class n. This gives us the important result:

THEOREM 3. *If K is a complex of order n, the lines of K which pass through a general point generate a cone of order n and the lines of K which lie in a general plane generate an envelope of class n.*

If K is of order 1, this theorem implies that the lines of K which pass through a point P all lie in a plane π through P, and the lines of K which lie in a plane π all pass through a point P in π. Thus a complex of this kind, which is known as a *linear complex*, consists of the lines of ∞^3 flat pencils, one for every point of space. There are only ∞^3 lines altogether, since each belongs to an infinity of the pencils.

If K is of order 2 it is called a *quadratic complex*. The lines of K through P generate a quadric cone, and the lines of K in π envelop a conic. If the cone is a plane-pair, P is a *singular point* of K; and if the conic is a point-pair, π is a *singular plane* of K.

In special cases it may happen that a complex K has *total points* or *total planes*. A total point is such that every line through it belongs to K, and a total plane is such that every line in it is a line of K.

The linear complex

The equation of a linear complex K may be written in the form $(\mathbf{a}, \mathbf{p}) = 0$; but it is often better to renumber the coefficients and write it as

$$\Omega_{ap} \equiv a_{01}p_{23}+a_{02}p_{31}+a_{03}p_{12}+a_{23}p_{01}+a_{31}p_{02}+a_{12}p_{03} = 0.$$

If $\Omega_{aa} = 0$, so that $a_{23},..., a_{03}$ are themselves coordinates of a line a, then K consists of all the lines of S_3 which meet a. We say in this case that the complex is *special* and that it has a as its axis. C_P is then the plane joining P to a, and E_π is the point in which π is met by a. Every plane through a and every point which lies on a is total for K.

If $\Omega_{aa} \neq 0$, K is a general linear complex. The equation of the complex plane C_P associated with the point P whose coordinates are $(\xi_0, \xi_1, \xi_2, \xi_3)$ is then

$$a_{01}(\xi_2 x_3-\xi_3 x_2)+...+a_{23}(\xi_0 x_1-\xi_1 x_0)+... = 0.$$

This plane is called the *polar plane* of P with respect to the complex, and its coordinates are obtained from the coordinates of P, if we now denote these by x_0, x_1, x_2, x_3, by the linear transformation

$$\begin{pmatrix} u_0 \\ u_1 \\ u_2 \\ u_3 \end{pmatrix} = \begin{pmatrix} 0 & a_{23} & a_{31} & a_{12} \\ -a_{23} & 0 & a_{03} & -a_{02} \\ -a_{31} & -a_{03} & 0 & a_{01} \\ -a_{12} & a_{02} & -a_{01} & 0 \end{pmatrix} \begin{pmatrix} x_0 \\ x_1 \\ x_2 \\ x_3 \end{pmatrix}$$

or, say, $\mathbf{u} = \mathbf{A}\mathbf{x}$, where $\mathbf{A}$ is a skew-symmetric matrix whose determinant Ω_{aa}^2 does not vanish. The transformation is therefore a non-singular null polarity ν. The polar plane of P in ν is its polar plane with respect to the linear complex, and two points P, Q are conjugate for ν if and only if PQ is a line of the complex. It follows that the pole of any plane π with respect to K is its pole in ν, and that K consists of all the lines of space which are self-polar for ν. This establishes the fundamental connexion between null polarities and linear complexes.

THEOREM 4. *Every linear complex K consists of the totality of lines that are self-polar for a null polarity ν, and conversely. If ν is singular, K is special and consists of all the lines which meet a fixed line.*

It follows from Theorem 4 that all the lines of K which meet a line l also meet the polar line l' of l. Conversely, if a line l is not self-polar for ν (i.e. not a line of K) then every line which meets both l and its polar line l' is a line of K. (See p. 361.)

Since K and ν determine each other uniquely, we may say that two lines are polar lines with respect to K when they are polar for ν.

EXERCISES

(i) If the reference lines X_0X_1 and X_2X_3 are polar lines for a linear complex K, show that the equation of K reduces to the form $p_{01} = \lambda p_{23}$. Find the polar line of any given line q with respect to this complex.

(ii) If four lines p_1, p_2, p_3, p_4 of a linear complex K have two (and only two) distinct transversals, prove that these are polar lines for K. If the lines have a unique transversal, prove that this belongs to K; and deduce that the range cut on it by the lines is homographic with the pencil of planes joining it to the lines. If the lines belong to a regulus, prove that this regulus is contained in K, and that the complementary regulus consists of pairs of polar lines for K.

(iii) Show that there exists a unique linear complex containing five general lines. Discuss the special case when the five given lines have a unique transversal line. Show also that if the five lines have two transversals they belong to an infinity of linear complexes.

Quadratic complexes. The tetrahedral complex

The geometry of the general quadratic complex is a very large subject which we shall not attempt to go into here. We remark only that it centres largely round the singular surface of the complex—a remarkable quartic surface, called the *Kummer surface*, possessing sixteen nodes—which is at the same time the locus of singular points and the envelope of singular planes. Our discussion will be limited to two special quadratic complexes which are of more immediate geometrical interest.

The first of these systems is the complex of tangent lines to a quadric ψ. To find its equation we may make use of the fact that a line touches ψ if and only if it meets its polar line with respect to ψ. Let the equation of ψ be

$$a_0 x_0^2 + a_1 x_1^2 + a_2 x_2^2 + a_3 x_3^2 = 0,$$

and let p be a general line, containing the points (y_i) and (z_i). Then the line-coordinates of p are given by

$$p_{ij} = y_i z_j - y_j z_i.$$

The polar line p' of p is the line of intersection of the polar planes of the points (y_i) and (z_i), i.e. the planes $(a_i y_i)$ and $(a_i z_i)$, and it therefore has dual line-coordinates,

$$\pi'_{ij} = a_i a_j (y_i z_j - y_j z_i) = a_i a_j p_{ij}.$$

We have then $$\Omega_{pp'} = \tfrac{1}{2} \sum a_i a_j p_{ij}^2,$$

and the condition for p to be a tangent line to ψ is

$$a_2 a_3 p_{23}^2 + a_3 a_1 p_{31}^2 + a_1 a_2 p_{12}^2 + a_0 a_1 p_{01}^2 + a_0 a_2 p_{02}^2 + a_0 a_3 p_{03}^2 = 0.$$

This equation is accordingly the equation of ψ regarded as a complex of tangent lines. The complex cone C_P is of course the enveloping cone with vertex P, and the complex envelope E_π is the section of ψ by π. The points and the tangent planes of ψ are all singular for the quadratic complex.

The second special quadratic complex that we wish to consider is the *tetrahedral complex*, which may be defined as the totality of lines which meet the faces of a given tetrahedron in a range of points with given cross ratio.

Suppose K is a tetrahedral complex, the defining tetrahedron being the tetrahedron of reference $X_0 X_1 X_2 X_3$. If p is a line of K, determined by two points (y_i), (z_i), the coordinates of a general point of p may be written as $(y_i - \lambda z_i)$, and the parameters of the points in which p meets the faces of the tetrahedron are then given by $\lambda_i = y_i/z_i$ $(i = 0,..., 3)$. If the value of the constant cross ratio is α, we then have

$$\alpha = \left\{\frac{y_0}{z_0}, \frac{y_1}{z_1}; \frac{y_2}{z_2}, \frac{y_3}{z_3}\right\} = \frac{y_0 z_2 - y_2 z_0}{y_1 z_2 - y_2 z_1} \bigg/ \frac{y_0 z_3 - y_3 z_0}{y_1 z_3 - y_3 z_1} = -\frac{p_{02}\, p_{31}}{p_{03}\, p_{12}}.$$

Since the expression on the right may also be written as $-\dfrac{\pi_{31}\,\pi_{02}}{\pi_{12}\,\pi_{03}}$ we see, incidentally, that the cross ratio of the four points in which p is met by the faces of the tetrahedron of reference is equal to the cross ratio of the four planes joining p to the vertices of the tetrahedron of reference—so that the tetrahedral complex is self-dual. The equation of the system may be written

$$p_{02}\, p_{31} + \alpha p_{03}\, p_{12} = 0,$$

and since the p_{ij} are connected by the identical relation

$$p_{01}\, p_{23} + p_{02}\, p_{31} + p_{03}\, p_{12} = 0,$$

the same complex may be represented in infinitely many ways by an equation of the form

$$a p_{01}\, p_{23} + b p_{02}\, p_{31} + c p_{03}\, p_{12} = 0.$$

Conversely, any equation of this form, in which a, b, c are all unequal, defines a tetrahedral complex.

The special peculiarity of the tetrahedral complex K, as a quadratic complex, is that it possesses four total points and four total planes, namely the vertices and faces of the basic tetrahedron. This is apparent at once from the form of the equation just obtained. The complex cone C_P consequently contains the four lines joining P to the vertices of the tetrahedron, and the complex envelope E_π contains the lines in which π is met by its four faces. The locus of singular points of K is the set of four faces of the tetrahedron, and the envelope of singular planes is the set of four vertices.

Tetrahedral complexes are met with in various connexions, and the following instances are especially important.

(i) If ϖ is a general space collineation, with united points A, B, C, D, then all the joins of corresponding points P, P' generate a tetrahedral complex based on $ABCD$, and the lines of intersection of corresponding planes π, π' also generate the same tetrahedral complex.

(ii) If (S) is a general pencil of quadrics the axes ϑ_P, relative to (S), of all points P of space generate a tetrahedral complex and the vertices of the tetrahedron are the vertices of the four cones of the pencil.

(iii) Dually, if (Σ) is a general range of quadrics the axes ϑ_π, relative to (Σ), of all planes π of space generate a tetrahedral complex; and the faces of the tetrahedron are the planes of the four disk quadrics of the range.

To prove the statement (i) we need only observe that if P, P' have coordinates (x_0, x_1, x_2, x_3), (ax_0, bx_1, cx_2, dx_3) respectively, the coordinates of the line PP' satisfy the equation

$$(a+b)(c+d)p_{01}p_{23}+(a+c)(d+b)p_{02}p_{31}+(a+d)(b+c)p_{03}p_{12} = 0.$$

It now follows immediately that the axes ϑ_P of a pencil of quadrics constitute a tetrahedral complex—for the polar planes π, π' of a point P with respect to two quadrics S, S' of the pencil are collinearly related; and, by duality, the axes ϑ_π of a range also make up a tetrahedral complex. An interesting special case of (iii) arises when the range (Σ) is a confocal system. In this case we can state the result in the following terms:

The normals of all the members of a confocal system of central quadrics generate a tetrahedral complex based on the three principal planes and the plane at infinity.

It follows from this general result that all the normals that can be drawn to the confocals from a fixed point generate a quadric cone, of which one generator passes through the centre of the system and three others are parallel to the axes. It follows further that all the normals to the confocals which lie in a plane envelop a parabola which touches the three principal planes.

EXERCISES

(i) If ψ is a central quadric, prove that the feet of those normals which belong to a given linear complex all lie on a quadric ψ' which passes through the centre of ψ and through the points at infinity on its axes, and conversely. Discuss the special case which arises when the linear complex is special and the intersection of ψ with ψ' is a pair of conics.

(ii) Prove that the normals at two points of a quadric ψ intersect if and only if the join of the points is a normal to some quadric confocal with ψ. Deduce from this that if a plane section of ψ contains one triad of points at which the normals to ψ are concurrent, then it contains ∞^1 such triads, and prove that the locus of the points of concurrence of the triads of normals is a line.

(iii) Form the equation of the complex of lines which meet the conic whose equations are $x_2^2-x_1x_3 = 0 = x_0$.

§ 4. Line Congruences

If a variable line is restricted in such a way as to reduce its freedom from 4 to 2, so that it depends effectively on the values of a set of parameters of which only two are independent, then it is said to describe a *line congruence* K. Thus, for instance, K might be the totality of lines common to two complexes—the complete intersection of the complexes; or it might be a two-dimensional component of a reducible system of this kind—a partial intersection of two complexes. In any case we expect there to be a finite number of lines of K which satisfy any restriction which imposes two further conditions on the two independent parameters. There will, in particular, be a finite number m of lines of K through a general point P; and this number m (which is independent of the choice of P if K is an algebraic congruence) is called the *order* of K. Similarly there will be a finite number of lines of K which lie in a general plane, and this number is called the *class* of K. A congruence of order m and class n will be called an (m, n) congruence.

As simple examples of congruences we have:

(i) the star, which is a $(1, 0)$ congruence;
(ii) the ruled plane, which is a $(0, 1)$ congruence;
(iii) the set of lines which meet two skew lines, which is a $(1, 1)$ congruence;
(iv) the $(1, 3)$ congruence of chords of a twisted cubic.

A point is said to be a *singular point* of the congruence K if infinitely many lines of K pass through it, and a plane which contains infinitely many lines of K is called a *singular plane.*

The linear congruence

The lines common to two distinct linear complexes L_1, L_2 evidently form a $(1, 1)$ congruence K, and such a system of lines is called a *linear congruence.* The equations of K may be written in the form $\Omega_{ap} = 0$, $\Omega_{bp} = 0$; and the lines of K clearly belong to every linear complex of the pencil of linear complexes given by $\Omega_{a+\lambda b,p} = 0$. This last complex is special if and only if λ satisfies the quadratic equation $\Omega_{a+\lambda b,a+\lambda b} = 0$, i.e.

$$\Omega_{aa}+2\lambda\Omega_{ab}+\lambda^2\Omega_{bb} = 0.$$

If this equation has distinct roots λ_1, λ_2, the pencil of linear complexes contains two complexes which are special, and these may be chosen as L_1 and L_2. Then K is the system of all lines which meet the axes of the two special complexes. This is the general type of linear congruence. Conversely, the lines which meet two general lines constitute a linear congruence with the two lines as axes.

If the quadratic in λ has coincident roots, K has only one axis, and it is then said to be special. It consists in this case of all the lines which meet a fixed line a and also belong to a linear complex L. The axis a is itself a line of L, since otherwise the lines of K would all meet the polar line of a with respect to L. The lines of K which pass through any point P of a lie in a plane π through a—namely the polar plane of P with respect to L—and the range (P) and the pencil (π) are homographically related.

An even more special case arises when the quadratic in λ is satisfied identically. This happens when L_1, L_2 are both special complexes and when their axes a, b intersect. K consists in this case of all the lines through the common point of a and b, together with all the lines coplanar with a and b.

Recapitulating, we have:

THEOREM 5. *The general linear congruence consists of all the lines which meet two fixed non-intersecting lines. A special linear congruence is generated by lines p which meet a fixed line a in such a way that the point ap and the plane ap correspond in a fixed homographic relation. A degenerate linear congruence is composed of a star* $[A]$ *and a ruled plane* $[\alpha]$, *with* A *in* α.

Other congruences

Among other types of congruence we may refer briefly to the following.

The quadratic congruence is a $(2, 2)$ congruence which is the complete intersection of a linear complex S and a quadratic complex T. If K is a general congruence of this type, it can be shown that K has sixteen singular points, each of which is the vertex of a pencil of lines of K; and the planes of the sixteen pencils so arising are singular planes of K. When, in particular, T happens to be a tetrahedral complex, four of the singular points must lie at the vertices of the tetrahedron and four more at the poles of its faces with respect to S; and in each face there lie two other singular points, whose polar planes with respect to S are the planes which form with that face the degenerate complex cones of T at the points in question.

The congruence of chords of a twisted cubic c is a $(1,3)$ congruence which has every point of c as a singular point. Dually, the congruence of axes of a cubic developable δ is a $(3,1)$ congruence.

The congruence of normals to a quadric ψ is a $(6,2)$ congruence whose lines are mapped on the points of ψ at which they are normal. The normals of ψ which belong to any general linear complex are mapped in this way by the points of the curve in which ψ is met by another quadric ψ' through the centre of ψ and the points at infinity on its axes (cf. p. 376, Exercise (i)). In this way the totality of sections of the congruence by all linear complexes is associated with the totality of sections of ψ by the ∞^5 quadrics ψ' through the four specified points.

§ 5. Ruled Surfaces

A *ruled surface R* is the line system generated by a variable line whose position depends effectively on the value of a single parameter. The *order n* of the surface, when this is algebraic, is the number of its lines which meet a general line of S_3.

A particularly simple ruled surface is the intersection of three linear complexes L_1, L_2, L_3. If the equations $L_i = 0$ $(i = 1,2,3)$ represent three general linear complexes, so that $L_i \equiv \Omega_{a_i,p}$, then the lines common to them all belong to every complex of the ∞^2 linear family $\lambda L_1+\mu L_2+\nu L_3 = 0$. The lines therefore belong, in particular, to the ∞^1 special complexes of this family, and they have the axes of these complexes as common transversals. But these axes are given by coordinate vectors of the form

$$\lambda \mathbf{a}_1+\mu \mathbf{a}_2+\nu \mathbf{a}_3;$$

and since we may take three of the special complexes as the L_i, it follows from Theorem 2 that the set of ∞^1 axes is a regulus, and the set of lines of which they are common transversals is the complementary regulus. Thus we have the theorem:

THEOREM 6. *The intersection of three linear complexes is, in general, a regulus, i.e. a ruled surface of order* 2.

Among ruled surfaces we distinguish as a special class those which are *developable*, i.e. those with the property that consecutive generators always intersect. Suppose that $\mathbf{p}(t)$ is the coordinate vector of a variable generator of a developable ruled surface R, t being a parameter, and suppose that $\mathbf{p}(t)$ is a differentiable function of t, with derivative $\dot{\mathbf{p}}$. Then, since R is developable, $\mathbf{p}(t)$ must satisfy the differential equation $\Omega_{p\dot{p}} = 0$.

The lines of a developable ruled surface R are, in general, the tangent lines of a curve C, the cuspidal edge of R, and they are also the focal lines of an envelope E, the envelope of osculating planes of C.

EXERCISES ON CHAPTER XV

1. Show that the line whose equations are $x_1/a = x_2/b = x_3/c$ is the axis of the special linear complex given by $ap_{23}+bp_{31}+cp_{12} = 0$ and that the line whose equations are $x_0 = 0 = \alpha x_1+\beta x_2+\gamma x_3$ is the axis of the special linear complex whose equation is $\alpha p_{01}+\beta p_{02}+\gamma p_{03} = 0$.

If $a_0 \neq 0$, show that a line belongs to the star with vertex (a_0, a_1, a_2, a_3) if and only if its coordinates satisfy the conditions

$$a_0 p_{23}+a_2 p_{30}+a_3 p_{02} = a_0 p_{31}+a_3 p_{10}+a_1 p_{03} = a_0 p_{12}+a_1 p_{20}+a_2 p_{01} = 0.$$

Find analogous equations for the ruled plane whose ordinary equation is $\alpha_0 x_0+\alpha_1 x_1+\alpha_2 x_2+\alpha_3 x_3 = 0$, where $\alpha_0 \neq 0$.

2. Show that the equation $p_{01} = \lambda p_{02}$ represents the special linear complex whose axis is the line $x_1-\lambda x_2 = 0 = x_0$.

3. Show that the lines which satisfy the equations $p_{01} = p_{02} = p_{03}$ are those which lie in the plane $x_0 = 0$ and those which pass through the point $(0, 1, 1, 1)$.

Find all the lines which satisfy the equations $p_{23} = p_{31} = p_{12}$.

4. Identify all types of line system obtained by taking one or more of the six line-coordinates to be zero.

5. Find the coordinate vectors of the two lines which meet all the lines of the linear congruence whose equations are

$$7p_{01}-5p_{02}+21p_{03}+2p_{23}+p_{31}+p_{12} = 0,$$
$$5p_{01}-2p_{02}+13p_{03}+2p_{23} = 0.$$

6. Show that the tetrahedron formed by the polar planes of the vertices of a given tetrahedron with respect to a null polarity is both circumscribed to and inscribed in the given tetrahedron.

Show also that any pair of tetrahedra related in this way are reciprocals of each other with respect to a null polarity.

7. Find the equation of the unique linear complex which contains the following five lines: the edges $X_0 X_1$, $X_0 X_2$, $X_3 X_1$, $X_3 X_2$ of the tetrahedron of reference and the line with coordinates $(1, 1, 2, 1, 1, -1)$.

8. If the coordinate vectors of three skew lines are $\mathbf{p} = (1, 0, 0, 0, 0, 0)$, $\mathbf{q} = (0, 0, 0, 1, 0, 0)$, and $\mathbf{r} = (1, 0, 1, 1, 0, -1)$, show that any line of the regulus containing these three has a coordinate vector of the form

$$(1-\lambda)\mathbf{p}+(\lambda^2-\lambda)\mathbf{q}+\lambda\mathbf{r}.$$

Show that the quadric surface which contains the regulus is $x_0 x_3 = x_1 x_2$.

9. If a, b, c are constants, show that the lines whose coordinates satisfy the equations $p_{02}/a = (p_{01}-p_{23})/b = p_{31}/c$ all meet two fixed generators of the quadric $x_0 x_1 = x_2 x_3$.

10. Five skew lines and a point O being given in general position, transversals are drawn from O to the pairs of transversals of sets of four of the lines. Show that the five transversals so obtained lie in a plane.

11. Find the line-coordinates of the lines joining the reference points to the poles of the opposite reference planes with respect to the quadric whose plane-equation is $\sum A_{ik} u_i u_k = 0$, and verify that these four lines are linearly dependent.

Deduce that the joins of corresponding vertices of two tetrahedra which are reciprocal for a quadric belong to a regulus.

12. A conic and a line are given, of general position in space, and a homographic correspondence is set up between the points of the one and the points of the other. Show that all the joins of corresponding points meet a second fixed line.

13. If p, q, r, s are tangents to the twisted cubic $c(\theta^3, \theta^2, \theta, 1)$ at the points whose parameters are θ_1, θ_2, θ_3, θ_4, show that $2\Omega_{pq} = (\theta_1-\theta_2)^4$, and deduce that if the signs of the fourth roots are chosen suitably then

$$(\Omega_{pq}\Omega_{rs})^{\frac{1}{4}} + (\Omega_{pr}\Omega_{sq})^{\frac{1}{4}} + (\Omega_{ps}\Omega_{qr})^{\frac{1}{4}} = 0.$$

14. The six coordinates p_{ik} of a variable line are quadratic polynomials in a parameter θ, which satisfy identically the fundamental relation $\Omega_{pp} = 0$, and the matrix of coefficients of the six polynomials is of rank 3. Show that in general the line generates a regulus, and find the complementary regulus.

15. Find the equation of the complex of lines which meet the conic $x_2^2 - x_1 x_3 = 0 = x_0$, and that of the complex of lines which meet the twisted cubic $c(\theta^3, \theta^2, \theta, 1)$.

16. If four skew lines, which do not lie on a quadric, meet a line l in points A, B, C, D and lie respectively in planes α, β, γ, δ through l, show that they have no second common transversal other than l if and only if

$$(A, B, C, D) \barwedge (\alpha, \beta, \gamma, \delta).$$

17. Show that if one quadric is transformed into another by the null polarity belonging to a linear complex, the two quadrics have four generators in common.

Show, conversely, that if two quadrics have four common generators there exists a null polarity which transforms the one into the other.

CHAPTER XVI

PROJECTIVE GEOMETRY OF n DIMENSIONS

In this concluding chapter we propose to give the reader some indication, through the medium of specific examples, of the natural extension of projective geometry to spaces of higher dimensionality, and of the important function of higher space in providing convenient and suggestive representations of variable geometrical entities in spaces of two and three dimensions. The systematic development of many-dimensional geometry, initiated in the second half of the nineteenth century by the Italian school of geometers, and by Veronese and the great Corrado Segre in particular, was a momentous undertaking and one which revealed unsuspected potentialities in the subject of projective geometry. Segre showed that to stop at three dimensions gives an unnatural and distorted picture even of much of three-dimensional projective geometry, since configurations in S_3 that appear to be very complicated can often be obtained as sections or projections of much simpler ones in higher space. Again, the introduction of higher space has proved of inestimable value in making possible new developments in ordinary projective geometry by allowing the construction of simple models of systems of geometrical entities in S_3. Quite apart from all this, the study of the geometry of S_n, both for its own sake and also with a view to solving purely algebraic problems, has opened up a vast and fruitful field of research.

Four-dimensional geometry

For the sake of simplicity we begin with four-dimensional space S_4, in which the points are given, in any allowable representation $\mathscr{R}$, by homogeneous coordinate vectors $\mathbf{x} = (x_0, x_1, \ldots, x_4)$. The representation may be specified by choosing five reference points $X_0, \ldots, X_4$, the vertices of a four-dimensional pentahedron of reference, and a unit point E.

The fundamental entities in S_4 are points, lines, planes, and solids. Lines, planes, and solids are linear subspaces of S_4, and they may be defined as totalities of points linearly dependent respectively on two, three, or four linearly independent points of S_4.

Alternatively, a solid may be characterized as the totality of points whose coordinates satisfy a single linear equation $\mathbf{u}^T\mathbf{x} = 0$;

and the vector $\mathbf{u}$ is then a coordinate vector of the solid. Planes, lines, and points are clearly the intersections of pairs, triads, and tetrads of linearly independent solids.

The solids through a plane form a pencil, those through a line form a line-star, and those through a point form a point-star.

The reader will have no difficulty in interpreting and verifying the following scheme indicating the normal intersections of linear subspaces in S_4:

	line	plane	solid
line	—	—	point
plane	—	point	line
solid	point	line	plane

It will be noted, in particular, that a line normally meets a solid in a point. This enables us to define an operation of projection from a point V on to a solid Π, by which any general point P of S_4 is transformed into the point of intersection of the line VP with Π. Further, two planes normally meet in a point; and so we can define projection from a line v on to a plane π, by which a point P goes into the point of intersection of the plane vP with π.

In S_4, as in S_2 and S_3, we have a principle of duality. This is indicated by the scheme

$$\begin{pmatrix} \text{point} & \text{line} & \text{plane} & \text{solid} \\ \text{solid} & \text{plane} & \text{line} & \text{point} \end{pmatrix}.$$

Thus, for example, to the proposition that two planes normally intersect in a point corresponds the dual proposition that two general (non-intersecting) lines have as their join a solid.

Exercises

(i) Show that a line and a plane do not meet unless they lie in a solid, and that two planes meet in a line if and only if they lie in a solid.

(ii) Verify that S_4 contains ∞^4 points and solids and ∞^6 lines and planes.

(iii) Show directly, and also by applying the principle of duality, that three general lines of S_4 have a unique transversal line. Explain the circumstances in which three lines of S_4 may have infinitely many transversals.

In S_4 there are three kinds of point-locus: (*a*) *primals*, which are three-dimensional loci, each given by a single homogeneous equation in $x_0, x_1, \dots, x_4$, (*b*) *surfaces*, which are loci of ∞^2 points, and (*c*) *curves*, which are loci of ∞^1 points.

The primals of order 1, given by linear equations, are of course the solids of S_4. Next in simplicity after these come the primals of order 2, or quadrics. Every quadric is represented by an equation of the form

$$S \equiv \sum_{i=0}^{4} \sum_{k=0}^{4} a_{ik} x_i x_k = 0 \quad (a_{ki} = a_{ik});$$

and, as was the case in S_3, the projective character of the locus depends only on the rank of the matrix (a_{rs}). According as the rank is 5, 4, or 3, the quadric is (i) a general quadric, (ii) a *point-cone*, generated by the ∞^2 lines which join a fixed point to all the points of an ordinary quadric surface lying in a solid, or (iii) a *line-cone*, formed by the ∞^1 planes which join a fixed line to the points of a conic.

One of the simplest surfaces which belong properly to S_4 (i.e. which do not lie in any solid in S_4) is the *Segre quartic surface* F, the surface of intersection of two general quadrics in S_4. This surface is known to have sixteen lines lying on it; it projects from any general point of itself into a general cubic surface in S_3; and it lies on each of the five cones belonging to the pencil defined by the given pair of quadrics. Another simple surface, the *cubic ruled surface* of S_4, is the residual intersection of two quadrics (necessarily cones) which pass through a common plane.

The simplest curve properly belonging to S_4 is the *rational normal quartic* curve C^4—the analogue of the conic in S_2 and the twisted cubic in S_3—whose parametric equations, referred to a suitably chosen system of reference, are

$$x_0 : x_1 : x_2 : x_3 : x_4 = \theta^4 : \theta^3 : \theta^2 : \theta : 1.$$

This curve is met by any solid in four points, whose parameters may be regarded as the roots of a quartic equation, and the resulting association between solids in S_4 and quartic equations $f(\theta) = 0$ provides us with a means of inferring algebraic properties of such equations from known geometrical properties of C^4.

The theorem of the fifth associated plane in S_4

In order to illustrate the working of four-dimensional geometry, we propose now to give a direct algebraic proof of an altogether remarkable and yet simple theorem which is the basis of much interesting geometry in S_4.

THEOREM 1. *The lines which meet four given general planes of S_4 all meet a fifth plane, which, together with the four original planes, makes up a symmetrical set of five associated planes.*

Proof. Let $\pi_1, \pi_2, \pi_3, \pi_4$ be the four given planes. If $A_{ij} = \pi_i \pi_j$, we can choose coordinate vectors $\mathbf{a}_{ij}$ for the six points A_{ij} in such a way that

$$\mathbf{a}_{12}+\mathbf{a}_{13}+\mathbf{a}_{14}+\mathbf{a}_{23}+\mathbf{a}_{24}+\mathbf{a}_{34} = 0. \tag{1}$$

Now let p be a line which meets π_1 and π_2, say in the points represented by the vectors $\mathbf{a}_{12}+x\mathbf{a}_{13}+y\mathbf{a}_{14}$ and $\mathbf{a}_{12}+z\mathbf{a}_{23}+t\mathbf{a}_{24}$ respectively. Then a general point of p is given by

$$(1+\lambda)\mathbf{a}_{12}+x\mathbf{a}_{13}+y\mathbf{a}_{14}+\lambda z\mathbf{a}_{23}+\lambda t\mathbf{a}_{24}, \tag{2}$$

and p meets π_3 provided that, for some value of λ, this vector is of the form $l\mathbf{a}_{13}+m\mathbf{a}_{23}+n\mathbf{a}_{34}$, and hence, by (1), of the form

$$-n\mathbf{a}_{12}+(l-n)\mathbf{a}_{13}-n\mathbf{a}_{14}+(m-n)\mathbf{a}_{23}-n\mathbf{a}_{24}.$$

This requires that, for some λ,

$$1+\lambda = y = \lambda t,$$

and so yields the condition

$$t = y/(y-1). \tag{3}$$

In the same way, the condition for p to meet π_4 is

$$z = x/(x-1). \tag{4}$$

Suppose, now, that conditions (3), (4) are satisfied, so that p meets all four planes; and consider the point of p for which

$$\lambda = -(x-1)(y-1).$$

Substituting this value for λ in (2), we obtain, as the coordinate vector of the point,

$$(x+y-xy)\mathbf{a}_{12}+x\mathbf{a}_{13}+y\mathbf{a}_{14}-x(y-1)\mathbf{a}_{23}-y(x-1)\mathbf{a}_{24},$$

i.e. $$(x+y-xy)(\mathbf{a}_{12}+\mathbf{a}_{23}+\mathbf{a}_{24})+x(\mathbf{a}_{13}-\mathbf{a}_{24})-y(\mathbf{a}_{23}-\mathbf{a}_{14}).$$

But this is the coordinate vector of a point lying in the plane π_5 of the points given by $\mathbf{a}_{12}+\mathbf{a}_{23}+\mathbf{a}_{24}$, $\mathbf{a}_{13}-\mathbf{a}_{24}$, and $\mathbf{a}_{23}-\mathbf{a}_{14}$. The same plane is also determined by the second and third of these points together with the point represented by

$$2(\mathbf{a}_{12}+\mathbf{a}_{23}+\mathbf{a}_{24})+(\mathbf{a}_{13}-\mathbf{a}_{24})-(\mathbf{a}_{23}-\mathbf{a}_{14}) = \mathbf{a}_{12}-\mathbf{a}_{34}.$$

We have proved, therefore, that any line which meets $\pi_1, \pi_2, \pi_3, \pi_4$ also meets the plane π_5 which is determined by the three points whose coordinate vectors are $\mathbf{a}_{23}-\mathbf{a}_{14}$, $\mathbf{a}_{13}-\mathbf{a}_{24}$, $\mathbf{a}_{12}-\mathbf{a}_{34}$; and this plane may be said to be *associated* with $\pi_1, \pi_2, \pi_3, \pi_4$.

In order to show that the five planes π_i are symmetrically related, we find coordinate vectors for the points $\pi_i \pi_5$ $(i = 1,2,3,4)$.

These may be taken as

$$\mathbf{a}_{15} = -(\mathbf{a}_{12}+\mathbf{a}_{13}+\mathbf{a}_{14}),$$
$$\mathbf{a}_{25} = -(\mathbf{a}_{12}+\mathbf{a}_{23}+\mathbf{a}_{24}),$$
$$\mathbf{a}_{35} = -(\mathbf{a}_{13}+\mathbf{a}_{23}+\mathbf{a}_{34}),$$
$$\mathbf{a}_{45} = -(\mathbf{a}_{14}+\mathbf{a}_{24}+\mathbf{a}_{34}).$$

It now follows that the plane associated with $\pi_5, \pi_2, \pi_3, \pi_4$ is determined by the three points whose coordinate vectors are

$$\mathbf{a}_{45}-\mathbf{a}_{23} = \mathbf{a}_{12}+\mathbf{a}_{13}, \qquad \mathbf{a}_{35}-\mathbf{a}_{24} = \mathbf{a}_{12}+\mathbf{a}_{14},$$

and

$$\mathbf{a}_{25}-\mathbf{a}_{34} = \mathbf{a}_{13}+\mathbf{a}_{14}$$

—i.e. it is the plane π_1; and this completes the proof of the theorem, since $\pi_1, \pi_2, \pi_3, \pi_4$ enter symmetrically into the problem.

The lines which meet the five associated planes π_i are subject to only four simple conditions, and they therefore form a doubly infinite system. This means that they generate a threefold locus, and this locus is called the *Segre primal*. In order to elucidate the nature of this primal we need to represent it algebraically, and this may be done in the following manner.

If $A_{12}, A_{13}, A_{14}, A_{23}, A_{24}$ are taken as vertices of a new pentahedron of reference, the coordinates of a general point of p are given, in accordance with (2), by

$$\frac{x_0}{1+\lambda} = \frac{x_1}{x} = \frac{x_2}{y} = \frac{x_3}{\lambda z} = \frac{x_4}{\lambda t} = \rho, \text{ say.}$$

Then $\rho = \dfrac{zx_0-x_3}{z} = \dfrac{x_1}{x}$, and hence $z = \dfrac{x_3 x}{x_0 x-x_1}$.

But, by (4), $z = \dfrac{x}{x-1}$, and hence $x = \dfrac{x_3-x_1}{x_3-x_0}$.

Similarly, by eliminating t instead of z, we obtain

$$y = \frac{x_4-x_2}{x_4-x_0}.$$

Since $\dfrac{x_1}{x} = \dfrac{x_2}{y}$, we have finally

$$\frac{x_1}{x_2} = \frac{x_3-x_1}{x_3-x_0}\bigg/\frac{x_4-x_2}{x_4-x_0}.$$

The locus generated by p is therefore the cubic primal whose equation is

$$x_1(x_0-x_3)(x_2-x_4)-x_2(x_0-x_4)(x_1-x_3) = 0.$$

This equation may be written in the form

$$XYZ = X'Y'Z',$$

where X, Y, Z, X', Y', Z' are six linear forms in the five point-coordinates which are connected by the identical relation

$$X+Y+Z = X'+Y'+Z'.$$

EXERCISES

(i) Show that the Segre cubic primal defined above contains fifteen planes in all, including the five original associated planes; also that it has ten nodes, of which four lie in each plane.

(ii) Verify that the equations

$$X = y(t-z), \qquad Y = z(t-x), \qquad Z = x(t-y),$$
$$X' = z(t-y), \qquad Y' = x(t-z), \qquad Z' = y(t-x),$$

give a $(1, 1)$ parametric representation of the Segre primal on the points (x, y, z, t) of a three-dimensional space; and find the surfaces in this space which correspond to the sections of the primal by solids of S_4.

(iii) Enunciate the dual of Theorem 1, i.e. the theorem of the fifth associated line.

Five-dimensional geometry

The points of S_5 are given, in any allowable representation $\mathscr{R}$, by means of vectors $\mathbf{x} = (x_0, x_1, \ldots, x_5)$; and any particular representation can be specified by means of the six vertices $X_0, \ldots, X_5$ of the simplex of reference and the unit point E.

The fundamental entities in S_5 are points, lines, planes, solids, and primes (four-dimensional linear subspaces). Five linearly independent points are joined by a unique prime, and five linearly independent primes intersect in a unique point.

The scheme of normal intersections of linear subspaces in S_5 is as follows:

	line	plane	solid	prime
line	—	—	—	point
plane	—	—	point	line
solid	—	point	line	plane
prime	point	line	plane	solid

It will be noted, in particular, that two planes of S_5 do not ordinarily intersect.

The principle of duality in S_5 is indicated by the scheme

$$\begin{pmatrix} \text{point} & \text{line} & \text{plane} & \text{solid} & \text{prime} \\ \text{prime} & \text{solid} & \text{plane} & \text{line} & \text{point} \end{pmatrix}.$$

The plane features here as a self-dual entity, and the ∞^2 points in a plane are dual to the ∞^2 primes passing through a plane.

In S_5 we have to consider loci V_k $(k = 4, 3, 2, 1)$ of four different dimensionalities: primals V_4 (each given by a single equation), threefolds V_3, surfaces V_2, and curves V_1.

A quadric primal V_4^2 is said to be general or non-singular if it is given by the vanishing of a quadratic form whose matrix is of rank 6. If the rank is 5, 4, or 3, the primal is said to be a point-cone, a line-cone, or a plane-cone. These three manifolds are obtained respectively by joining the points of a quadric threefold V_3^2 by lines to a point, the points of a quadric surface V_2^2 by planes to a line, or the points of a conic V_1^2 by solids to a plane.

EXERCISES

(i) If two planes have a point in common, show that they lie in a prime.

(ii) Show that through any general point there passes a unique transversal line of two given non-intersecting planes. State the dual result.

(iii) Prove that in S_5 there are ∞^5 points and primes, ∞^8 lines and solids, and ∞^9 planes.

(iv) Discuss projection from a point, line, or plane in S_5.

Representation of conics in S_2 by points in S_5

We have already referred in an earlier chapter (see p. 106) to the fact that when considering systems of conic loci k in a plane π it is natural to represent all such loci by points of S_5, taking as coordinates x_i' of the representative point in S_5 the six coefficients which appear in the equation $\sum_{i=0}^{2}\sum_{k=0}^{2} a_{ik}x_i x_k = 0$. When this is done, the four types of linear system of conics, those of freedom 1, 2, 3, and 4 respectively, are represented by the four different types of linear subspace of S_5. Thus, in particular, the conics of a pencil are mapped on the points of a line in S_5.

We shall now give a few properties of this useful representation in order to exhibit some of its more typical features. The first thing to observe is that while a general conic in π is represented simply by a general point of S_5, various special kinds of conic are represented by points belonging to certain loci in S_5. Thus we have the following results.

The points of S_5 which represent repeated lines of the plane form a surface ϕ, called the Veronese surface; and those which represent line-pairs form a cubic primal M_4^3, called the cubic symmetroid. M_4^3 is the locus generated by all chords of ϕ.

To prove this we observe first that if k is the repeated line given by $(\lambda x_0+\mu x_1+\nu x_2)^2 = 0$, the coordinates x_i' of the image point K are given by

$$\frac{x_0'}{\lambda^2} = \frac{x_1'}{\mu^2} = \frac{x_2'}{\nu^2} = \frac{x_3'}{\mu\nu} = \frac{x_4'}{\nu\lambda} = \frac{x_5'}{\lambda\mu}; \tag{1}$$

and these equations constitute a parametric representation of the surface ϕ—defining a mapping of the points of the surface on the points (λ, μ, ν) of a plane α. If T is any general solid of S_5, defined as the intersection of the two primes whose equations are

$$\sum_{i=0}^{5} u_i x_i' = 0, \qquad \sum_{i=0}^{5} v_i x_i' = 0,$$

then the points of intersection of ϕ with T are represented in α by the four points (λ, μ, ν) that are common to the two conics whose equations are

$$u_0\lambda^2+u_1\mu^2+...+u_5\lambda\mu = 0$$

and

$$v_0\lambda^2+v_1\mu^2+...+v_5\lambda\mu = 0.$$

Thus ϕ is met by a general solid in four points, and we accordingly describe it as a quartic surface.

The points of S_5 which represent line-pairs of π are evidently those whose coordinates satisfy the equation

$$\begin{vmatrix} x_0' & x_5' & x_4' \\ x_5' & x_1' & x_3' \\ x_4' & x_3' & x_2' \end{vmatrix} = 0,$$

and this is the equation of the cubic symmetroid M_4^3.

Finally, since a quadratic form $\sum_{i=0}^{2}\sum_{k=0}^{2} a_{ik}x_i x_k$ is a product of linear factors if and only if it can be expressed as a linear combination $\lambda L^2+\mu M^2$ of the squares of two linear forms in x_0, x_1, x_2, it follows that a point of S_5 lies on M_4^3 if and only if it lies on a chord of ϕ; i.e. M_4^3 is the locus of chords of ϕ.

We can now establish the following additional property of the manifolds ϕ and M_4^3:

The surface ϕ contains a doubly infinite system of conics, one conic passing through any two given points of the surface. M_4^3 is the locus *of planes of these conics, and it has the tangent planes of ϕ as a second system of generating planes.*

To prove this, we consider the curve c on ϕ whose points represent repeated lines through a fixed point V in π. This curve is represented by a straight line in the plane α of the parameters λ, μ, ν. Its parametric equations are obtained therefore by replacing λ, μ, ν in (1) by linear functions of a parameter θ, and this procedure clearly leads to equations of a curve of the second order, i.e. a conic. Also the points of the plane of c represent all line-pairs with V as vertex, and this plane therefore lies on M_4^3. Since, furthermore, any two lines of π meet in a unique point V, any two points of ϕ are joined by a unique conic c on the surface. Any two conics c, c' on ϕ meet in a unique point.

Finally, it is easy to verify that all the line-pairs of π with a fixed line d as component are mapped by points of the tangent plane to ϕ at the point which represents the repeated line d; and this shows that the tangent planes to ϕ form a second doubly infinite system of generating planes of M_4^3.

EXERCISES

(i) Show that the three line-pairs of a pencil of conics in π are represented in S_5 by the three points of intersection with M_4^3 of the image line of the pencil. Deduce that all the conics which touch a fixed conic k in π are represented by points of the tangent cone to M_4^3 from the image point of k.

(ii) Prove that a double-contact pencil of conics is represented by a line which meets ϕ; also that any trisecant plane of ϕ represents the net of conics for which a fixed triangle is self-polar.

(iii) Show that all the conics of π which touch a fixed line are represented in S_5 by the points of a quadric primal through ϕ.

Representation of the lines of S_3 by the points of a quadric primal Ω of S_5

Our last illustration of the use of geometry of higher dimensionality is one of a kind that has proved to be extraordinarily useful and suggestive, as well as being capable of very wide application. It concerns a representation in which a system of geometrical entities—in this case the lines of S_3—is mapped not on a whole space S_n but on a certain algebraic manifold in such a space.

We have already discussed the elements of line-geometry in Chapter XV, and the reader must have been struck by the fact that the geometry of line systems of one, two, and three dimensions is much harder to visualize than the more familiar geometry of point-loci. A further awkward feature of line-geometry, to which we drew specific attention, is the fact that a linear combination of coordinate vectors of lines is not in general itself the coordinate vector of a line—so that the coordinate vectors of the lines of S_3 do not constitute a vector space. What we now propose to do is to show briefly how these and other similar difficulties may be avoided by transforming the whole theory of line systems in S_3 into the theory of ordinary point-loci on a four-dimensional manifold Ω in S_5, whose points map unexceptionally the lines of S_3.

In terms of any fixed allowable representation of S_3, every line p has a set of six line-coordinates which make up the coordinate vector

$$\mathbf{p} = (p_{23}, p_{31}, p_{12}, p_{01}, p_{02}, p_{03});$$

and this vector satisfies the identical relation

$$\Omega_{pp} \equiv p_{01}p_{23}+p_{02}p_{31}+p_{03}p_{12} = 0.$$

If, now, we interpret $\mathbf{p}$ as the coordinate vector $(x_0,\ldots,x_5)$ of a point in S_5, it is obvious that a one-one correspondence is set up between the lines of S_3 and the points of S_5 which lie on the quadric primal Ω whose equation is

$$\Omega \equiv x_0x_3+x_1x_4+x_2x_5 = 0.$$

This is the representation to which we have already alluded. Complexes, congruences, and ruled surfaces of S_3 are represented respectively by threefold loci, surfaces, and curves cut on Ω by algebraic manifolds in S_5; and the condition $\Omega_{pq} = 0$ for two lines p, q in S_3 to intersect clearly expresses the conjugacy of their representative points in S_5 with respect to the quadric Ω.

Plainly, in order to profit from this representation, we need to be familiar with the properties of Ω. Our knowledge of such properties comes in the first instance partly from such algebraic results as we already know concerning quadratic forms and symmetrical polarities and partly from the results that we have already obtained by direct study of the line-geometry of S_3.

To begin with, Ω is a non-singular quadric of S_5; for a simple non-singular transformation of coordinates reduces $x_0x_3+x_1x_4+x_2x_5$ to a sum of the six squares of the new coordinates.

Then again, by obvious generalization of the polar properties of conics and quadric surfaces, we may infer the following properties of Ω.

(i) Ω defines a symmetrical (non-singular) point-prime polarity of S_5, in which the uniquely associated pairs of polar spaces are (*a*) point and prime, (*b*) line and solid, (*c*) plane and plane. Any point of either space of a polar pair is conjugate to all points of the other, and any prime through either space of a pair is conjugate to every prime through the other. (The definitions of conjugacy are, of course, immediate generalizations of the definitions of conjugacy of points and of planes with respect to a quadric in S_3.)

(ii) The points of Ω are those points of S_5 which lie on their polar primes, and the tangent primes of Ω are those primes which contain their poles.

(iii) Through any solid T there pass two tangent primes to Ω, and the polar line of T is the line which joins their points of contact. Through any general plane π there pass ∞^1 tangent primes of Ω, and these touch Ω at points of the conic in which Ω is met by the polar plane of π.

Analogy with the quadric in S_3 leads us to expect that Ω will contain certain systems of generating linear spaces which lie wholly on it, and such spaces do in fact exist. In order to find them we can use the representation of lines; for any generating space of Ω must be the image on Ω of a linear line space in S_3 (see Chapter XV, § 2). Now we know exactly what special line systems of this kind exist in S_3, and we accordingly infer that Ω contains the following generating spaces:

(i) ∞^5 lines, representing pencils of lines in S_3;

(ii) ∞^3 planes of one system, α-planes say, representing point-stars of S_3;

(iii) ∞^3 planes of another system, β-planes say, representing ruled planes of S_3.

Every line on Ω is clearly the intersection of a unique α-plane with a unique β-plane.

Since two point-stars of S_3 have a unique ray in common, it follows that two α-planes meet always in a single point; and similarly two β-planes meet in a single point. A point-star and a ruled plane, on the other hand, have either no line in common or else contain a common pencil of lines (when the vertex of the star lies in the plane); and therefore an α-plane and a β-plane in general do not intersect, but if they have an intersection this must be a line.

We are now in a position to discuss the representation on Ω of various types of line system in S_3 which we have already met in Chapter XV.

In the first place, any linear complex—obtained by imposing a fixed linear condition on the p_{ij}—is represented by the threefold locus V_3^2 in which Ω is met by a prime Π. If Π touches Ω at a point P, then the section V_3^2 is a point-cone with P as vertex, and the corresponding linear complex is then the special complex whose axis is the line corresponding to P.

A quadratic complex K is represented by the quartic threefold V_3^4 in which Ω is met by a second quadric Ω'; and the geometry of K is accordingly related to that of the pencil of quadrics defined by Ω and Ω'. The singular points and planes of K correspond to α-planes and β-planes of Ω which touch Ω', i.e. which meet it in line-pairs.

A linear congruence is represented by the quadric surface F in which Ω is met by a solid T, and its two directrices are represented by the points of intersection of Ω with the polar line of T.

A quadratic congruence is represented by the Segre quartic surface common to two quadrics Ω, Ω' and a prime Π.

A regulus is represented by the conic in which Ω is met by a plane π; and the complementary regulus is represented by the conic in which Ω is met by the polar plane π' of π. This follows from the fact that if P, Q are two points, one on each of the conics, representing lines p, q respectively, then P and Q are conjugate points for Ω, and the condition $\Omega_{pq} = 0$ for p and q to intersect is therefore satisfied.

For further details and applications of the representation just considered, in addition to those few given in the exercises below, we refer the reader particularly to the excellent account given in Baker, *Principles of Geometry* (Cambridge, 1925), volume iv; and also to the accounts given in Edge, *Ruled Surfaces* (Cambridge, 1931), chapter i, and Semple and Roth, *Introduction to Algebraic Geometry* (Oxford, 1949), chapter x.

Conclusion

Our main concern in the preceding chapters has been with some of the simplest manifolds in two and three dimensions, especially the conic, quadric, and twisted cubic, and we have been able to give a tolerably complete account of these three manifolds by relying on a few general notions such as that of homographic correspondence. We have also touched lightly upon certain powerful methods of proving geometrical theorems, of which the methods of transformation and representation are typical. We have met various examples of birational transformation, including Cremona transformations of a plane into itself and projection of a quadric on to a plane; and we have seen how conics in S_2 and lines in S_3 may be represented by points in higher space.

In order to illustrate Segre's discovery that figures in S_2 or S_3 can sometimes be derived by projection or section from simpler figures in higher space, we should have to study a number of other special manifolds in addition to those already mentioned, and to do this we should need to go more fully into the theory of birational correspondences. This would take us beyond the confines of elementary projective geometry into the realm known as algebraic geometry. References to a number of books on this subject have already been given.

A yet more ambitious project may just be discerned upon the horizon—to discuss, in full generality, the properties of algebraic manifolds as such in projective space S_n. This is the ultimate objective in algebraic geometry, and one towards which some progress has been made from various directions. Thus the geometers of the Italian school rely largely upon correspondence methods, while other algebraic geometers are bringing the full resources of modern algebra to bear upon their problems. The reader who wishes to learn more of this side of geometry should turn, say, to *Methods of Algebraic Geometry* by Hodge and Pedoe.

Our task has been a much more elementary one; but it has its justification in the intrinsic interest of the particular manifolds we have discussed in detail, and also in the fact that one has to be thoroughly acquainted with the properties of a variety of special manifolds in order to be in a position to tackle more general geometrical problems.

EXERCISES ON CHAPTER XVI

1. Show that any quadric of S_4 that contains a plane is necessarily a cone (or a pair of solids).

2. If two quadric cones of S_4 have a plane π in common, show that in general they meet residually in a cubic ruled surface, and that the generators of this surface meet π in the points of a conic.

Show also that the equations of such a cubic ruled surface can be written in the form $u/u' = v/v' = w/w'$, where u, v, w, u', v', w' are linear forms in the point-coordinates.

3. Deduce from Exercise 2 that the cubic ruled surface lies on a quadric line-cone whose vertex meets all the generating lines of the surface.

Show also that the surface is generated by the lines which join corresponding points of homographic ranges on a line and on a conic respectively, and that any general surface so generated is a cubic ruled surface.

[*Hint.* Write

$$u/u' = v/v' = w/w' = (\lambda u+\mu v+\nu w)/(\lambda u'+\mu v'+\nu w')$$
$$= (\lambda' u+\mu' v+\nu' w)/(\lambda' u'+\mu' v'+\nu' w'),$$

and choose λ, μ, ν, λ', μ', ν' so that $\lambda' u+\mu' v+\nu' w \equiv \lambda u'+\mu v'+\nu w'$.]

4. Show that all the planes of S_4 which meet three general lines meet a fixed solid in the lines of a tetrahedral complex, and that any tetrahedral complex can be generated in this way. [*Hint.* If the unique transversal line of the given lines meets them in A, B, C and meets the solid in D, and if π is any one of the planes, then the pencil of solids $\pi(A, B, C, D)$ meets the fixed solid in a pencil of planes of fixed cross ratio.]

5. If C is the rational normal quartic curve whose parametric equations are

$$x_0 : x_1 : x_2 : x_3 : x_4 = \theta^4 : \theta^3 : \theta^2 : \theta : 1,$$

show that the solid which joins the four points of the curve whose parameters are θ_i $(i = 1, 2, 3, 4)$ has the equation

$$x_0 - s_1 x_1 + s_2 x_2 - s_3 x_3 + s_4 x_4 = 0,$$

where s_1, s_2, s_3, s_4 are the elementary symmetric functions of θ_1, θ_2, θ_3, θ_4.

Deduce from this the equations of a trisecant plane and of a chord of the curve.

6. Show that the chords of the curve C in the previous exercise generate the cubic primal whose equation is

$$\begin{vmatrix} x_0 & x_1 & x_2 \\ x_1 & x_2 & x_3 \\ x_2 & x_3 & x_4 \end{vmatrix} = 0.$$

Show that three chords of C meet a general line of S_4.

Show that four osculating solids of C pass through a general point of S_4. If P is this point and Π is the solid containing the points of contact of the four osculating solids, show that Π is the polar solid of P with respect to a fixed non-singular quadric containing all the tangents of C.

7. Show that the trisecant planes of a rational normal quartic curve which pass through a general point generate a quadric cone with vertex at the point.

8. The quartic equation $a_0 x^4 + 4a_1 x^3 + 6a_2 x^2 + 4a_3 x + a_4 = 0$ is represented by the point $(a_0, \dots, a_4)$ of S_4. Show that (i) the locus of points of S_4 which represent equations with four equal roots is a rational normal quartic curve C, (ii) the locus of points which represent equations with three equal roots is a sextic surface, locus of tangent lines of C, (iii) the locus of points which represent equations with two equal roots is a sextic primal, locus of osculating planes of C, and (iv) the locus of points which represent equations with a given root is an osculating solid of C.

Find the locus of points which represent equations with two pairs of equal roots.

9. Prove, by the methods of Exercise 8, that a necessary (and in general sufficient) condition for the quartic polynomial $(a_0,\ldots,a_4\between x,1)^4$ to be expressible as a sum of two fourth powers is

$$\begin{vmatrix} a_0 & a_1 & a_2 \\ a_1 & a_2 & a_3 \\ a_2 & a_3 & a_4 \end{vmatrix} = 0.$$

In what special circumstances is this condition insufficient?

10. A *rational* quartic curve in the plane or in space being defined as a quartic curve given by parametric equations in which the coordinates are proportional to linearly independent quartic polynomials in a parameter θ with no common factor, show that every such quartic curve is a projection of the rational normal quartic curve in S_4 from a line or from a point.

Show that the general rational plane quartic has three double points.

Show also how to choose the line-vertex of projection so that the rational normal quartic shall project into a tricuspidal plane quartic; and interpret in this way the fact that the three cuspidal tangents of such a curve are concurrent.

11. Prove that the general rational quintic curve in S_3 has one quadrisecant line, and that the general rational quintic curve in S_4 has one trisecant line.

12. Prove that the general quintic polynomial in one variable x is expressible in one and only one way as a linear combination of three fifth powers $(x-\alpha)^5$, $(x-\beta)^5$, $(x-\gamma)^5$.

13. Show that in S_5 a unique transversal plane of three given lines (of general position) can be drawn through any general point P.

If P describes a fourth line, show that the transversal plane cuts homographic ranges on all four lines.

14. Establish the following results for the representation of the conics of S_2 by points of S_5 (p. 388).

(i) The surface ϕ is double on M_4^3, i.e. any general line through a point P of ϕ has two of its three intersections with M_4^3 coincident at P.

(ii) Any inflexional tangent line to M_4^3 represents a pencil of conics with three-point contact.

(iii) Any plane which does not meet ϕ but meets M_4^3 in the sides of a triangle represents a net of conics which pass through three fixed points.

(iv) A solid which contains a conic of ϕ meets ϕ in general in one further point not on the conic, and it represents the system of conics which have a given point and a given line as pole and polar.

(v) The cone which projects ϕ from any general point of S_5 represents the system of all conics which have double contact with a fixed conic.

(vi) Any prime of S_5 represents the system of conics which are outpolar to a fixed conic envelope, and this envelope is a proper conic, a point-pair, or a repeated point according as the prime meets ϕ in an irreducible quartic curve, a pair of conics, or a single conic counted twice.

15. Establish the following properties of the representation of the lines of S_3 by the points of a quadric Ω in S_5.

(i) The section of Ω by the tangent prime at any point P is a quadric cone

V_3^2, generated by the ∞^1 α-planes and the ∞^1 β-planes through P, and it represents the lines of S_3 which meet a fixed line.

(ii) A twisted cubic curve C on Ω represents a cubic ruled surface R^3 in S_3, and in general the lines of R^3 meet two fixed lines, two of them passing through any general point of one of the lines and one of them through any point of the other.

(iii) The chords of a twisted cubic c in S_3 are represented on Ω by the points of a Veronese surface which meets any general α-plane in one point and any general β-plane in three points, and the axes (lines which lie in two osculating planes) of c are represented by another Veronese surface which meets any general α-plane in three points and any general β-plane in one point.

16. Show that the vertices and faces of a tetrahedron in S_3, regarded as point-stars and ruled planes respectively, are represented on Ω by a double-four

$$\begin{matrix} \alpha_1 & \alpha_2 & \alpha_3 & \alpha_4 \\ \beta_1 & \beta_2 & \beta_3 & \beta_4 \end{matrix}$$

of α-planes and β-planes such that α_i meets β_j in a line $(i, j = 1, 2, 3, 4)$ except when $i = j$.

Show also that the section of Ω by a quadric Ω' through such a double-four of planes represents a tetrahedral complex in S_3.

APPENDIX

TWO BASIC ALGEBRAIC THEOREMS

MUCH of the argument of the preceding chapters rests upon two theorems in pure algebra, one on linear transformations and the other on quadratic forms, and we shall conclude our account of algebraic projective geometry with a formal statement of these basic theorems.

The application of linear algebra to projective geometry is complicated by the fact that the coordinates used are homogeneous and, in consequence, the correspondence between points and coordinate vectors is not one–one. For this reason we have usually to take scalar factors of proportionality into account, as in the lemma which follows.

LEMMA. *If* $\mathbf{x}_0,\ldots,\mathbf{x}_{n+1}$ *are* $n+2$ *vectors in* $V_{n+1}(K)$, *no* $n+1$ *of which are linearly dependent, and if* $\mathbf{e}_0,\ldots,\mathbf{e}_n,\mathbf{e}_{n+1}$ *are respectively the vectors* $(1,0,\ldots,0),\ldots,(0,0,\ldots,1)$, $(1,1,\ldots,1)$, *there exists a non-singular linear transformation* $\mathbf{x}\rightarrow\mathbf{A}\mathbf{x}$ *of* V_{n+1} *into itself such that* $\mathbf{A}\mathbf{e}_i=\lambda_i\mathbf{x}_i$ $(i=0,\ldots,n+1)$, *where the* λ_i *are non-zero scalars; and the matrices of any two transformations with this property differ at most by a scalar factor.*

Proof. Let $\mathbf{x}_i$ have components $(x_{i0},x_{i1},\ldots,x_{in})$. The matrix $\mathbf{A}$ satisfies the $n+1$ conditions

$$\mathbf{A}\mathbf{e}_i=\lambda_i\mathbf{x}_i \quad (i=0,\ldots,n)$$

if and only if the elements of its $(i+1)$th column are the components of the vector $\lambda_i\mathbf{x}_i$, i.e. if $\mathbf{A}\equiv(a_{rs})=(\lambda_s x_{sr})$. We have to show, therefore, that the values of $\lambda_0,\ldots,\lambda_n$, and λ_{n+1}, can be so chosen that the remaining condition

$$\mathbf{A}\mathbf{e}_{n+1}=\lambda_{n+1}\mathbf{x}_{n+1}$$

is also satisfied. For this to be the case,

$$\begin{pmatrix}\lambda_0x_{00} & \lambda_1x_{10} & \cdot & \cdot & \cdot & \lambda_nx_{n0}\\ \lambda_0x_{01} & \lambda_1x_{11} & \cdot & \cdot & \cdot & \lambda_nx_{n1}\\ \cdot & \cdot & \cdot & \cdot & \cdot & \cdot\\ \lambda_0x_{0n} & \lambda_1x_{1n} & \cdot & \cdot & \cdot & \lambda_nx_{nn}\end{pmatrix}\begin{pmatrix}1\\1\\ \cdot\\1\end{pmatrix}=\begin{pmatrix}\lambda_{n+1}x_{n+1,0}\\ \lambda_{n+1}x_{n+1,1}\\ \cdot\quad\cdot\quad\cdot\\ \lambda_{n+1}x_{n+1,n}\end{pmatrix}.$$

The λ_i must accordingly be chosen to satisfy the equations

$$\sum_{k=0}^{n}x_{ki}\lambda_k-x_{n+1,i}\lambda_{n+1}=0 \quad (i=0,\ldots,n);$$

and since, by the hypothesis concerning the linear independence of the vectors $\mathbf{x}_i$, all the $(n+1)$-rowed determinants formed from the matrix of coefficients are non-zero, the ratios of the λ_i are uniquely determined. Furthermore, none of the λ_i is zero. The matrix $\mathbf{A}$ is thus uniquely determined apart from a scalar factor, and it is clearly non-singular.

THEOREM 1. *If* $\mathbf{x}_0,\ldots,\mathbf{x}_{n+1}$ *and* $\mathbf{y}_0,\ldots,\mathbf{y}_{n+1}$ *are two sets of* $n+2$ *vectors in* $V_{n+1}(K)$, *no* $n+1$ *vectors in either set being linearly dependent, there exists a non-singular linear transformation* $\mathbf{x} \rightarrow \mathbf{Px}$ *of* $V_{n+1}(K)$ *into itself such that* $\mathbf{Px}_i = \rho_i \mathbf{y}_i$ $(i = 0,\ldots,n+1)$, *where the* ρ_i *are scalars; and the matrix* $\mathbf{P}$ *is uniquely determined apart from a scalar factor.*

Proof. By the lemma, we can choose a non-singular matrix $\mathbf{A}$ and a set of non-zero scalars $\lambda_0,\ldots,\lambda_{n+1}$ so that

$$\mathbf{Ae}_i = \lambda_i \mathbf{x}_i \quad (i = 0,\ldots,n+1),$$

and in the same way we can choose $\mathbf{B}$ and $\mu_0,\ldots,\mu_{n+1}$ so that

$$\mathbf{Be}_i = \mu_i \mathbf{y}_i \quad (i = 0,\ldots,n+1).$$

Then

$$\mathbf{BA}^{-1}\mathbf{x}_i = \frac{\mu_i}{\lambda_i}\mathbf{y}_i \quad (i = 0,\ldots,n+1),$$

and we need only put $\mathbf{BA}^{-1} = \mathbf{P}$, $\dfrac{\mu_i}{\lambda_i} = \rho_i$.

If, further, $\mathbf{Px}_i = \rho_i \mathbf{y}_i$ and $\mathbf{Qx}_i = \sigma_i \mathbf{y}_i$, then $\mathbf{PAe}_i = \rho_i \lambda_i \mathbf{y}_i$ and $\mathbf{QAe}_i = \sigma_i \lambda_i \mathbf{y}_i$; and hence, by the lemma, $\mathbf{PA} = \tau\mathbf{QA}$, i.e. $\mathbf{P} = \tau\mathbf{Q}$ for some scalar τ.

Theorem 1 supplies the foundation for our whole geometrical system. Interpreting $\mathbf{x}$ as a homogeneous coordinate vector of a point in $S_n(K)$ we can infer at once (i) that an allowable representation of S_n is determined uniquely when $n+2$ points, no $n+1$ of which are linearly dependent, are chosen as reference points and unit point respectively, and (ii) that a self-collineation of S_n is determined uniquely by $n+2$ corresponding pairs, provided that the usual restrictions as to linear independence are satisfied.

The second fundamental theorem alluded to above is required in the theory of conics and quadrics, and it may be stated in the following terms.

THEOREM 2. *If* $Q(\mathbf{x}, \mathbf{x}) \equiv \mathbf{x}^T\mathbf{A}\mathbf{x}$ *is a quadratic form in* $x_0,\ldots, x_n$, *and* $\mathbf{A}$ *is of rank* r, *a non-singular linear transformation* $\mathbf{x} = \mathbf{P}\mathbf{y}$ *and non-zero constants* $d_0,\ldots, d_{r-1}$ *can be found such that*

$$Q \equiv d_0 y_0^2 + \ldots + d_{r-1} y_{r-1}^2.$$

If the elements of $\mathbf{A}$ *belong to a field* K, *the elements of* $\mathbf{P}$ *and the coefficients* d_i *may all be chosen in* K.

The proof of this theorem is well known, and we do not need to give it here. The reduction of Q to diagonal form may be carried out in any given case by Lagrange's method.

INDEX

(Most of the numbers refer to the pages on which the topics are first mentioned, but some additional page references have been given where it seemed that they might be useful.)

Printed in the United Kingdom by
Lightning Source UK Ltd., Milton Keynes
137621UK00001BA/6/A